KT-443-576
05/07
UNIVERSITY OF
WOLVERHAMPTON
3e

*To Anselm*
*December 1916 - September 1996*

*Scholar and Humanist*
*Who touched the minds and lives of all who came into contact with him*

# Basics of QUALITATIVE RESEARCH

## Techniques and Procedures for Developing Grounded Theory 3e



Juliet Corbin
Anselm Strauss

*For information:*

Sage Publications, Inc.
2455 Teller Road
Thousand Oaks,
California 91320
E-mail: order@sagepub.com

Sage Publications Ltd.
1 Oliver's Yard
55 City Road
London EC1Y 1SP
United Kingdom

Sage Publications India Pvt. Ltd.
B 1/I 1 Mohan Cooperative
Industrial Area
Mathura Road, New Delhi 110 044
India

Sage Publications Asia-Pacific Pte. Ltd.
33 Pekin Street #02-01
Far East Square
Singapore 048763

Printed in the United States of America.

*Library of Congress Cataloging-in-Publication Data*

Corbin, Juliet M., 1942-
Basics of qualitative research : techniques and procedures for developing grounded theory/Juliet Corbin and Anselm Strauss. — 3rd ed.
p. cm.
Rev. ed. of: Basics of qualitative research/Anselm Strauss, Juliet Corbin. 2nd ed.
The 1990 and 1998 eds. present Strauss as the first author and Corbin as second.
Includes bibliographical references and index.
ISBN 978-1-4129-0643-2 (cloth)
ISBN 978-1-4129-0644-9 (pbk.)
1. Social sciences—Statistical methods. 2. Grounded theory. I. Strauss, Anselm L. II. Strauss, Anselm L. Basics of qualitative research. III. Title.

HA29.S823 2008
300.72—dc22 2007034189

This book is printed on acid-free paper.

07 08 09 10 11 10 9 8 7 6 5 4 3 2 1

*Acquisitions Editor:* Vicki Knight
*Associate Editor:* Sean Connelly
*Editorial Assistant:* Lauren Habib
*Production Editor:* Karen Wiley
*Copy Editor:* Gretchen Treadwell
*Typesetter:* C&M Digitals (P) Ltd.
*Proofreader:* Kristen Bergstad
*Indexer:* Kay Dusheck
*Cover Designer:* Candice Harman
*Cover Photo:* Michael Henkin
*Marketing Manager:* Stephanie Adams

# Contents

to build theory. There is a chapter in this book devoted to theory construction, but many of the analysis chapters are designed to be useful to researchers who are interested in thick and rich description, concept analysis, or simply pulling out themes. Fourth, in this edition, rather than just talking about analysis, I am actually doing analysis—taking the reader through the steps from concept identification to theoretical development. Fifth, and absent from previous editions, there are exercises at the end of each chapter to reinforce learning. Sixth, explanations of how to integrate computer programs into analysis are included.

In every seminar that I've taught over the past years, there is the inevitable question about the use of computer programs for qualitative analysis. Though the use of computer programs in qualitative research is debatable and outright rejected by some researchers, computer programs for analysis are here to stay and their ability to support the research process increases with each improvement in the many programs that are available. Notice that I say "support" and not "take over" or "direct" the research process. I think one of the most interesting aspects of this edition is that it demonstrates that the analytic process remains a researcher-driven thinking and feeling process, even with the supplementation of a computer program. This is a very important point. Though users of computer programs sometimes rigidify the analytic process, this need not be. The evolving analysis should determine how the researcher will use the computer program and not the reverse. There is no reason to restrict analysis to the limits of a program's capabilities. Computer programs are tools, like the many other analytic tools presented in this book. They can enhance the ability of the researcher to search for, store, sort, and retrieve materials. They help a researcher keep track of his or her codes, provide easy access to memos, and facilitate the making of diagrams. Furthermore, the researcher need not be committed to an analytic scheme too early in the analytic process because computer programs allow the researcher to move materials around and think about them in other ways. Everything is at the analyst's fingertips. There is no more rummaging through boxes or notebooks looking for that important memo. Finally, computer programs provide for transparency of the research process. The researcher can retrace the analytical process, an option that didn't exist twenty years ago. For researchers who are interested in "reliability" and being able to provide an "audit trail," the ability to retrace the analytic process makes it easier both during and at the end of analysis to evaluate the research process. Always keep in mind that findings are only as good as the work that the researcher is willing to put into the analysis. The researcher has to think and feel his or her way through the process. The computer program is an option, a tool, one meant to facilitate and not distract from the analytic process. Computer programs are not integral to this method or necessary for doing the exercises included in the book, but the option is now there.

The computer program utilized in this book is MAXQDA (Kuckartz, 1988/2007). This author does not advocate the use of one computer program over another and acknowledges that there are many excellent programs out there, including N-vivo, Atlas.ti, and Ethnograph, among others. While I happen to use MAXQDA, I use it because it does in a very clear and well-organized way what I want a computer program to do and it is relatively easy to learn and use. And with my nontechnological mind I am able to understand it. In certain places in the book there are details of how to use this software and what it would look like in specific phases and steps of the analysis. Moreover, the data and the analysis presented in this book are prepared as a MAXQDA project, which is provided as a free download from the Sage Web site at www.sagepub.com/corbinstudysite or alternatively from the MAXQDA Web site at www.maxqda.com/Corbin-BasicsQR.

Thus, with the software, readers will have the opportunity to work "live" with the data, do additional coding, add codes, write your own memos, and so on. Readers may download a free demo version of the MAXQDA software together with the project, which is named "JC-BasicsQR.mx3." There is also a step-by-step tutorial available that introduces you, in a clear and easy way, to the basic functions of the program. Moreover, you will find detailed information about how to handle the project.

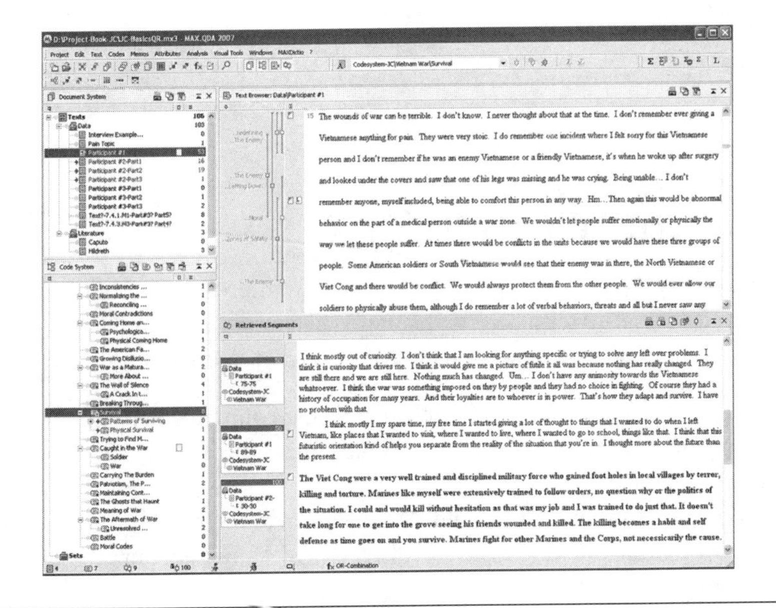


**Screenshot 0**

The pictures shows the project "JC-BasicsQR.mx3," which contains all memos and the interview data of this book. The project can be downloaded at www.maxqda.com/; there you will also find all necessary information to work with the project. The screenshot displays the workspace of MAXQDA 2007: The four-window main screen is structured along the four major areas of qualitative data analysis: The Data Set (window: "Document System"), the codes/categories (window: "Code System"), the results of Retrievals (window: "Retrieved Segments") and the text work space, where codes are assigned, Memos are written and attached (window: "Text Browser"). Most of the management options are integrated into the four windows as a context menu, accessible via the right mouse button. The basic selection principle of MAXQDA is the action of activating, which allows completely free, one-click selections of any number and combination of codes and texts. In the screenshot you see that the one text of Participant #1 and the three texts of Participant #2 are activated and the code "Survival" together with its subcodes. This selection means that all coded segments of the activated texts that have been assigned with the activated codes are displayed in the "Retrieved Segments" window. The currently opened text is the one of Participant #1, displayed in the Text Browser window. All assigned codes and it exact position is displayed in the code margin by the colored code stripes (colors are freely chosen by the researcher). Memos are created and displayed in the margin beside it, they can be opened by double clicking a memo icon.

# Acknowledgments

I want to thank my husband Dick for all his computer help and support during the writing process. He certainly made up for my deficits in the use of computers.

I also want to thank my friend Anne Kuckartz for her support and help while writing this book. Her interest in the project and encouragement kept me going.

I am indebted to following reviewers of this text for all their helpful comments: T. Gregory Barrett, University of Arkansas at Little Rock; J. Randy McGinnis, University of Maryland, College Park; J. Randall Koetting, University of Nevada, Reno; Anthony N. Maluccio, University of Connecticut and Boston College; and Kathleen Slobin, North Dakota State University. Their input helped to make this a better book.

Finally, I want to thank the Vietnam veterans who participated in this research both through the interviews and through their written memoirs. They fought for the freedoms that we living in a democratic society enjoy and so take for granted.

# Preface

> *Also, at my intellectual core perhaps is the sense that—however naïve you think this—the world of social phenomena is bafflingly complex. Complexity has fascinated and puzzled me much of my life. How to unravel some of that complexity, to order it, not to be dismayed or defeated by it? How not to avoid the complexity nor distort interpretation of it by oversimplifying it out of existence? This is of course, an old problem: Abstraction (theory) inevitably simplifies, yet to comprehend deeply, to order, some degree of abstraction is necessary. How to keep a balance between distortion and conceptualization?* (Strauss, 1993, p. 12)

Whenever an author is asked to write a revision of a text there are always those persons, including this author, who say, "Is another revision necessary? Wasn't everything said in past editions?" I thought so, yet when I looked at the 2nd edition of this book I realized how much both the field of qualitative research and I had changed since its publication.

I grew up intellectually in the Age of the Dinosaurs, or so it seems when I read the literature pertaining to qualitative research today. I carried within me the values, beliefs, attitudes, and knowledge of my profession and the times. I believed what I was told and wrote about it. But one day I looked about and found that I had been labeled a "post-positivist" (Denzin, 1994). "Oh dear," I thought, "I've been classified and labeled just like we do in qualitative research!" It seems that while I was going about business as usual, a Qualitative Revolution was taking place. As part of that revolution the word "interpretation," the *byword* of qualitative research in the old days, became passé. The new qualitative jargon centered on letting our respondents talk for themselves. Also, it was now considered okay to "go native," a dreaded accusation in the "old days." It gets worse. I knew my research world, like that of Humpty Dumpty, had tumbled down when the notion of "objectivity" was dismissed as impossible to achieve. Instead of being the "objective researcher," the postmodern movement put the researcher right into the center of the study. But the final assault on my research identity came when the notion of being able to capture "reality" in data was deemed a fantasy. All is relative. There are "multiple perspectives." The postmodern era had arrived. Everything was being "deconstructed" and re-"constructed"

It's safe to assume that I was just a little exasperated and concerned as I heard about these new ideas. I feared that researchers would become so concerned with "examining their own navels" and "telling nice stories" that they would lose sight of the purpose of doing research (at least from my perspective) and that is to *generate a professional body of empirical knowledge*. Most of all, I feared that qualitative methods would lose whatever credibility they had accrued within the "scientific world." However, the more I thought about it, the more I realized that there were some valid points being made by the "postmodern," "deconstructionist," and "constructionist" schools of thought. With my original research "bubble" burst, I wondered what was left. I have to add to this "confession" that during these past years I was doing a lot of teaching in various countries on how to do analysis, and the interactions with students also helped shaped my new understanding of qualitative research.

It wasn't until I was asked to write the 3rd edition of *Basics* that I started to think about putting my thoughts together. As I drafted an outline for the book, I was confronted with a series of questions. Questions such as: What are methods? Are they merely sets of procedures? Or are they philosophical approaches with few, if any, procedures? What role do procedures play in research? Are they guides, or just a broad set of ideas? What and how much structure is necessary to give students? And what is the role of the researcher? How can the researcher be acknowledged while still telling the story of participants? How much or how little interpretation should be involved?

Part of the challenge I faced in writing this new edition was determining who I was as a researcher. I was trained as a grounded theorist. At the time of my training, supposedly there was one "grounded theory" approach, though this point is open to debate. Throughout the years, what was initially grounded theory has evolved into many different approaches to building theory grounded in data. Each evolution has been an attempt to modernize or to extend the original method, bringing it more in line with contemporary thought. Yet, I also wanted to hold on to the methodological vision of Anselm Strauss, now deceased, who continued to believe until the end of his life in the value of theory and its importance to the development of any professional body of knowledge. Complicating this last point was the fact I no longer believed that theory construction is the only way to develop new knowledge.

Thick and rich description, case analysis, bringing about change in a difficult situation, and telling a story are all valid reasons for doing research. Each form of research is powerful in its own way. I came to realize that in order to remain true to Strauss's vision, yet still hold on to my own beliefs, I would have to find a way in this book of accommodating other research goals in addition to theory building. Then, too, there is the whole issue of complexity as pointed out in the quotation by Strauss introducing this Preface. Since complexity was so important to Strauss, there is no doubt that the method presented in this book would have to provide a way of capturing some of that complexity. In other words, I would have to find a way of blending art with science and interpretation with complex storytelling—qualities that certainly characterized Strauss's writing. Strauss was, to those who knew him well, the master storyteller, though no one can deny the scientific contribution of his work.

Needless to say, with all things considered, I wondered if I could live up to the challenge of writing the 3rd edition, and I felt rather daunted when I first sat down to write. I procrastinated, wrote and rewrote as one does when trying out ideas. But once I got into the "groove" of writing I found myself enjoying the process. I discovered that I wasn't delineating a whole new method. I was modernizing the method I had grown up with, dropping a lot of the dogma, flexing some of the procedures, and even thinking about how computers might enhance the research process.

In this 3rd edition of *Basics*, I try to keep Anselm's vision in mind as I write. My aim is not to recreate his approach to analysis, but to combine what was good about the old editions of *Basics* with some aspects of contemporary thinking. I don't wish to be labeled as a "this" or a "that" because once labels are applied they tend to stick. Labels don't take into account that times change, the state of knowledge changes, and, most of all, people change along with these.

This book is based upon the belief that though there are multiple interpretations that can be constructed from one set of data (I've done this myself), generating concepts is a useful research endeavor. It is useful for two reasons. First, it increases understanding of persons in their every day lives—their routines, habits, problems, and issues—and how they handle or resolve these. Second, concepts provide a language that can be used for discussion and debate leading to the development of shared understandings and meanings. The understandings can then be used to build a professional body of knowledge and enhance practice.

*The Basics of Qualitative Research*, Third Edition, is not a recipe for doing qualitative research and I would be offended if it is viewed as such. Rather, it presents a set of analytic techniques that can be used to make sense out of masses of qualitative data. Researchers are encouraged to use the procedures in their own way. There is one thing about which I feel strongly, however. Researchers should be very clear at the beginning of a study what it is they are setting out to do. If the goal is to do description, then fine, do so. I just want researchers to do "quality" description, and using this book should help them do so. However, if the goal is to develop theory, the findings should be integrated to form an overarching theoretical explanatory scheme. Too often persons do description and call it theory, leaving the reader confused about what is theory and what is not.

How a person does qualitative analysis is not something that can be dictated. Doing qualitative research is something that a researcher has to feel him- or herself through. A book only provides some ideas and techniques. It is up to the individual to make use of procedures in ways that best suit him or her.

In the first part of the book readers will notice the use of the pronoun "we." In the second half of the book, the pronoun changes to "I." Please don't be confused. There is a reason for this. The first part of the book, which includes all of the methodological procedural chapters, is based on materials that Anselm Strauss and I worked on together—many of which were published in previous editions of this text. The second half of the book is devoted to demonstrating how to do analysis using materials from the Vietnam War. These are new chapters and I take full responsibility for them. Though Anselm Strauss has been dead for some years, he remains a strong part of this book. In the fifteen years we worked together, it became difficult to separate my views from his. Times have changed and so have I, yet the words contained in this book are grounded in all that he taught me. I hope that in this 3rd edition I am true to him as well as true to the person I have become. For Anselm, the techniques and procedures were more than just a way of doing research. They were his way of learning about life.

In this 3rd edition, there are some new features. First, the book is a more open, analytically, reflecting changes that have occurred in myself. Second, the first chapter begins by explicating the theoretical foundation underlying the approach to research presented in this book. Though this chapter was written several years before Anselm's death and was meant to be part of the 2nd edition, when it came time for publication, the theoretical material in the chapter was deleted by the editor. It was thought that perhaps it was too theoretical for a basics book. This time, the chapter is being included. Third, the book is not limited just to persons who want to build theory. Theory construction is a long process made up of many analytic steps. Persons using this book can do quality research without going on to the final step of theory building as long as they make it clear that they are not out to build theory. There is a chapter in this book devoted to theory construction, but many of the analysis chapters are designed to be useful to researchers who are interested in thick and rich description, concept analysis, or simply pulling out themes. Fourth, in this edition, rather than just talking about analysis, I am actually doing analysis—taking the reader through the steps from concept identification to theoretical development. Fifth, and absent from previous editions, there are exercises at the end of each chapter to reinforce learning. Sixth, explanations of how to integrate computer programs into analysis are included.

In every seminar that I've taught over the past years, there is the inevitable question about the use of computer programs for qualitative analysis. Though the use of computer programs in qualitative research is debatable and outright rejected by some researchers, computer programs for analysis are here to stay and their ability to support the research process increases with each improvement in the many programs that are available. Notice that I say "support" and not "take over" or "direct" the research process. I think one of the most interesting aspects of this edition is that it demonstrates that the analytic process remains a researcher-driven thinking and feeling process, even with the supplementation of a computer program. This is a very important point. Though users of computer programs sometimes rigidify the analytic process, this need not be. The evolving analysis should determine how the researcher will use the computer program and not the reverse. There is no reason to restrict analysis to the limits of a program's capabilities. Computer programs are tools, like the many other analytic tools presented in this book. They can enhance the ability of the researcher to search for, store, sort, and retrieve materials. They help a researcher keep track of his or her codes, provide easy access to memos, and facilitate the making of diagrams. Furthermore, the researcher need not be committed to an analytic scheme too early in the analytic process because computer programs allow the researcher to move materials around and think about them in other ways. Everything is at the analyst's fingertips. There is no more rummaging through boxes or notebooks looking for that important memo. Finally, computer programs provide for transparency of the research process. The researcher can retrace the analytical process, an option that didn't exist twenty years ago. For researchers who are interested in "reliability" and being able to provide an "audit trail," the ability to retrace the analytic process makes it easier both during and at the end of analysis to evaluate the research process. Always keep in mind that findings are only as good as the work that the researcher is willing to put into the analysis. The researcher has to think and feel his or her way through the process. The computer program is an option, a tool, one meant to facilitate and not distract from the analytic process. Computer programs are not integral to this method or necessary for doing the exercises included in the book, but the option is now there.

The computer program utilized in this book is MAXQDA (Kuckartz, 1988/2007). This author does not advocate the use of one computer program over another and acknowledges that there are many excellent programs out there, including N-vivo, Atlas.ti, and Ethnograph, among others. While I happen to use MAXQDA, I use it because it does in a very clear and well-organized way what I want a computer program to do and it is relatively easy to learn and use. And with my nontechnological mind I am able to understand it. In certain places in the book there are details of how to use this software and what it would look like in specific phases and steps of the analysis. Moreover, the data and the analysis presented in this book are prepared as a MAXQDA project, which is provided as a free download from the Sage Web site at www.sagepub.com/corbinstudysite or alternatively from the MAXQDA Web site at www.maxqda.com/Corbin-BasicsQR.

Thus, with the software, readers will have the opportunity to work "live" with the data, do additional coding, add codes, write your own memos, and so on. Readers may download a free demo version of the MAXQDA software together with the project, which is named "JC-BasicsQR.mx3." There is also a step-by-step tutorial available that introduces you, in a clear and easy way, to the basic functions of the program. Moreover, you will find detailed information about how to handle the project.

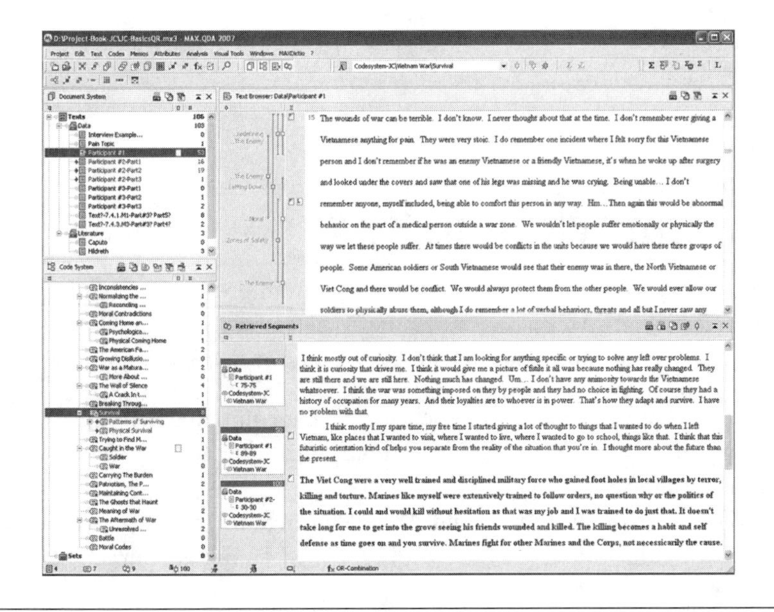


**Screenshot 0** The pictures shows the project "JC-BasicsQR.mx3," which contains all memos and the interview data of this book. The project can be downloaded at www.maxqda.com/; there you will also find all necessary information to work with the project. The screenshot displays the workspace of MAXQDA 2007: The four-window main screen is structured along the four major areas of qualitative data analysis: The Data Set (window: "Document System"), the codes/categories (window: "Code System"), the results of Retrievals (window: "Retrieved Segments") and the text work space, where codes are assigned, Memos are written and attached (window: "Text Browser"). Most of the management options are integrated into the four windows as a context menu, accessible via the right mouse button. The basic selection principle of MAXQDA is the action of activating, which allows completely free, one-click selections of any number and combination of codes and texts. In the screenshot you see that the one text of Participant #1 and the three texts of Participant #2 are activated and the code "Survival" together with its subcodes. This selection means that all coded segments of the activated texts that have been assigned with the activated codes are displayed in the "Retrieved Segments" window. The currently opened text is the one of Participant #1, displayed in the Text Browser window. All assigned codes and it exact position is displayed in the code margin by the colored code stripes (colors are freely chosen by the researcher). Memos are created and displayed in the margin beside it, they can be opened by double clicking a memo icon.

# Acknowledgments

I want to thank my husband Dick for all his computer help and support during the writing process. He certainly made up for my deficits in the use of computers.

I also want to thank my friend Anne Kuckartz for her support and help while writing this book. Her interest in the project and encouragement kept me going.

I am indebted to following reviewers of this text for all their helpful comments: T. Gregory Barrett, University of Arkansas at Little Rock; J. Randy McGinnis, University of Maryland, College Park; J. Randall Koetting, University of Nevada, Reno; Anthony N. Maluccio, University of Connecticut and Boston College; and Kathleen Slobin, North Dakota State University. Their input helped to make this a better book.

Finally, I want to thank the Vietnam veterans who participated in this research both through the interviews and through their written memoirs. They fought for the freedoms that we living in a democratic society enjoy and so take for granted.

# Overview of the Contents

Chapter 1 introduces the reader to this book and presents the philosophical belief underlying the Corbin/Strauss approach to analysis. Chapter 2 discusses practical considerations when doing qualitative research. This is similar to a chapter in past editions. Chapter 3, titled "Prelude to Analysis," is a new chapter. It explains what is meant by analysis. Chapter 4 combines several chapters from the previous edition. It presents a series of analytic procedures and techniques that can be used to analyze data. Chapter 5 uses familiar materials on context and process, but includes them early in the text and presents them as additional analytic tools. It also includes a section on integration. Chapter 6 then focuses on memos and diagrams. What is different about this chapter in the present edition is its placement early in the text. Chapter 7 is about theoretical sampling, again an earlier placement of this chapter. Chapter 8 is the first chapter in a series of five new chapters that demonstrate the "doing of analysis" and uses materials from the Vietnam War as the research project. The focus of Chapter 8 is on concept identification. Chapter 9 moves on to concept elaboration. Chapter 10 is about contextualizing data. Chapter 11 brings process into the analysis. Chapter 12 is devoted to integration and theory development. Chapter 13 offers slight revisions of an earlier chapter on writing theses, monographs, and giving talks about research. Chapter 14 is about evaluating qualitative research and has some new components to bring it more in line with contemporary thinking. Finally, Chapter 15 is on student questions and answers to these questions—a chapter that remains popular from previous editions

This book remains a joy to write and a testament to my teacher and mentor, Anselm Strauss. I hope that through it we can inspire a new generation of qualitative researchers in both the art and science of doing qualitative research analysis.

*If the artist does not perfect a new vision in his process of doing, he acts mechanically and repeats some old model fixed like a blueprint in his mind.*

—John Dewey, *Art as Experience,* 1934, p. 50

# 1

# Introduction

*If what is designated by such terms as doubt, belief, idea, conception, is to have any objective meaning, to say nothing of public verifiability, it must be located and described as behavior in which organism and environment act together, or* inter-*act. (Dewey, 1938, p. 32)*

**Table 1.1** Definition of Terms

*Grounded Theory*: A specific methodology developed by Glaser and Strauss (1967) for the purpose of building theory from data. In this book the term grounded theory is used in a more generic sense to denote theoretical constructs derived from qualitative analysis of data.

*Methodology*: A way of thinking about and studying social phenomena.

*Methods:* Techniques and procedures for gathering and analyzing data.

*Philosophical Orientation*: A worldview that underlies and informs methodology and methods.

*Qualitative Analysis*: A process of examining and interpreting data in order to elicit meaning, gain understanding, and develop empirical knowledge.

## Dewey and Mead: Pragmatist Philosophy of Knowledge

Every **methodology** rests on the nature of knowledge and of knowing, and so does ours. We don't mean to frighten away the novice researcher by placing an abstract discussion of the philosophy of knowledge at the beginning of the book. Rest assured, the remainder of the book is much more concrete. Our purpose here is to lay the groundwork for the methodology that will follow and provide insight for why, methodologically, we do the things that we do.

This methodology's epistemology has come to it in a two-step evolution, involving both the tradition of Chicago Interactionism and the philosophy of Pragmatism inherited largely from John Dewey and George Mead (Fisher & Strauss, 1978, 1979a, 1979b; Strauss, 1991). When later Interactionists trained in that particular tradition write about or even touch on epistemological matters, they draw at least implicitly on these sources (see for example: Baszanger, 1998; Becker, 1982, 1986a; Blumer, 1969; Charmaz, 1983, 1991; Clarke, 1991, 2005; Davis, 1991; Denzin, 1989; Fagerhaugh & Strauss, 1977; Fisher, 1991; Fujimora, 1987, 1991; Gerson, 1991; Hughes, 1971; Lofland, 1991; Schatzman, 1991; Shibutani, 1991; Star, 1989; Suczek & Fagerhaugh, 1991; Wiener, 1991).

For those readers who might not be familiar with Interactionism or Pragmatism we present the following two quotes, which though not very inclusive do bring out very important elements of the two philosophies. According to Blumer (1969), "symbolic interaction" refers to a particular form of interaction that occurs between persons. He says:

> The peculiarity consists in the fact that human beings interpret or "define" each other's actions instead of merely reacting to each other's actions. Their "response" is not made directly to the actions of one another but instead is based on the meaning which they attach to such actions. (p.19)

Mead (1956) tells us something about pragmatism and its origins. He says:

> It is out of interest in the act itself and the relationship of thought to the act itself that the last phase of more recent philosophy dealt with above, that is, pragmatism, arises. Out of the type of psychology which you may call "behavioristic" came a large part of the stimulus for a pragmatic philosophy. There are several sources, of course; but that is one of the principal ones. (p. 404)

The influential Pragmatist writings were published mainly in the first three decades of the twentieth century (Dewey, 1917, 1922, 1929, 1938; Mead, 1917, 1956, 1959). The writings present an innovative philosophy of knowledge, easily recognizable as the framework for our own methodology. Both Dewey and Mead assume, for instance, that knowledge is created through action and interaction. Dewey (1929) states, "ideas are not statements of what is or has been but of acts to be performed" (p. 138). Or, more properly speaking, knowledge arises through (note the verbs) acting and interacting of self-reflective beings. Typically the activity is precipitated by a problematic situation, where one can't just act automatically or habitually. According to Dewey (1929), "All reflective inquiry starts from a problematic situation, and no such situation can be settled in its own terms" (p. 189). And Mead (1938) states, "Reflective thinking arises in testing the means which are presented for carrying out some hypothetical way of continuing an action which has been checked" (p. 79). The issue before the actor is the resolution of a problem. Its answer is uncertain, and judgment of it can be made only in terms of further action (consequences) directed by the provisional answer. According to Dewey (1929), "The test of ideas, of thinking generally, is found in the consequences of the acts to which the ideas lead, that is in the new arrangement of things which are brought into existence" (p. 136).

The activity of thinking, even at its quickest and most spontaneous, has temporal aspects. The envisioned end of the action affects whatever action is actually taken—and often it is altered in midstream as the actor reassesses its effectiveness. Past memories and recollections also enter directly or indirectly into the actions. Dewey (1929) says, "reflective thought, thinking that involves inference and judgment, is not originative. It has its test in antecedent reality as that is disclosed in some non-reflective immediate knowledge" (p. 109). Because of this temporality of action—in a world of contingencies—the Pragmatists were also concerned with processes. "Because we live in a world in process, the future, although continuous with the past is not its bare repetition" (Dewey, 1929, p. 40).

Furthermore, the Pragmatists did not subscribe to a then popular duality of person and group (or collectivity). So, even if it is single person, rather than a team or an organization, who discovers or creates some new understanding of reality, he or she does this only because already socialized to the perspectives that have been inherited. "Neither inquiry nor the most abstractly formal set of symbols can escape from the cultural matrix in which they live, move and have their being" (Dewey, 1938, p. 20). So the Pragmatists believed in the accumulation of collective knowledge. (Though this point may seem obvious to us today, there are still philosophers of knowledge who give unquestioned primacy to the individual knower.)

In this assumption, the Pragmatists were looking closely at natural science (mainly biology) as a model. They believed that new knowledge was also provisional until checked out empirically by peers. A summarizing paragraph from Mead's (1917) "Scientific Method and the Individual Thinker" ends in this way:

> In both of these processes, that of determining the structure of experience which will test by experiment [that is, controlled inquiry] the legitimacy of any new hypothesis, and that of formulating the problem and the hypothesis for its

> solution, the individual functions in his full particularity, and yet in organic relationship with the society that is responsible for him. (p. 227)

The experiences of whoever is engaged in an inquiry are vital to the inquiry and its implicated thought processes. Dewey (1929) states, "Insofar, we have the earnest of a possibility of human experience, in all its phases, in which ideas and meanings will be prized and will be continuously generated and used. But they will be integral with the course of the experience itself, not imported from the external source of a reality beyond" (p. 138).

There is still the question of what is called "validity," or what philosophers call "truth," and the Pragmatists were centrally interested in that passionately contested question. The answer for them lay in the consequences. "A definition of the nature of ideas in terms of operations to be performed and the test of the validity of the idea by the consequences of these operations establishes connectivity within concrete experience" (Dewey, 1929, p. 114). They are careful to emphasize that acts of knowing embody perspectives. Thus, what is discovered about "reality" cannot be divorced from the operative perspective of the knower, which enters silently into his or her search for, and ultimate conclusions about, some event. This Pragmatist position does not at all lead to radical relativism (as currently in one version of postmodernism). Radical relativism reasons that since no version or interpretation can be proven, therefore no certainty about any given one can be assumed. Instead the Pragmatists, like any practicing scientist in their day or ours, must make a couple of key assumptions. One is that truth is equivalent to "for the time being this is what we know—but eventually it may be judged partly or even wholly wrong." Another assumption is that despite that qualification, the accumulation of knowledge is no mirage. The world is not flat nor the Milky Way the center of the universe; neither is the discovery of electricity and all its theoretical and practical implications to be disregarded. True, some people may not believe in evolution or in a spherical earth, but generally those matters are part of accumulated belief. Perhaps knowledge about societies and human activity is less cumulative, but the Pragmatists and undoubtedly most social scientists believe some social knowledge certainly is cumulative and provides the basis for the evolution of thought and society.

One last Pragmatist position vis-à-vis the philosophy of knowledge is that knowledge can be useful for practice or practical affairs. They saw no necessity for assuming a yawning gap—another false dualism—between knowledge and everyday action. Dewey (1929) says, "Our discussion has for the most part turned upon an analysis of knowledge. The theme, however, is the relation of knowledge and action; the final import of the conclusions as to knowledge resides in the changed idea it enforces into action" (p. 245). Indeed, they (knowledge and action) both feed into each other. Knowledge leads to useful action, and action sets problems to be thought about, resolved, and thus is converted into new knowledge. In a continuously changing world, generating one contingency after another, this interplay of practice and inquiry is also continual. (This is the philosophical equivalent to enunciating the scientist's interplay between data and theory.) This is why the Pragmatists drew no hard and fast line between everyday ("commonsense") thinking and the more systematically controlled scientific types. They didn't address—but would certainly have applauded—views of organizational action, for instance, much of which is hardly haphazard but carefully planned and evaluated. Yet the Pragmatists were not just interested in practice or the practical. They addressed matters like those of aesthetics and ethics, of language and meaning, and others equally abstract. Mead (1938) states, "pragmatism holds no brief against aesthetic experience. It is an activity to be acknowledged like all other human activities, and like these faces its own problems, those of appreciation, and solves them by reflection" (p. 98).

Some of these assumptions, but probably not all, about knowledge and the world are shared by other methodologies of social research. However, it is important to know that still others are (and have been) supported by quite different assumptions. Thus, methodologies may firmly rule out personal experience from inquiry in the name of "objectivity." Then, too, they may undervalue, from our standpoint, the importance of self-reflection both in its relation to what reality "is" and to its role in "knowing" it. As is apparent to you, action and interaction are crucial to Pragmatists' and our own conceptions of the world and knowledge.

## Ontology: Assumptions About the World

Probably most researchers who use our methodology (and certainly those who use only its procedures) have not reflected upon the assumptions that underlie the method presented in this book. Perhaps they assume that methodology evolves strictly from practice. Though it does to some degree, it is also considerably influenced by worldview, or the beliefs and attitudes about the world we live in.

What is the nature of this world that we wish to study? Here is a quotation that expresses our version of this world:

> We are confronting a universe marked by tremendous fluidity; it won't and can't stand still. It is a universe where fragmentation, splintering, and disappearance are the mirror images of appearance, emergence, and coalescence. This is a universe where nothing is strictly determined. Its phenomena should be partly determinable via naturalistic analysis, including the phenomenon of

men [and women] participating in the construction of the structures which shape their lives. (Strauss, 1993, p. 19)

What kind of theory might fit the nature of the universe assumed in the quotation above? A world that is complex, often ambiguous, evincing change as well as periods of permanence; where action itself although routine today may be problematic tomorrow; where answers become questionable and questions ultimately produce answers.

Where we are leading is to a set of assumptions, which in turn lie behind the method and the research strategies explicated in this book. For instance, our assumptions about the inevitability of contingencies, the significance of process, and the complexity of phenomena direct us to examine problematic as well as routine situations and events. Important to us are the great varieties of human action, interaction, and emotional responses that people have to the events and problems they encounter. The nature of human responses creates conditions that impact upon, restrict, limit, and contribute toward restructuring the variety of action/interaction that can be noted in societies. In turn, humans also shape their institutions; they create and change the world around them through action/interaction.

Next we shall present some working axioms that lie quite specifically behind our conception of methodology. Most of them rest on the Pragmatist and Interactionist philosophies presented above. As readers become more familiar with this book they should easily grasp the relevance of the assumptions to the version of the methodology. We reproduce some of the assumptions here.[1]

**Assumption 1.** The external world is a symbolic representation, a "symbolic universe." Both this and the interior worlds are created and recreated through interaction. In effect, there is no divide between external or interior world (Blumer, 1969).

**Assumption 2.** Meanings (symbols) are aspects of interaction, and are related to others within systems of meanings (symbols). Interactions generate new meanings . . . as well as alter and maintain old ones (Mead, 1934).

**Assumption 3.** Actions are embedded in interactions—past, present and imagined future. Thus, actions also carry meanings and are locatable within systems of meanings. Actions may generate further meanings, both with regard to further actions and the interactions in which they are embedded (Mead, 1934).

**Assumption 4.** Contingencies are likely to arise during a course of action. These can bring about change in its duration, pace, and even intent, which may alter the structure and process of interaction (Dewey, 1929).

**Assumption 5.** Actions are accompanied by temporality, for they constitute courses of action of varying duration. Various actors' interpretations of the temporal aspects of an action may differ according to the actors' respective perspectives; these interpretations may also change as the action proceeds (Mead, 1959).

**Assumption 6.** Courses of interaction arise out of shared perspectives, and when not shared, if action/interaction is to proceed, perspectives must be negotiated (Blumer, 1969).

**Assumption 7.** During early childhood and continuing all through life, humans develop selves that enter into virtually all their actions and in a variety of ways (Mead, 1959).

**Assumption 8.** Actions (overt and covert) may be preceded, accompanied, and/or succeeded by reflexive interactions (feeding back onto each other). These actions may be one's own or those of other actors. Especially important is that in many actions the future is included in the actions (Dewey, 1929).

**Assumption 9.** Interactions may be followed by reviews of actions, one's own and those of others, as well as projections of future ones. The reviews and evaluations made along the action/interaction course may affect a partial or even complete recasting of it (Dewey, 1929).

**Assumption 10.** Actions are not necessarily rational. Many are nonrational or, in common parlance, "irrational." Yet rational actions can be mistakenly perceived as not so by other actors (Dewey, 1929).

**Assumption 11.** Action has emotional aspects: To conceive of emotion as distinguishable from action, as entities accompanying action, is to reify those aspects of action. For us, there is no dualism. One can't separate emotion from action; they are part of the same flow of events, one leading into the other (Dewey, 1929).

**Assumption 12.** Means-ends analytic schemes are usually not appropriate to understanding action and interaction. These commonsense and unexamined social science schemes are much too simple for interpreting human conduct (Strauss, 1993).

**Assumption 13.** The embeddedness in interaction of an action implies an intersection of actions. The intersection entails possible, or even probable, differences among the perspectives of actors (Strauss, 1993).

**Assumption 14.** The several or many participants in an interactional course necessitate the "alignment" (or articulation) of their respective actions (Blumer, 1969).

**Assumption 15.** A major set of conditions for actors' perspectives, and thus their interactions, is their memberships in social worlds and subworlds. In contemporary societies, these memberships are often complex, overlapping, contrasting, conflicting, and not always apparent to other interactants (Strauss, 1993).

**Assumption 16.** A useful fundamental distinction between classes or interactions is between the routine and the problematic. Problematic interactions involve "thought," or when more than one interactant is involved then also "discussion." An important aspect of problematic action can also be "debate"—disagreement over issues or their resolution. That is, an arena has been formed that will affect the future course of action (Dewey, 1929; Strauss, 1993).

### Methodological Implications

The methodological implications of the above can be summarized as follows. The world is very complex. There are no simple explanations for things. Rather, events are the result of multiple factors coming together and interacting in complex and often unanticipated ways. Therefore any methodology that attempts to understand experience and explain situations will have to be complex. We believe that it is important to capture as much of this complexity in our research as possible, at the same time knowing that capturing it all is virtually impossible. We try to obtain multiple perspectives on events and build variation into our analytic schemes. We realize that, to understand experience, that experience must be located within and can't be divorced from the larger events in a social, political, cultural, racial, gender-related, informational, and technological framework and therefore these are essential aspects of our analyses.

Process is integral to our studies because we know that experience, and therefore any action/interaction that follows, is likely to be formed and transformed as a response to consequence and contingency. We don't necessarily want to reduce understanding of action/interaction/emotion to one explanation or theoretical scheme; however, we do believe that concepts of various levels of abstraction form the basis of analysis. Concepts provide ways of talking about and arriving at shared understandings among professionals. If you don't have a language, you can't talk—and if you can't talk, you can't do, and the basis of many professions is still doing (Blumer, 1969).

## Impact of Recent Trends on This Methodology

Though the above reflects the thinking of both Strauss and Corbin, what is about to be written now reflects more of the thinking of Corbin. Therefore, in this next section, the first person pronouns will be used. Though the above philosophical and epistemological assumptions laid the foundation for this methodology, there is no doubt that contemporary thought has had some influence on my thinking about methodology. Much of what has been written in recent years has given me invaluable insight, shown me the error of some of my past ways of doing, and has made me wonder at times how I could have been so "misinformed." But then that is the nature of knowledge. It does progress and change with time and so does methodology. Some researchers have simply walked away from the more traditional approaches to doing qualitative research, while others, like me, have tried to hold on to what is good about the past while updating it to bring it more in line with the present. Like most persons, I have chosen parts of both past and present and rejected others from this smorgasbord of ideas, based upon who and what I am.

There is not doubt that I, Corbin, have been influenced to some degree by the writings of contemporary feminists, constructionists, and postmodernists. I especially admire the works of both Clarke (2005) and Charmaz (2006) and how they have applied postmodernist and postconstructivist paradigms to **grounded theory** methodology, thus taking up the challenge of Denzin (1994, p. 512) to move interpretative **methods** more deeply into the regions of postmodern sensibility. In this section, I want to explain how my approach to analysis has been affected by recent directions in qualitative research, while still retaining most of Strauss's basic approach to doing analysis.

Readers of this text should remember that this is a basic book about analysis. It's an attempt to take an extremely complicated process and make it understandable to beginning qualitative researchers. This author knows full well that something occurs when doing analysis that is beyond the ability of a person to articulate or explain. She agrees with Denzin (1998) when he says, "Interpretation is an art that cannot be formalized" (p. 338). Yet, without some formalization of method, how would one teach it in a text? I would not say that this new edition is as much oriented toward formalizing method as it is teaching persons how to think more self-consciously and systematically about data; that is why I have included the demonstration chapters on analysis that follow later in this book. I acknowledge that I do not know it all and this approach to analysis will not solve every methodological problem or respond to every contemporary philosophical argument. I believe that method evolves to handle the methodological problems that we as researchers face out in the field. The changes that have occurred in me will be noticeable to persons using this text, in the language that I use and in how procedures are implemented, especially in the chapters where I demonstrate analysis.

I have no simple term to classify the person I've become methodologically, or a simple term to describe the method presented here. Strauss and I and this book are a mixture of many things. As Denzin (1998) says so well when talking about research and qualitative researchers today, "Clearly simplistic classifications do not work. Any given qualitative researcher-as-bricoleur can be more than one thing at the same time, can be fitted into both the tender- and the tough-minded categories" (p. 338). I would say that this describes me—tender and tough minded. To be more specific, here are some of the changes I've undergone.

I realize there is no one "reality" out there waiting to be discovered (Geertz, 1973); however, I do believe there are external events, such as a full moon, a war, and an airplane crashing into a building. As Schawndt (1998) states, "One can reasonably hold that concepts and ideas are invented (rather than discovered) yet maintain that these inventions correspond to something in the real world" (p. 237). However, it is not the event itself that is the issue in our studies, because each person experiences and gives meaning to events in light of his or her own biography or experiences, according to gender, time and place, cultural, political, religious, and professional backgrounds. To see the validity of this statement, one only has to turn on the television and listen to a group of people discussing an event, such as a president's speech. There is much discourse and sometimes outright conflict about what was said, especially if politics are involved, but rarely a total agreement about the meaning of the event. What a viewer sees and hears are multiple viewpoints on the same topic with no apparent consensus. Add to this picture the notion that what is being seen and heard on the television is filtered through the viewer's interpretation of the event based upon his or her personal history and biography and we get a very complicated picture, one that at best can never be fully understood or reconstructed by the researcher.

I agree with the constructivist viewpoint that concepts and theories are *constructed* by researchers out of stories that are constructed by research participants who are trying to explain and make sense out of their experiences and/or lives, both to the researcher and themselves. Out of these multiple constructions, analysts construct something that they call knowledge. Schawndt (1998) says:

> In a fairly unremarkable sense, we are all constructivists if we believe that the mind is active in the construction of knowledge. Most of us would agree that knowing is not passive—a simple imprinting of sense data on the mind—but active; mind does something with these impressions, at the very least forms abstractions of concepts. In this sense, constructivism means that human beings do not find or discover knowledge so much as construct or make it. We invent concepts, models, and schemes to make sense of experience and, further, we continually test and modify these constructions in light of new experience. (p. 237)

Though I realize that knowledge is constantly evolving in light of new experience, perhaps it is the nurse in me that is talking, but I believe analytic work necessitates some degree of conceptual language to talk about "findings." Without a conceptual language, there is no basis for discussion, conflict, negotiation, or the development of a knowledge based practice. We can't have practitioners walking around doing things without having a disciplined body of knowledge, along with experience, as the basis for their actions. Knowledge may not mirror the world but it does help us to understand it. If you were a patient in the intensive care ward, would you want just anyone coming in off the street to take care of you? Or would you prefer a nurse working from sound theoretical principles, a nurse who understands that no theory should be applied dogmatically, but rather theory should be reevaluated and adjusted to meet the situation at hand.

I am practical in what I want to accomplish with my research. Coming from a nursing background, I want to develop knowledge that will guide practice. In drawing upon my Pragmatist and Interactionist (Hughes, 1971; Park, 1967; Thomas, 1966) theoretical orientations and keeping with the social justice aim of feminist research (Oleson, 1998), I want to bring about social change and make persons' lives better. But you are not likely to find me on a street corner carrying a sign or leading a protest. I recently attended a conference where I made mention of the small demonstration research project that I did as part of this book. One member in the audience responded, "But I don't hear enough outrage in you." Well, I am indeed outraged, and in fact I was very disturbed while doing the mini-study on veterans of the Vietnam War. But I can't let the outrage take over my life. I am of the type that would rather educate with words than take my arguments to the streets. I leave the antiwar "marching" to persons much younger than I. My hope is that in telling these veterans' stories, people will understand the physical, emotional, and moral problems young soldiers face. Maybe their stories will generate social outrage. Maybe if there is sufficient social outrage, there will be no more sending young men and women off to war. Or if war is inevitable, then society should welcome veterans home as heroes and provide the care and support structures that they need to "fit" back into civilian life.

I agree with the feminists in that we don't separate who we are as persons from the research and analysis that we do. Therefore, we must be self-reflective about how we influence the research process and, in turn, how it influences us. Hamberg and Johansson (1999) explain what they did to be self-reflective and I too try to carry this out. They say:

> For this reflexive analysis, we have reread the coded interviews to scrutinize parts featuring tension, contradictions, or conflicting codes—passages that had often been discussed when we were striving to find reasonable and legitimate interpretations. We have also read our memos to recall our instant reactions during, and after, the interviews and our discussions when we compared our coding. (p. 458)

I think that I make it very clear in the memos that I wrote while doing the mini-research project on Vietnam in the later chapters of this book that I tried to keep a record of how the research was affecting me. Actually, I felt a strong need to write my experiences and feelings down because I was often so disturbed. I truly identified with the stories that I was told and read. I was involved. But I also was concerned with my role as investigator and the need to tell my participants' stories. I certainly would never want to exploit my participants and did give them an opportunity to read and give input into chapters that involved them. I told them that the interviews would be used in a methodology book and they agreed to this. I also worried about ethics and made certain that they were agreeable to putting their words into print as some of them were very graphic—but their words tell the story far better than I—a woman who has never been to war—could. What was not feasible for participants and myself was to construct these findings together, though in a way it is a co-construction because it does present their words along with mine.

Though readers of research construct their own interpretations of findings, the fact that these are constructions and reconstructions does not negate the relevance of findings nor the insights that can be gained from them. I believe that we share a common culture out of which common constructions are arrived at through discourse. Concepts give us a basis for discourse and arriving at shared understandings. Therefore, I will continue to believe in the power of concepts and advocate their use.

At the same time, I want to emphasize that techniques and procedures are tools, not directives. No researcher should become so obsessed with following a set of coding procedures that the fluid and dynamic nature of **qualitative analysis** is lost. The analytic process, like any thinking process, should be relaxed, flexible, and driven by insight gained through interaction with data rather than being overly structured and based only on procedures. With all of this reflection behind me, I'm ready to move on to the purpose of this book: teaching students how to do research.

## Why Do Qualitative Research

Why do qualitative research? The most frequently given, and probably the most accurate, response to this question is that the research question should dictate the methodological approach that is used to conduct the research. Other reasons given include: qualitative research allows researchers to get at the inner experience of participants, to determine how meanings are formed through and in culture, and to discover rather than test variables. But I think there are additional reasons why some persons make a career out of doing qualitative research. Committed qualitative researchers tend to frame their research questions in such a way that the only manner in which they can be answered is through qualitative research.

Committed qualitative researchers lean toward qualitative work because they are drawn to the fluid, evolving, and dynamic nature of this approach in contrast to the more rigid and structured format of quantitative methods. Qualitative researchers enjoy serendipity and discovery. Statistics might be interesting, but it is the endless possibilities to learn more about people that qualitative researchers resonate to. It is not distance that qualitative researchers want between themselves and their participants, but the opportunity to connect with them at a human level. Qualitative researchers have a natural curiosity that leads them to study worlds that interest them and that they otherwise might not have access to. Furthermore, qualitative researchers enjoy playing with words, making order out of seeming disorder, and thinking in terms of complex relationships. For them, doing qualitative research is a challenge that brings the whole self into the process. This is not to denigrate quantitative researchers or to imply that they do not share many of the same traits but merely to make the comment that committed qualitative researchers tend to be of a certain type.

Though not specific to qualitative researchers, those who do "good" qualitative research tend to share the following characteristics:

- A humanistic bent
- Curiosity
- Creativity and imagination
- A sense of logic
- The ability to recognize diversity as well as regularity
- A willingness to take risks
- The ability to live with ambiguity
- The ability to work through problems in the field
- An acceptance of the self as a research instrument
- Trust in the self and the ability to see value in the work that is produced

The researchers that these authors have trained tend to really enjoy working with data. They enjoy the mental challenge of working with data. They are unafraid to draw on their own experiences when analyzing materials, having rejected more traditional ideas of "objectivity" and the dangers of using personal experience. Our former students regard their ideas as provisional. Even after publication, they view their work as modifiable and open to negation as new knowledge is accrued. In the work itself, researchers trained by us certainly tend to be flexible, a characteristic enhanced in seminars and occasional team research where they are open to criticism and can enjoy the play of ideas in the give and take of group discussion. For example, consider the following statement:

> I'm part of a writing group that has met about once a month for a couple of years. We pass around work in progress and criticize it, sometimes help with analytic rough spots. Recently an old member of the group returned and described to us her unsuccessful attempt to start a similar group in another location. Participants in her group had followed the same procedures we had, in form, but had gotten very harsh with each other's work and focused more on competitive speeches than genuine collaboration. Our group tried to analyze why we'd been successful, and realized that it had a lot to do with the fact that four of us had been through the grounded theory [seminar]. It isn't just that we shared an analytic focus, though, because in fact we're very different. The striking thing was that we had learned to work together in a collaborative and supportive way. (Leigh Star as cited in Strauss, 1987, pp. 303–304)

Flexibility and openness are linked with having learned to sustain a fair amount of ambiguity. The urge to avoid uncertainty and to get quick closure on one's research is tempered with the realization that phenomena are complex and their meanings not easily fathomed or just taken for granted. Research itself is a process, one that our former students are likely to be self-reflective about. In doing their research, they enjoy the flow of ideas, but not merely the substantive ones since they have learned that theoretical ideas have their own precious value. Yet, they are skeptical of theories, however enticing they seem, unless these are eventually grounded through active interplay with data.

There are two additional important points. The first is that probably most researchers hope that their work also has some relevance for nonacademic audiences. This is because the qualitative researchers take with great seriousness the words and actions of the people studied. Or as poignantly expressed, "I saw that being an intellectual didn't have to be removed from people's lives, that it could be connected directly to where people were in the world and what they thought about it" (B. Fisher as cited in D. Maines, 1991, p. 8). Our second point is that almost inevitably researchers trained in qualitative analysis become completely "absorbed in the work," which though not always "in the foreground [of our lives] is never gone" (A. Clarke, personal communication, March 21, 1990).

That sense of absorption in, and devotion to, the work *process* as such, and in consequence a sense of enhanced integrity, was reflected in a description written by another student. We quote her at length, because her words eloquently emphasize so many of our assertions about the characteristics of qualitative researchers and their work. Trained in public health, she had worked for three or four years on a Sioux Indian reservation, becoming engrossed with the question: What are these people's basic conceptions of health, for their conceptions are so different than ours? Returning to the research seminar after several months in the field, she commented soon after in a memo to the instructor:

> These concerns and fears [that the class would misread her non-Western, cross-cultural data] were systematically and carefully dispelled over the course of the two-hour session. I watched very carefully and listened intently to what people said and how they worked their ideas and images through the data, carefully questioning me when more information was needed and not jumping to conclusions in advance of important additions. The students seemed to search carefully for the richness in the data, picking out critical issues and playing them off against one another for more meaning, noting several possible interpretations to many situations. I was quite overjoyed at the degree of fit between what these analysts were identifying and what I had heard and seen while doing the work. Both the integrity and precision aspects of these sessions were spared by and sustained by the pedagogical style, which is to say (for it cannot be separated from) the formulations of Interactionist epistemology and the conceptual and analytic framework of qualitative research. (K. Jurich as cited in Strauss, 1987, p. 304)

## In Conclusion

> Like Coleridge and Kublai Khan I woke up dreaming, but since it isn't a complete dream but only the germ I thought out the words and here they are. . . . (Anselm Strauss)

Persons choose to do research because they have a dream that somehow they will make a difference in the world through the insights and understandings they arrive at. But it is not enough to dream. Dreams must be brought to fruition. The purpose of this book is to offer qualitative researchers a means of fulfilling their research dreams. It is not a perfect means, and we acknowledge this. However, we do provide some words of wisdom derived from our years of experience as researchers and analysts. We present a few analytic procedures and techniques—ones that we have found useful, and the idea behind them is to provide students with something that they can turn to when they feel overwhelmed with data. We give insight into how we personally analyze data by taking students through a demonstration research project. The authors recognize how difficult it is to have collected masses of qualitative research and not know what to do with that data.

Though we wish we could reach across the world and train everyone who is interested in learning qualitative analysis, we know that this is not possible. Therefore, we have written this book with the hope that we can become "teachers/mentors in absentia." Like all good teachers, our aim is not to provide a recipe book. Rather, our purpose is to build a solid foundation in data analysis, a foundation that will enable students to pursue their careers and reach the research heights that they dream of. Though students often need direction and crave structure when it comes to doing qualitative analysis, there is no one right way of doing things.

Qualitative analysis is many things, but it is not a process that can be rigidly codified. What it requires, above all, is an intuitive sense of what is going on in the data; trust in the self and the research process; and the ability to remain creative, flexible, and true to the data all at the same time. Qualitative analysis is something that researchers have to feel their way through, something that can only be learned by doing. Some persons using this book will be interested in developing a "grounded theory," while other researches will aim for thick and rich description or perhaps just delineating basic themes. Regardless of the research aim, we think this book will prove useful.

With this introduction behind us, we want to take our readers on a journey of knowledge acquisition—a journey that we hope will enlighten, empower, and inspire our readers for years to come.

## Summary of Important Points

There are many reasons for choosing to do qualitative research, but perhaps the most important is the desire to step beyond the known and enter into the world of participants, to see the world from their perspective and in doing so make discoveries that will contribute to the development of empirical knowledge. A qualitative researcher should be curious, creative, and not afraid to trust his or her instincts. Though there are different styles and approaches to doing qualitative research, in this book we present a methodological approach and procedures that we have found useful for analyzing qualitative data. Our approach was derived from a combination of Chicago style Interactionism and Pragmatism. These traditions guide our way through the data collection and analysis. During data collection and analysis we look for many things and ask many questions, but the foundation upon which our analysis rests is concepts.

Though we are interested in how persons experience events, and the meanings that they give to those experiences, at the same time we consider that any explanation of experience would be *incomplete* without (a) locating experience within the larger conditional frame or context in which it is embedded; and (b) describing the process or the ongoing and changing forms of action/interaction/emotions that are taken in responses to events and the problems that arise to inhibit action/interaction. We also look for consequences because these come back to be part of the next sequence of action. Users of this book need not adopt our theoretical stance to find the book useful. Many of the procedures, such as making comparisons, asking generative questions, and theoretical sampling, are not theoretically based but come out of Strauss's earlier work with Glaser (Glaser & Strauss, 1967) and can be used by anyone regardless of whether their research aim is theory building; rich, thick description; or case study analysis.

## Activities for Thinking, Writing, and Group Discussion

1. Take some time to think about your own personal and professional **philosophical orientation** and beliefs about the world. Write a paragraph or two describing how you think these might influence your approach to doing research. Share your thoughts and discuss these with a classmate or colleague.

2. What attracts you to doing qualitative research? How do you think your personal characteristics will enhance your ability to do good qualitative research?

3. In a group, discuss what you think are the qualities of a good qualitative researcher and how these qualities might be fostered through proper mentorship and the teaching/learning situation.

## Note

1. The introduction to this chapter was originally written for the 2nd edition of this book. At that time, Sage deleted the section because it was considered too complicated for a beginning text on qualitative research. However, this author (Corbin) believes that the elimination of that section explicating the philosophy underlying this method was a mistake because it does locate the method for those unfamiliar with Strauss's other work. In light of all the discussion and controversy that is now taking place about methods, grounded theory in particular, Corbin believes that it is essential to incorporate the philosophic background materials into this 3rd edition. What is noticeable about this section is that it is very similar, in fact, to what Strauss wrote in his book *Continual Permutations of Action* (1993) and I refer readers to this text for a fuller description of what is written here. However, this chapter definitely has my (Corbin) stamp on it also.

# 2

# Practical Considerations

*You are desperate to communicate, to edify or entertain, to preserve moments of grace or joy or transcendence, to make real or imagined events come alive. But you cannot will this to happen. It is a matter of persistence and faith and hard work. So you might as well just go ahead and get started. (Lamott, 1994, p. 7)*

**Table 2.1** Definition of Terms

*Nontechnical Literature*: Biographies, diaries, documents, memoirs, manuscripts, records, reports, catalogues, and other materials that can be used as primary data or to supplement interviews and field observations.

*Research Problem*: The general issue or focus of the research.

*Research Question*: The specific query to be addressed by this research. The question(s) sets the perimeters of the project and suggests the methods to be used for data gathering and analysis.

*Sensitivity*: The ability to pick up on subtle nuances and cues in the data that infer or point to meaning.

*Technical Literature*: Reports of research studies, and theoretical or philosophical papers characteristic of professional and disciplinary writing.

## Introduction

For the inexperienced qualitative researcher, doing qualitative analysis can be a daunting process. It is intimidating because there are the overriding concerns: "Am I doing it correctly?" "Am I being true to the data?" Naturally, researchers want to do the best research that they can. Even experienced researchers ask those questions. However, we want to assure our readers that there is no need for trepidation. Qualitative analysis builds upon natural ways of thinking. To quote Schatzman (1986), "Underlying this paper is the contention that analysis is a natural, generic process of thinking learned very early in social life along with language and almost constantly in experience" (p. 1).

Much of what we'll be talking about in this book will sound familiar. For example, during analysis it is suggested that analysts think in terms of concepts because concepts form the basis for generating common understandings. Think of the word "chair." Immediately, something to sit on comes to mind. Also during analysis, it is suggested that analysts ask questions of the data. Consider your first thought when you come across something you have never experienced before. You are likely to ask, "What is this?" Or, "What's going on here?" Additionally, analysts are encouraged to make comparisons between different pieces of data in order to determine what is the same and what is different about them. When you go out to purchase new tires, you compare brands and ask, "Which tire is the best value for the money?" Most of the time conceptualizing, asking questions, and making comparisons occur quite unconsciously. They are the tools that persons use to become acquainted with and understand the worlds they live in. The difference between everyday life and doing analysis is that in analysis researchers take a more self-conscious and systematic approach to knowing.

The notion that analysis builds upon everyday ways of thinking becomes evident to students when we sit down to work on their data. Invariably our graduate students say, "Doing this together is so different from reading about it in a book." Indeed it is! It is not until we stop the action during our analytic sessions and say, "See, you are asking questions, making comparisons, talking in terms of concepts" that students become aware that they are doing analysis. Our experience has taught us that analysis is much easier to teach in a classroom than it is to teach in a book.

The discussion about analysis is getting ahead of ourselves. Before a researcher can do analysis he or she must have a research project and some data. So, let's put aside the discussion about analysis for a few minutes. The purpose of this chapter is to offer a few practical suggestions for getting started on a research project. The chapter begins with a discussion about choosing the **research problem** and stating the **research question.** Next comes a short section on data collection. Later in the chapter there is a discussion on the uses of the **technical** and **nontechnical literature.** The chapter ends with a short section on theoretical frameworks.

## Choosing a Research Problem

One of the most difficult aspects of doing research is deciding upon a topic for investigation. The topic is something that the researcher will have to live with for some time, so it has to be something of interest. The two major questions related to deciding upon a topic are (a) How do I identify a problem that I would like to research? (b) How then do I narrow the problem down sufficiently to make it into a workable project?

Choosing a topic and defining its perimeters may seem especially difficult for the novice because the research problem in qualitative research is not as easily structured as it is in quantitative inquiries. Qualitative research begins with a broad question and often no preidentified concepts. Concepts are identified in and constructed from data. It is this very openness to discovery that makes doing a qualitative research so interesting and yet somewhat daunting for the novice researcher.

### Sources of Problems

The *sources* of problems in qualitative inquiries are not much different from those of other forms of research. There are four main areas:

- Problems that are suggested or assigned by an advisor or mentor
- Problems derived from technical and nontechnical literature
- Problems derived from personal and professional experience
- Problems that emerge from the research itself

Each of these will be discussed in turn.

First there are the suggested or assigned research problems. One way to arrive at a problem is to ask for suggestions from a professor doing research in an area of interest. Often he or she has ongoing research projects and welcomes having a graduate student do a small part of a project. This way of finding a problem tends to increase the possibility of getting involved in a doable and relevant research problem. This is because the more experienced researcher already knows what needs to be done in a particular substantive area. On the other hand, a choice arrived at in this manner may not be the most interesting to the student. It is important to remember that whatever problem is selected, the researcher will have to live with it for quite a while, so the final choice should be something that engages his or her curiosity.

A variant on the assigned or suggested problem source is to follow up on a professional or collegial remark that an inquiry into "such and such" would be useful and interesting. This is often a more palatable source of a research problem, especially if the researcher has some inclination toward that substantive area. For example, the interest of a woman who is athletic might be sparked by a remark such as, "I notice that women who exercise tend to feel more comfortable with their bodies." This broad and open statement can lead to all sorts of questions. How are women's images of body and exercise formed? What impact do school athletics, health beliefs, media, and cultural attitudes have on women's willingness to exercise? What is the process through which women who regularly exercise come to know their body and its strengths and limitations? What is the range of athletic activities that women are most likely to engage in? Why these activities and not others? Is there a difference between women who exercise and women who do not in how they experience their bodies? How does body experience with exercise translate into other aspects of women's lives?

Still another variation on the assigned problem is whether or not funds are available for research on certain topics. In fact, faculty sponsors may steer students in directions where funds are available. This is quite a legitimate suggestion, as often those are problem areas of special need.

A second source of problems is the technical and nontechnical literature. The literature can be a stimulus to research in several ways. Sometimes it points to a relatively unexplored area or suggests a topic in need of further development. At other times there are contradictions or ambiguities among the accumulated studies and writings. The discrepancies suggest the need for a study that will help to resolve those uncertainties. Alternatively, a researcher's reading on a subject may suggest that a new approach is needed to solve an old problem even though it has been well studied in the past. Something about the problem area and the phenomena associated with it remain elusive, and those unknowns, if discovered, might be used to reconstruct understanding. Also, while reading the literature, a researcher might come across a finding that is dissonant with his or her own experience that can lead to a study resolving that dissonance. Finally, reading may simply stimulate curiosity about a subject. The minute a potential researcher asks the question, "What if?" and finds there is no answer, there is a problem area.

A third source of problems is personal and professional experience. A person may undergo a divorce and wonder how other women or men experienced their own divorces. Or, someone may come across a problem in his or her profession or workplace for which there is no known answer. Professional experience frequently leads to the judgment that some feature of the profession or its practice is less than effective, efficient, humane, or equitable. So, it is believed that a good research study might help to correct that situation. Some professionals return to school to work for higher degrees because they are motivated by a reform ambition. The research problems that they choose are grounded in that motivation. Choosing a research problem through the professional or personal experience route may seem more hazardous than the suggested or literature routes. This is not necessarily true. The touchstone of a potential researcher's experience may be a more valuable indicator of a potentially successful research endeavor than another more abstract source.

A fourth source is the research itself. A researcher might enter the field having a general notion about what is desired to study but no specific problem area. A good way to begin is to do some initial interviews and observations. If the researcher is carefully listening to or observing the speech and actions of respondents, analysis should lead the researcher to discover the issues that are important or problematic in the respondents' lives. This acid test of paying attention to respondents' concerns is the key to where the focus of a research project should be. While, admittedly, there is no one and only relevant focus, the particular focus arrived at through respectful examination of respondents' concerns reduces the risks of being irrelevant or merely trivial. Consider the following example.

A student from Botswana, who was taking a class in fieldwork, grew desperate when studying "older Americans" in a senior resident home. To begin with, the ideas she had when she entered the field didn't seem to fit what she was hearing and observing. But, if that was so, then what, then, were the "real" issues? What she carried initially into this research situation were assumptions derived probably from three different sources. She was young and had some incorrect and even stereotypical conceptions about older people. Also, she was from a foreign country and thought in terms of her own culture. Then again, she was a beginning researcher and had not yet learned how to pick up cues from the subjects themselves about their concerns, and she was unfamiliar with how to let this information guide her choice of research problem. In the instance of this particular student, there was an additional difficulty that she faced. She was working voluntarily for a social work agency that had its own agenda, which included an evaluation of its work with these elders. So, the agency was urging her to obtain particular information that she discovered had little or nothing to do with the elders' lives or interests. Yet she was responsible to the agency. Finally, by listening closely to the elders, she formulated a significant research problem.

Certainly, anyone who is curious or concerned about the world around himself or herself—and anyone who is willing to take risks—should not, after some deliberation, have too much trouble finding a problem area to study. The next step is asking the proper research question.

## The Research Question

All research inquiries necessitate a question of some sort to guide the inquiry. However, qualitative research questions tend to be different from quantitative ones.

### Defining Issues

The manner in which a researcher asks the research question(s) is important because it determines to a large extent the research methods that are used to study it. Herein lies a dilemma. Does a researcher choose qualitative analysis because the problem area, and the question that stems from it, suggest that this form of research will be most productive? Or, does a committed qualitative researcher frame the question to fit the method? Is it a conscious or unconscious process that determines the research approach, as Pierce (1995) suggests? This issue is difficult to respond to because the answer is not clear. Although the basic premise is that the research question(s) dictates the method, it is our belief that persons tend to be more disposed toward either quantitative or qualitative research. For example, coming from these authors' background in an attraction to Pragmatism and Interactionism, it seems only natural that our preference would be for qualitative methods. Therefore, even when a problem area suggests that either qualitative or quantitative methods might be used, we tend to frame the question in a manner that enables us to carry out the project using qualitative methods. There is no reason for us to belabor this point; we only want to emphasize that some problems clearly suggest one form of research over another and that investigators should be true to the problem but also to themselves and their research preferences.

Furthermore, even when a researcher decides to use a qualitative approach, there remains the question of which method among the many qualitative options should be used by the investigator. There are now so many different qualitative approaches that just to present an introduction to each of them is beyond the scope of this book. To read about the various methods we refer readers to Berg (2006), Creswell (1998), Denzin and Lincoln (2005), Flick (2002), Gilgun, Daly, & Handel (1992), Marshall & Rossman (2006),

Morgan (1996), Morse (1994), Morse and Field (1995), Silverman (2004), Somekh and Lewin (2005). For three recent books that take a different approach to doing grounded theory than that presented in this book see: Charmaz (2006), Clarke (2005), Goulding (2002). Charmaz takes what she calls a "constructionist" approach, and Clarke refers to her method as "situational analysis," while Goulding orients her book to management, business, and market researchers.

Another important aspect of the research question is that it helps to establish the boundaries of what will be studied. It is impossible for any investigator to cover all aspects of a problem. Therefore, designing the question appropriately, even in qualitative studies, is very important. Sometimes a research problem requires the use of mixed methods or qualitative and quantitative approaches. This presents another whole set of methodological issues that are beyond the scope of this book. (We refer readers who are interested in mixed methods to Creswell, 2003; Greene, Kreider, & Mayer 2005.)

## Framing the Research Question

What do questions look like in qualitative studies? How do they differ from those of quantitative studies, and why? Qualitative studies are usually exploratory and more hypothesis generating rather than testing. Therefore, it is necessary to frame the research question(s) in a manner that provides the investigator with sufficient flexibility and freedom to explore a topic in some depth. Also underlying the use of qualitative methods is the assumption that all of the concepts pertaining to a given phenomenon have not been identified, or aren't fully developed, or are poorly understood and further exploration on a topic is necessary to increase understanding. While research questions in qualitative studies tend to be broad, they are not so broad as to give rise to unlimited possibilities. The purpose of the question is to lead the researcher into the data where the issues and problems important to the persons, organizations, groups, and communities under investigation can be explored.

The research question in a qualitative study is a statement that identifies the topic area to be studied and tells the reader what there is about this particular topic that is of interest to the researcher. Here is an example of how one might write a qualitative research question. "How do women with a pregnancy complicated by a chronic illness manage their pregnancy and life in a way to secure a positive pregnancy outcome?" This question, while it may be considered too general and nonspecific for a quantitative study, is a perfectly good one for a qualitative research study (Corbin, 1993). The question tells the reader that the study will investigate women during pregnancy, and that the pregnancy will be complicated by a chronic illness. Furthermore, the study will be looking at management of the pregnancy and everyday life from the women's perspective; that is, not from a doctor's or any other person's perspective. And, most important, the women in the study desire a positive outcome from their pregnancy—that is, they wish to have the baby.

Of course, in a qualitative inquiry, it is important to obtain as many perspectives on a topic as possible. In the study above, the researcher might also want to obtain some data on what the doctors, nurses, spouses, and significant others do and say about chronic illness, pregnancy, and all the issues involved because these interactions may influence how women view and manage their pregnancies. However, as indicated by the question, the focus remains on the women. Keeping that focus prevents the researcher from becoming distracted by unrelated and unproductive issues rather than trying to obtain data on the universe of possibilities. For example, rather than studying the entire world of high-risk obstetrics, only those tests and treatments that make their way into the study because they are actual or potential parts of the obstetrical care of participants become part of the investigation and are followed up on. The same holds true for the chronic conditions. It is only logical that not every chronic condition and its range of treatments can be examined in the proposed study. Only those aspects of chronic illness, or its treatment as these aspects enter into and affect the pregnancy of participants, become part of the investigation.

## Other Relevant Points

There are a few other points about questions in qualitative research that we want to make. A qualitative study need not be confined to individuals. The investigation can be focused on families, organizations, industries, and other fruitful lines of endeavor. Here is an example taken from the literature of questions pertaining to an interactional and organizational study. Shuval and Mizrahi (2004) in their study of boundaries of institutional structures, the dynamics of configuration, and the nature of permeability asked the following questions: "How do organizational and cognitive boundaries relate to each other? Why do biomedical practitioners allow the invasion of competitors? How do alternative practitioners 'fit' into the social and geographic space of clinic and hospital structures? What mechanisms or rituals of acceptance or rejection are visible in practice settings?" (p. 680).

In their biographical study of three generational families, Rosenthal and Völter (1998) asked the following questions: "How do three generations of families live today with the family and collective past during the Nazi period? What influences does this past of the first generation, and their own ways of dealing with it, have upon the lives of their offspring and on the ways in which the latter come to terms with their family history?" (p. 297).

Notice how broad the above questions are, and how they address the topics of interest, while at the same time limit the scope of the research. The interesting aspect of qualitative research is that though a researcher begins a study with a general question, questions arise during the course of the research that are more specific and direct further data collection and analysis. This point will seem clearer as readers come to the chapters on analysis later in this book.

## Data Collection

One of the virtues of qualitative research is that there are many alternative sources of data. The researcher can use interviews, observations, videos, documents, drawings, diaries, memoirs, newspapers, biographies, historical documents, autobiographies, and other sources not listed here. In any study, the researcher can use one or several of these sources alone or in combination, depending upon the problem to be investigated. Other considerations are the desire to triangulate or obtain various types of data on the same problem, such as combining interview with observation, then perhaps adding documents for the purpose of verifying or adding another source of data. Since there are many excellent texts that present in-depth discussions on data collection techniques, such as how to do interviews or observations, we will not go into detail about the procedures themselves. Instead, we will confine our discussion to more general matters of data collection as related to analysis. (For excellent texts on interviewing see Gubrium & Holstein, 2001, and Weiss, 1994. For texts that discuss doing fieldwork or observation see Lofland, Snow, Anderson, & Lofland, 2006; Patton, 2002; and Schatzman & Strauss, 1973—an old classic.)

Though many factors contribute to the quality of analysis, one of the most important factors is the quality of the materials that one is analyzing. Persons sometimes think that they can go out into the field and conduct interviews or observations with no training or preparation. Often these persons are disappointed when their participants are less than informative and the data are sparse, at best. Interviewing and observing are skills that take training and practice to acquire.

Our experience has demonstrated that perhaps the most data dense interviews are those that are unstructured; that is, they are not dictated by any predetermined set of questions (Corbin & Morse, 2003). It takes practice to sit with an open mind and an open agenda and not let nervousness get in the way of the free flow of information. For example, one might ask, "Tell me about your experience with cancer? I want to hear the story in your own words.

After you have completed your storytelling, then if I have further questions or something is not clear I will ask you. But for now just talk freely."

The use of the unstructured interview format does not mean that the researcher has no influence over the course of an interview. Mishler (1986) views interviews as a form of discourse between a researcher and the person being interviewed. He says, "Questioning and answering are ways of speaking that are grounded in and depend on culturally shared and often tacit assumptions about how to express and understand beliefs, experiences, feelings, and intentions" (p. 7). He goes on to explain how the interview is shaped both in its construction and meaning through the questions that are asked, the pauses, facial expressions, and other verbal and nonverbal communications that occur between the respective parties.

One of the most difficult aspects of interviewing for beginning researchers is facing periods of silence in the interview. Two German biographical researchers, Riemann (2003) and Schütze (1992a, 1992b), have developed a style of interviewing and analysis that takes silences into account. (For still another example of how to do and analyze biographical interviews, see Rosenthal, 1993.)

It is not unusual for qualitative researchers to come across persons who agree to be interviewed but have little to say once the interview begins, leaving the researcher uncertain about where to go next. At these times, it is good to have backup questions. Often the problem is that the person just does not know what to say, or is uncomfortable with the interview situation. Asking a few questions often relaxes the study participant and stimulates his or her memory so that he or she becomes more talkative and spontaneous. Sometimes a person has not thought about the issue for a while. Or, it may be that a topic generates a lot of emotion and the participant has to retreat into silence for a while to regain composure.

A sensitive interviewer knows when to step aside and let the interviewee guide when to resume the interview. What this researcher has found most interesting is that participants often offer some of the most interesting data as soon as the tape recorder has been turned off. I suppose there are many reasons for participants waiting until the recorder is turned off to present the last "tidbits" of information. One reason for "revelations" coming at the end of the interview might be because the interview process provides participants an opportunity to talk in depth about issues that they hadn't talked much about before, giving them additional insights into their own behavior. The final words are afterthoughts that they want to share. Another reason, and probably the more plausible explanation, is that many persons feel uncomfortable revealing what they consider "sensitive information" when the tape recorder is on. They don't mind the interviewer using the material or they would not reveal it, but the thought of possible identification through a voice recording makes them uncomfortable despite assurances that the tape will be destroyed after transcription. Since I always bring pencil and paper with me, in addition to a tape recorder, I usually ask for permission to write the added information down. Interviewees have always agreed, though perhaps their agreement is because of a possible power differential between interviewer and interviewee and because of politeness.

A researcher can never be certain why persons agree to be research participants; all a researcher can do if there is a question is ask for permission and try to be sensitive to nonverbal as well as verbal responses from research participants. Perhaps this is the point at which to make a few comments about the ethics of fieldwork.

Most institutional ethics committees are in place to ensure that safeguards exist to protect the anonymity and confidentiality of research participants and to protect their health and well-being in biomedical or potentially socially/psychologically disturbing research. In addition, the researcher has a responsibility during the research process to treat participants in a manner that he or she would like him- or herself and/or family members to be treated. A safe rule is if you don't think you would like it, than the participants probably wouldn't like it either. There is another point to be made also. People have the right to let their voices be heard. Sometimes a researcher feels uncomfortable or awkward with interview material or something that is observed. However, participants are not. In fact, they want their stories out there. This point will become evident when readers get to Chapters 8–12.

Additionally, a researcher can't make judgment on the words of others, unless the words have the potential to cause undue harm to someone. These authors acknowledge that ethics is a very relevant topic when it comes to research. Since a lengthy discussion of ethics is beyond the scope of this book, we refer our readers to Long and Johnson (2007) and Piper and Simons (2005).

Doing observations or fieldwork is often more difficult for novice researchers. Perhaps this is one reason why interviewing is used more often than observation for data collection by many qualitative researchers. Another reason might be that some researchers think of fieldwork as specific to anthropologists. Also, doing observations is more time consuming and can be intrusive, so researchers themselves are reluctant to use this mode of data collection. But observations have a lot to offer the qualitative researcher and should be considered as an option when deciding upon data collection methods. The reason why observation is so important is that it is not unusual for persons to say they are doing one thing but in reality they are doing something else. The only way to know this is through observation. Also, persons may not be consciously aware of, or be able to articulate, the subtleties of what goes on in interactions between themselves and others. Observations put researchers right where the action is, in a place where they can see what is going on. Patton (2002) states, "Creative fieldwork means using every part of oneself to experience and understand what is happening. Creative insights come from being directly involved in the setting being studied" (p. 302).

Observations have their potential drawbacks. A researcher may give meaning to action/interaction based on observation without checking out that meaning with participants. It is always beneficial to combine observation with interview or leave open the possibility to verify interpretations with participants. Patton (2002) states, "Nonverbal behaviors are easily misinterpreted, especially cross-culturally. Therefore, whenever possible and appropriate, having observed what appear to be significant nonverbal behaviors, some effort should be made to follow up with those involved to find out directly from them what the nonverbal behaviors really meant" (p. 291).

What is it that researchers look for when doing observations? The researcher begins by sitting or standing back and letting the scene unfold. Eventually, something interesting will catch the researcher's eye. The observation then focuses in on that. If the incident proves to be significant, the researcher begins taking notes on what is happening, what is being said and done, and by whom. It is, however, impossible to capture every bit of what is going on in a setting. From the perspective of these authors, the important thing to keep in mind when doing interviews and/or observations is that concepts drive the data collection and analysis. But where do these concepts come from? Let us give an example from our research on head nurses (unpublished study). Corbin began her first fieldwork session by meeting a head nurse as she prepared for her day and following her throughout the day, taking notes on just about everything the head nurse did or said, as well as recording notes on the context, and finally following up the observations with questions to obtain the head nurse's explanation about events. (It was impossible with everything happening in a busy hospital unit to write down everything, so I confined my observations to the head nurse and those activities and interactions that she concerned herself with.) Corbin then met with Strauss to analyze the notes. The concepts derived from that analysis became the basis for subsequent observations, though not entirely. Each additional day of observation offered opportunities to follow up on previously identified concepts as well as to discover new ones. If a subsequent observation yielded no data on a concept, at the end of the day the researcher would ask about the concept.

*Confidentiality* is an important issue when doing interviews or observations and later when writing. Lofland et al. (2006) state, "One of the central obligations that field researchers have with respect to those they study is the guarantee of anonymity via the 'assurance of confidentiality'—the promise that the real names of persons, places, and so forth will not be used in the research report or will be substituted by pseudonyms" (p. 51).

*Reflexivity* during data collection and analysis is another important consideration in qualitative research. We always knew in the past that researchers felt sad, angry, happy, and responded with approval or disapproval when collecting and analyzing data, yet we never thought much about the researcher's feelings or responses. There is no doubt that such emotions are conveyed to participants, and, in turn that participants react to researchers' responses by continually adjusting their stances as the interview or observation continues. Much of this occurs on an unconscious level. One might even say, due to this reciprocal influence, that researcher and participants co-construct the research (at least data collection) together (Finlay, 2002). Thus, examining the researcher's influence on the research process is important as Chesney (2001) states:

> I support the autobiographical analysis of self, not as separate from or in competition with the ethnographic words of the women but as a nurturing bed to place the research finding in and as part of the transparency of the research process. Reflecting honestly and openly has helped me retain some integrity and develop insight and self-awareness, and it has given me a certain self confidence. (p. 131)

Though now considered essential to the research process, the meaning that a researcher gives to reflexivity and the extent to which it is carried out is variable, depending upon the researcher's philosophical orientation and the degree of relevance accorded to the process. Each researcher must consider how much, when, and how. Though there is agreement upon the necessity of reflexivity, there is still some debate about its feasibility. Cutcliffe (2003) makes an interesting point when he asks how we can completely account for ourselves in the research since so much of what transpires takes place within the deeper levels of consciousness. Nevertheless, reflexivity remains as Finlay (2002) states, "a valuable tool to

- examine the impact of the position, perspective, and presence of the researcher;
- promote rich insight through examining personal responses and interpersonal dynamics;
- empower others by opening up a more radical consciousness;
- evaluate the research process, method, and outcomes; and
- enable public scrutiny of the integrity of the research through offering a methodological log of research decisions" (p. 532).

I, Corbin, found self-reflection to be a very natural and necessary process when doing the mini-project on the Vietnam veterans presented in the chapters on analysis later in this book. Self-reflection was cathartic and it helped me to see how I was slanting the data. I noticed that as I reviewed and thought about what I wrote in the memos, some were more reflective of my emotional response to the data than a conceptualization of what my respondents were telling me. I rewrote those memos, but I could certainly see myself in the analysis.

## Sensitivity

Data collection and analysis have traditionally called for "objectivity." But today we all know that objectivity in qualitative research is a myth. Researchers bring to the research situation their particular paradigms, including perspectives, training, knowledge, and biases; these aspects of self then become woven into all aspects of the research process (Guba & Lincoln, 1998). The questions that confront us include: "Is this so bad?" And, "How can we use what we as investigators bring to the research process in order to increase our sensitivity to what our participants are telling us?" Perhaps the answer to these questions is to focus on **sensitivity** (Glaser, 1978; Glaser & Strauss, 1967; Strauss, 1987).

### The Nature of Sensitivity

Sensitivity stands in contrast to objectivity. It requires that a researcher put him- or herself into the research. Sensitivity means having insight, being tuned in to, being able to pick up on relevant issues, events, and happenings in data. It means being able to present the view of participants and taking the role of the other through immersion in data. Sensitivity is a characteristic that comes more easily to some researchers than to others. Mostly it is a trait that develops over time through close association and work with both data and people. Through alternating processes of data collection and analysis, meanings and significance of data, often illusive at first, become clearer and the researcher begins to see the issues and problems from the perspectives of participants.

But insights into data do not just occur haphazardly, they happen to prepared minds during interplay with the data. Theories, professional knowledge that we carry within our heads, inform our research in multiple ways, even if quite subconsciously (Sandelowski, 1993). To quote Dey (1993), "In short, there is a difference between an open mind and an empty head. To analyze data researchers draw upon accumulated knowledge. They don't dispense with it. The issue is not whether to use existing knowledge, but how" (p. 63). As researchers move along in the analysis, it is their knowledge and experience (professional, gender, cultural, etc.) that enables them to respond to what is in the data. When we speak about what we bring to the research process, we are not talking about forcing our ideas on the data. Rather, what we are saying is that our backgrounds and past experiences provide the mental capacity to respond to and receive the messages contained in data—all the while keeping in mind that our findings are a product of data *plus* what the researcher brings to the analysis.

Sensitivity is a fascinating interplay of researcher and data in which understanding of what is being described in the data slowly evolves until finally the researcher can say, "Aha, that is what they are telling me" (at least from my understanding). Forcing (Glaser, 1992) the researcher's ideas on data is more likely to happen when the researcher ignores the relevance of self in the interpretation process and thinks that it is only the data talking when it is data talking through the "eyes" of the researcher. The more we are aware of the subjectivity involved in data analysis, the more likely we are to see how we are influencing interpretations.

Professional experience can enhance sensitivity. Though experience can prevent analysts from reading data correctly, experience can also enable researchers to understand the significance of some things more quickly. That is because researchers do not have to spend time gaining familiarity with surroundings or events. While a "fresh outlook" is often important, sometimes it takes a new researcher two to three weeks in an area just to feel comfortable, and during that period much time—and data—can be lost. Three things are important to remember. The first is to always compare knowledge and experience against data, never losing sight of the data themselves. The second is to always work with concepts in terms of their properties and dimensions, because it keeps the researcher focused on the similarities and differences in events and prevents being overwhelmed by descriptive data.

A third point is that it is not the researcher's perception of an event that matters. Rather, it is what participants are saying or doing that is important. For example, Corbin might know that a certain piece of equipment in a hospital is used to take X-rays. But in doing a study on patients' responses to, or experience with, hospital equipment, it is not the researcher's understanding of this equipment that is relevant. What is relevant is the meaning given to this equipment by the participant and how those meanings are formed and transformed. Does the participant describe the equipment as an outdated machine, a physical threat, something beneficial or life saving and therefore to be endured? Is the experience with it painful, frightening, or uncomfortable? What helps to keep the researcher focused on the data is having a comparative base from which to work. A researcher might say to him- or herself, "To me this is a piece of diagnostic equipment," an "inanimate object," "a useful medical tool"—all properties of the equipment. But I am seeing this from the perspective of a nurse. Do patients describe their experiences in the same way or do they see it differently? What meanings do they assign the equipment, what emotional responses does contact with this piece of equipment generate in them? The descriptions given by a participant tend to stand out when they are contrasted against descriptions given by other participants or the researcher.

Here is another example. Though the researcher may never experience an unwanted divorce, having undergone the death of a loved one does help him or her understand the meaning of grief and loss. Experience provides a comparative base for asking questions about grief and loss in divorce. Once a researcher has developed a list of general properties of loss and grief, the generated properties can be used as a comparative base to examine the data. In the end only the data themselves are significant, but it helps to have a little insight to start with. We do not reinvent life each day, otherwise we would never get anywhere. Rather, we build upon the foundation of knowledge that we have, comparing what we don't know against what we do know. It is the same with research.

It is amazing how sensitivity builds when a researcher is working with data. Sometimes analysts come upon a piece of datum and are stuck, unable to discern its meaning. What these authors have discovered is that researchers often carry their analytic problems around in their heads as they go about their daily activities. Then perhaps while reading the paper, talking with a colleague on the phone or via e-mail, or awakening from a dream, an insight occurs and the analyst is able to make sense out of data that up until this time had little or no meaning. Technically these insights pertain to the data, even though that insight was stimulated by another experience.

Background, knowledge, and experience not only enable us to be more sensitive to concepts in data, they also enable us to see connections between concepts. As the famous biologist Selye (1956) once wrote, "It is not to see something first, but to establish solid connections between the previously known and hitherto unknown that constitutes the essence of specific discovery" (p. 6). In other words, we have to have some background, either through immersion in the data or through personal experience, in order to know what we are "seeing" in data is significant and to be able to discern important connections between concepts.

This section on sensitivity is a good place to bring up cross-cultural research. Though these authors are not "experts" in this area, we do know that sensitivity is especially important when dealing with other cultures or even other genders. There is a wonderful book by Eva Hoffman titled *Lost in Translation* (1989). Hoffman, born in Poland, immigrated to Canada at the age of seven. One of the problems she confronted when she came to Canada was the lack of language to express her experiences in this new and strange country. The complexity was lost as she attempted to express her thoughts in English. Researchers should carry that message with them, especially when doing cross-cultural research. Something of the complexity will probably be lost. As teachers, these authors have been struck by the fact that foreign students doing research in their own countries often encounter concepts for which there are no specific English equivalents. For example, one of our students, Noriko Yamamoto, identified in her data two Japanese concepts used to express changes in the level and quality of care given by Japanese family caregivers as their parents' dementia increased. The two concepts were *amaeru*, used to describe the younger caregiver seeking indulgent love from a care recipient who is still able to respond; and *amayaksu* to describe the offering of indulgent love by the caregiver when the care recipient was no longer able to respond as an adult (Yamamoto & Wallhagen, 1998).

There are techniques that researchers can use to increase sensitivity in cross-cultural studies. For example, Chesney (2001) used Pakistani advisors to help her understand what her participants were telling her. She expresses that at times she wished she had been Pakistani herself to bridge the cultural and language barrier. See Green, Creswell, Shope, and Plano Clark (2007) for an excellent discussion on handling diversity in research.

## The Literature

Researchers bring to the inquiry a considerable background in professional and disciplinary literature. This background may be acquired while studying for examinations or simply to "keep up" with the field. During the research itself, analysts often discover biographies, memoirs, manuscripts, reports, or other materials that seem pertinent to the area under investigation. The question as it applies to this knowledge gained from the literature is how it can be used to enhance analysis.

Of course, the discipline, school, and perspective of the researcher will greatly influence how much literature is acquired and how it is used. To begin with, readers can be assured that there is no need to review all of the literature in the field beforehand, as is frequently done by researchers using quantitative research approaches. It is impossible to know prior to the investigation what salient problems or what relevant concepts will be derived from this set of data. There is always something new to discover. If everything about a topic is known beforehand, there is no need for a qualitative study. Also, the researcher does not want to be so steeped in the literature that he or she is constrained and even stifled by it. It is not unusual for students to become so enamored with a previous study or theory, either before or during their own investigation, that they become literally paralyzed. Becker (1986b) makes a good point when he says, "Use the literature, don't let it use you" (p. 149).

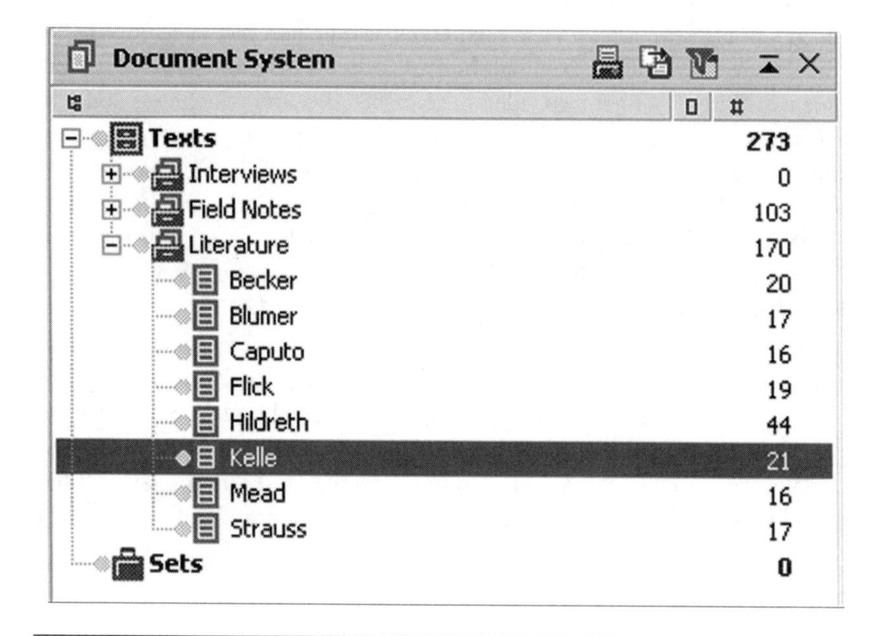


**Screenshot 1** This picture shows one of the four main windows of MAXQDA: The Document System Window, where all your data is stored and managed. It may be organized by text groups, enabling you to group your data in a meaningful way. The Document System is handled similar to Windows Explorer. In this example, "interviews" and "field notes" each have their own text group. Moreover, the text group "literature" has been created in order to integrate important articles, abstracts, and so on, into the project. This allows having immediate access to relevant material, thus you may copy and paste quotes of your literature directly into a memo.

## Making Use of the Technical Literature

Though the following list is by no means exhaustive, it does describe how the technical literature may be used:

- It can be a source for making comparisons.
- It can enhance sensitivity.
- It can provide a cache of descriptive data with very little interpretation.
- It can provide questions for initial observations and interviews.
- It can be used to stimulate questions during the analysis
- It can suggest areas for theoretical sampling (see Chapter7).
- It can be used to confirm findings, and just the reverse, findings can be used to illustrate where the literature is incorrect, simplistic, or only partially explains a phenomenon.

Each of these will be discussed below.

Concepts derived from the literature can provide a source for making comparisons with data as long as the comparisons are made at the property and dimensional level, and are not used as data per se. If a concept emerges from the data that seems similar or opposite to one recalled from the literature, then the researcher can examine both concepts for similarities and differences. For example, suppose a researcher was studying "coping" with the loss of a spouse due to an accident at work. Certainly coping with the loss of a spouse under those conditions will have similarities to other types of coping, such as coping with the loss of a spouse associated with divorce, but it will also have differences. Sometimes comparing the two conceptually similar but different situations (the concept in question is "loss") will delineate important features of each. This is especially so if one is thinking in terms of properties and dimensions, such as timing of loss, previous experience with loss, feelings about the spouse, and so on.

Familiarity with relevant literature can enhance sensitivity to subtle nuances in data. Though a researcher does not want to enter the field with an entire list of concepts, some may turn up over and over again in the literature and also appear in the data, thus demonstrating their significance. The important question for the researcher to ask when this happens is, "Are these concepts truly derived from data or am I imposing these concepts on the data because I am so familiar with them?" For example, it is not unusual for students in areas such as nursing and psychology to label everything as "coping" because this concept is relevant professionally. However, "coping" may not be the best term to describe what is going on in this particular research. An analyst has to learn to think "outside the box" and get away from professionally overused concepts, such as coping. However, if a concept is truly relevant, the question to ask is how the concept is the same and/or different from that in the literature

There is a special sense in which published descriptive materials can be useful to a researcher. Writings often provide illustrations of some concept or findings that include very descriptive data on a relevant topic with very little interpretation. Reading such literature is almost like reading field notes collected by another researcher for the same or another purpose. Such largely uninterpreted findings can stimulate thinking and make an analyst more sensitive to what is in his or her own data. It can also suggest questions that a researcher can ask of his or her data. Also, themes or concepts from a study may have relevance to a researcher's investigation. However, to repeat what we said earlier, researchers must be very careful to look for examples of incidents in their data and to identify the form that the concept takes in their study.

Before beginning a project, a researcher can turn to the literature to formulate questions for initial observations and interviews. After the first interview(s) or observation(s), the researcher will turn to questions and concepts derived from analysis of the data. Initial questions derived from the literature can also be used to satisfy human subjects committees by providing them with a list of conceptual areas to be investigated. Though new areas will emerge, at least the initial questions demonstrate overall intent of the research.

The technical literature can also be used to stimulate questions during the analytic process. For example, when there is a discrepancy between a researcher's data and the findings reported in the literature, that difference should stimulate the researcher to ask, "What is going on? Am I overlooking something important? Are conditions different in this study? If so, how, and how does this affect what I am seeing?"

Areas for theoretical sampling (Chapter 7) can be suggested by the literature, especially in the first stage of the research. The literature can provide insights about where (what place, time, papers) a researcher might go to investigate certain relevant concepts. In other words, it can direct a researcher to situations that he or she might not otherwise have thought of.

When an investigator has finished his or her data collection and analysis and is in the writing stage, the literature can be used to confirm findings, and just the reverse, findings can be used to illustrate where the literature is incorrect, simplistic, or only partially explains phenomena. Bringing the literature into the writing not only demonstrates scholarship, but also allows for extending, validating, and refining knowledge in the field. A researcher who has done a thorough job of investigating his or her topic should avoid being insecure about his or her discoveries, even though they do not match the published literature. Such discrepancies can point the way to new discoveries.

## Making Use of the Nontechnical Literature

Nontechnical literature consists of letters, biographies, diaries, reports, videotapes, memoirs, newspapers, catalogues, memos (scientific and otherwise), and a variety of other materials. The nontechnical literature can be used for all of the purposes listed above. In addition it has the following uses:

- It can be used as primary data
- It can be used to supplement interviews and observations.

The nontechnical literature can be used as primary data, especially in historical or biographical studies. Since it is often difficult to authenticate and determine the veracity of some historical documents, letters, memoirs, and biographies, it is very important to cross-check data by examining a wide variety of documents and supplementing these if possible with interviews and observations. In the chapters on analysis, beginning with Chapter 8, you will see how Corbin makes use of memoirs and historical documents as primary data.

The nontechnical literature can be used to supplement interviews and observations. For example, much can be learned about an organization, its structure, and how it functions (that may not immediately be visible in observations or interviews) by studying its reports, correspondence, and internal memos.

## Theoretical Frameworks

Before closing off this chapter it seems appropriate to say a few words about the use of theoretical frameworks. Theoretical frameworks are very common in quantitative research. They provide a conceptual guide for choosing the concepts to be investigated, for suggesting research questions, and for framing the research findings. For example, Patricia Vanhook (2007) used Corbin and Strauss's concepts of "Chronic Illness Trajectory" and "Comeback" (Corbin & Strauss, 1991a, 1991b) to guide her research on women with strokes. The concepts provide the structure for the study, down to her choice of measurement tools. Each tool was chosen to measure a component of comeback: physical recovery and rehabilitation, psychological adjustment to loss, and re-adaptation to the life course. In qualitative research, the use of theoretical frameworks is not so clear. There seems to be some controversy as to whether or not frameworks should be used and how. We know that many qualitative researchers do use of theoretical frameworks. In fact, a colleague, Jane Gilgun, and I (Corbin) have had many lengthy discussions on this topic.

Though it is these authors' preference not to begin our research with a predefined theoretical framework or set of concepts, we acknowledge in some instances theoretical frameworks can be useful. For example:

- After studying a topic the researcher finds that a previously developed framework is closely aligned to what is being discovered in the researcher's present study, and therefore can use it to complement, extend, and verify the findings.

- A framework from the literature can also be used to offer alternative explanations, for we all know that there is always more than one explanation for things—though we have to be careful here because a framework from the literature may explain part of the findings but not all, leaving the researcher trying to bend the findings to fit the theory.

- If the researcher is building upon a program of research or wants to develop middle-range theory, a previously identified theoretical framework can provide insight, direction, and a useful list of initial concepts. However, a researcher should remain open to new ideas and concepts and be willing to let go if he or she discovers that certain "imported" concepts do not fit the data. The importance of "remaining open" is essential even for experienced researchers working on their own program of research.

- A theoretical framework can help the researcher determine the methodology to be used.

A perusal of journal articles reveals that researchers often use theoretical frameworks to justify the use of a particular methodology or approach to a study, especially in nursing. For example, Holroyd (2003), in her study of care giving in China, used a framework drawn from a cognitive anthropological paradigm that focused "on shared, recognized, and transmitted internal representations of culture with the contention that cultural guidelines exist in dominant political and social structures of a particular society" (p. 306). By using that framework, Holroyd could expect to find that obligations and duties would be consistent with gender and generation guidelines of Chinese society. In another example, Cannaerts, Dierckx de Casterlé, and Grypdonck (2004) used symbolic interactionism to explain their use of grounded theory as the methodology of choice in their study of the nature of "palliative care."

A little different take on the use of theoretical frameworks occurs when a researcher comes to the research with a Marxist, or feminist, or interactionist, philosophical orientation. In this case the philosophical orientation is likely to influence the researcher's whole approach to doing research. Reid (2004), at the onset of her research monograph titled "The Wounds of Exclusion," positions herself as a feminist conducting feminist action research. In explaining what this means, she says, "In the rendered account that follows I have attempted to hold myself accountable back to the research participants and to myself for my critical analysis and responsible use of power" (p. 6).

Moving on to another related topic, if a qualitative researcher is interested in extending a substantive theory or raising a substantively derived theory to the level of middle-range theory, he or she can begin with a concept. Take, for example, the concept of "awareness" derived from Glaser, Strauss, and Benoliel's study of dying and reported in Glaser and Strauss's book *Awareness of Dying* (1965). The concept was used to explain how the various inter/actants (health professionals, family, and the dying person) managed information about the patient's dying. A researcher interested in developing a middle-range theory of "information management" could begin with "awareness" as described by Glaser and Strauss and use it as a basis for researching how information is revealed or kept secret in marital infidelities, in spies, and "in the closet" gays. New categories would most likely be discovered. There would be elaboration of previously identified categories. Finally, "awareness" would be raised to an even greater abstraction because it is now being applied across situations. See Strauss's (1995) article titled "Notes on the Nature and Development of General Theories" for a discussion on developing, checking, and linking general theories.

## Summary of Important Points

This chapter covered four major areas: (a) choosing a research problem and stating the questions, (b) developing sensitivity to what is in the data, (c) use of the literature, and (d) theoretical frameworks. Each of these areas must be considered before beginning the research inquiry.

*The Research Problem and Question.* The original research question and the manner in which it is phrased lead the researcher to examine data from a specific perspective and to utilize certain data collection techniques and modes of data analysis. The question(s) sets the tone for the research project and helps the researcher to stay focused even when there are masses of data. The original question in a qualitative study is often broad and open-ended. It tends to become more refined and specific as the research progresses and the issues and problems of the area under investigation are identified. The original research question(s) may be suggested by a professor or colleague or be derived from the literature or from a researcher's experience. Whatever the source of the problem, it is important that a researcher have an enthusiasm for the subject because he or she will have to live with it for some time.

*Sensitivity.* If one cannot achieve objectivity in qualitative research, then perhaps one can have sensitivity. Sensitivity, or insight into data, is derived through what the researcher brings to the study as well as through immersion in the data during data collection and analysis. Sensitivity enables a researcher to grasp meaning and respond intellectually (and emotionally) to what is being said in the data in order to be able to arrive at concepts that are grounded in data. Later, when it comes time to write findings, that same sensitivity enables researchers to present participants' stories with an equal mix of abstraction, detailed description, and, just as important, feeling.

*Using the Literature.* The technical and nontechnical literature tends to be useful in somewhat different and specific ways. Ingenious researchers, besides using the usual technical literature, will sometimes use various other types of published and unpublished materials to supplement their interviews and field observations. Though reports and biographies often come to mind, there are many useful types of nontechnical literature.

Technical literature can provide initial questions, initial concepts, and ideas for theoretical sampling. Nontechnical literature can be used as both primary and supplemental data, for making comparisons, and can act as the foundation for developing general theory. The important point for a researcher to remember is that the technical literature can hinder creativity if it is allowed to stand between the researcher and the data. But, if it is used for comparative purposes it can foster identification of properties and dimensions of relevant concepts.

Theoretical frameworks are a form of technical literature. They are often used in qualitative research; however, their use is different in quantitative studies. That is, they do not define the variables to be studied nor do they structure the research in the same manner as they do in quantitative studies. They tend to be used more as justification for the use of a particular methodology or as a guiding approach to the research. On the other hand, the theoretical orientation of the researcher plays a significant role in qualitative research. While the theoretical orientation does not determine the concepts to be studied, it does determine a committed approach to doing research. Also, a researcher interested in developing a middle-range theory from a substantive theory can use a substantive theory as a theoretical base for exploring the core concept across different groups, thereby increasing the theory's depth, breadth, and level of abstraction.

## Activities for Thinking, Writing, and Group Discussion

1. Write a paragraph describing the source of the problem for your research topic. Or, if you don't have a topic, explain how you might go about finding one based on the information provided in this book.

2. Take your research topic and write two questions from it, one qualitative and one quantitative. Then, describe how the questions would lead to different methods of data collection and analysis.

3. As a group, explore the notion of sensitivity. What does it mean to the various group participants, how can its development be enhanced, and how do the group members see its relevance to their research?

4. Discuss how the technical literature can enhance or hinder the qualitative research process.

5. Peruse the research journals in your field and note how the qualitative studies make use of theoretical or conceptual frameworks. Bring several examples to the group for discussion.

# 3

# Prelude to Analysis

*In reading of scientific discoveries one is sometimes struck by the simple and apparently easy observations which have given rise to great and far-reaching discoveries making scientists famous. But in retrospect we see the discovery with its significance established. Originally the discovery usually has no intrinsic significance; the discoverer gives it significance by relating it to other knowledge, and perhaps by using it to derive further knowledge. (Beveridge, 1963, p. 141)*

**Table 3.1** Definition of Terms

*Analysis*: Analysis involves examining a substance and its components in order to determine their properties and functions, then using the acquired knowledge to make inferences about the whole.

*Analytic Tools*: Analytic tools are thinking devices or procedures that if used correctly can facilitate coding.

*Concepts*: Words that stand for groups or classes of objects, events, and actions that share some major common property(ies), though the property(ies) can vary dimensionally.

*Dimensions*: Variations of a property along a range.

*"Feeling right"*: Indicates that after being immersed in the data for some time the researcher believes that the findings arrived at through reflective analysis express what participants are trying to convey through word and action and emotions, as seen through the "eyes" of the analyst.

*(Continued)*

(Continued)

*Microanalysis:* Detailed coding around a concept. A form of open coding used to break data apart and to look for varied meanings of a word or phrase.

*Properties*: Characteristics or components of an object, event, or action. The characteristics give specificity to and define an object, event, and/or action.

## Introduction

A researcher cannot continue to collect data forever. Sooner or later "something" has to be done with that data to give it significance. That something is termed **analysis.**

What is analysis? Analysis is a process of examining something in order to find out what it is and how it works. To perform an analysis, a researcher can break apart a substance into its various components, then examine those components in order to identify their **properties** and **dimensions**. Finally, the researcher can use the acquired knowledge of those components and their properties to make inferences about the object as a whole. In making inferences, analysts rely upon experience and training to recognize and give meaning as stated above in the quote from Beveridge (1963). Without some background, either from immersion in the data or professional/experiential knowledge, the ability to recognize and give meaning is not there.

Another approach to analysis would be to begin with the whole, observe to see what the substance does and how it seems to work, then take it apart to determine its various components, studying the makeup and function of the components and their relationship to the whole. Take blood, for example. A researcher could examine it for its general properties such as color, viscosity, apparent functions, and distribution within the body. Though helpful, this approach would not tell the whole story of blood. A researcher would have to dig deeper and examine blood more closely in order to determine its components, such as red blood cells, white blood cells, and plasma; and explore their properties and function. The final step would be to determine how the different components relate to each other and the whole. Still another approach would be to take something like blood, hypothesize its various functions, test each hypothesis to determine if any one or more are correct, and finally eliminate those that are wrong. Research is often both an inductive and deductive process.

Analysis is a very dynamic process. The analyst has to brainstorm, try out different ideas, eliminate some, and expand upon others before arriving at any conclusions. To make this point, let us provide the following example. A sculptor friend of Dr. Strauss once invited us to his workshop to see how he worked, and there we started talking about creativity. In his workshop were all types of metals of various shapes and forms. He explained how he works. First, he studies the different pieces of metal to see what possibilities lie within them, letting his imagination run free. Then the imaginary piece of sculpture is given form to see how the actual form holds up to his vision. If the resulting piece doesn't "work" aesthetically, it is dismantled, and the process repeated again until the piece of sculpture "looks and feels right."

It is the same with analysis. There are many different stories that can be constructed from data. How an analyst puts together the **concepts** often requires many tries before the story or findings "feel right" to him or her. **Feeling right** is a gut feeling. It means that after being immersed in the data the researcher believes that the findings reflect the "essence" of what participants are trying to convey, or represent one logical interpretation of data, as seen through the eyes of this particular analyst.

In this chapter we take analysis apart and examine its various components, a kind of mini-analysis of analysis. The purpose of the chapter is to provide the novice researcher with a strong foundation for what will follow in the remainder of the book.

## Some Properties of Qualitative Research

Qualitative research has many different properties. In the remainder of the chapter we'll explore each in greater depth.

- Analysis is an art and a science.
- Analysis is an interpretive act.
- More than one story can be created from data.
- Concepts form the basis of analysis.
- Concepts vary in levels of abstraction.
- There are different levels of analysis.
- Analysis can have different aims.
- Delineating context is an important aspect of analysis.
- Analysis is a process.
- Analysis begins with the collection of the first pieces of data.
- A researcher can do **microanalysis** or more general analysis as the analytic situation demands.

### Analysis is an Art and Science

Analysis is both an art and a science (Patton, 1990). The "art" aspect has to do with the creative use of procedures to solve analytic problems and the ability to construct a coherent and explanatory story from data, a story that "feels right" to the researcher. To bring the art aspect into analysis, the researcher must remain flexible in his or her use of procedures. He or she must learn to think "outside the box," be willing to take risks, and be able to spin "straw" into "gold"; that is, turn raw data into something that promotes understanding and increases professional knowledge. The art aspect of research transcends all forms of research and it is doubtful that any significant piece of research could be accomplished without it. Beveridge (1963) explains it this way:

> New knowledge very often has its origins in some quite unexpected observation or chance occurrence arising during an investigation. . . . Interpreting the clue and realizing its significance requires knowledge without fixed ideas, imagination, scientific taste and a habit of contemplating all unexplained observations. (p. 147)

Though qualitative research has its art aspect, it comes with certain responsibilities. There must also be the science part to call the product research. Sandelowski (1994) says:

> Celebrating the art in qualitative research is not an imprimatur for anarchy or for ignorance. Qualitative researchers are not free to make wild forays into fancy; they make, but cannot fake. Nor are they free to be ignorant of the logic and aesthetic of the varieties of research strategies encompassed by the label *qualitative research*. (p. 58)

The science aspect of qualitative research is not "science" in the traditional sense. The science comes from "grounding" concepts in data. Then, it systematically develops concepts in terms of their properties and dimensions and at the same time validates interpretations by comparing them against incoming data (Blumer, 1969, pp. 25–26; Glaser & Strauss, 1967). When we use the term "validate," we don't mean to imply that we are testing hypotheses in a quantitative sense. Validating here refers more to a checking out of interpretations with participants and against data as the research moves along.

In all qualitative research, there has to be some sort of balance between the art and science. Though data and findings are constructed and might be considered "stories" (Denzin, 1989), they are not "novels" in the traditional sense of imaginative yarns or tales meant to entertain. There can be no flights of fancy. Nor is qualitative research controlled laboratory science. How far the analysis varies dimensionally from art to science depends upon the philosophic background of the researcher, his or her discipline, and the qualitative method he or she is using.

## Analysis Involves Interpretation

Analysis involves interpretation (Blumer, 1969). Interpretation implies a researcher's understanding of the events as related by participants. As Denzin (1998) states:

> Interpretation is a productive process that sets forth the multiple meanings of an event, object, experience, or test. Interpretation is transformation. It illuminates, throws light on experience. It brings out, and refines, as when butter is clarified, the meanings that can be sifted from a text, an object, or slice of experience. (p. 322)

But Denzin (1998) does not stop there. He goes on to say, "So conceived, meaning is not in a text, nor does interpretation precede experience, or its representation. Meaning, interpretation, and representation are deeply intertwined in one another" (p. 322). Though interpretations are not exact replications of data, but rather the analyst's impressions of that data, it does not mean that researchers should give up doing research. Interpretation is not exact science. It can never be, nor should it be. But doing qualitative research with all its flaws remains an important endeavor. Qualitative research has made a major difference in my (Corbin's) nursing practice. It brought me out of the role of an "authority on health care" to one of cocreator or negotiator of care with my patients (Corbin & Cherry, 1997). Through qualitative research, I learned that patients knew more about their illnesses, their body's responses to it, and the regimens designed to control the illnesses than I could ever know because they lived with these every day.

Researchers are translators of other persons' words and actions. Researchers are the go-betweens for the participants and the audiences that they want to reach. As every language translator knows, it is not easy to convey meaning. Words can have different meanings from one language to another and from one situation to another. I (Corbin) learned this from experience. On more than one occasion, I have worked with translators while teaching in a foreign country. Invariably, students will laugh at something I say because the direct translation of a word conveys something other than what I intended. Or, students will tell me that they are not certain how much of my presentation they understood because the translation was so poor. Obviously, something was lost in translation. Though it is discouraging to me when I hear students say the translation of a book or presentation was "not very good," it does provide a lesson for all of us who are attempting to bring the words of our participants to life through research. Interpretations are often not exact and sometimes researchers are a "bit off" and, furthermore, some interpreters are better at it than others. Yet, the possibility of being a "bit off" at times should not discourage researchers from trying. Qualitative researchers have to push forward with analysis. With it we have more to gain than we have to lose.

One additional point about interpretation, before moving on, is that analysis is never quite finished, no matter how long a researcher seems to work on a study. Since researchers are always thinking about their data, they are always extending, amending, and reinterpreting interpretations as new insights arise and situations change. Such revisions are part of the qualitative process. As Denzin and Lincoln (1998) state in the introduction to Part II of *Collecting and Interpreting Qualitative Materials,* "Part II explores the art and politics of interpretation and evaluation, arguing that the processes of analysis, evaluation, and interpretation are neither terminal nor mechanical. They are always ongoing, emergent, unpredictable, and unfinished" (pp. 275–276).

### More Than One Story Can Be Derived From Data

Qualitative data are inherently rich in substance and full of possibilities. It is impossible to say that there is only one story that can be constructed from the data. Though participants speak through data, the data themselves do not wave flags denoting what is important and what is not. Different analysts focus on different aspects of data, interpret things differently, and identify different meanings. Also, different analysts arrive at different conclusions even about the same piece of data. Furthermore, the same analyst might look at the same data differently at different times. It all depends upon the angle or perspective that the analyst brings to the data. For example, interviews with persons who have chronic illnesses can be examined from the angle of illness management (Corbin & Strauss, 1988), identity and self (Charmaz, 1983), and of suffering (Morse, 2001, 2005; Riemann & Schütze, 1991). If a person examined the interviews conducted by these respective researchers, especially if they are unstructured interviews, that person would find that the interviews are not that much different in substance. What is different is the prism through which the analyst viewed the data. Management, identity, and suffering can all be found in data about chronic illness and all are valid interpretations. Each of these interpretations presents a more rounded picture of what chronic illness is all about. But different researchers tend to focus on different aspects. In other words, data talk to them in different ways. What is different about each study is the level of significance accorded to each of the different phenomena and how they are put together in a study.

## Levels of Analysis

Analysis can range from superficial description to theoretical interpretations. Superficial description tends to skim the top of data and looks more like journalism than research. It does not challenge thinking, present new understandings, or tell us anything we probably don't already know. A more in-depth analysis tends to dig deeper beneath the surface of data (and many journalists are now doing interpretation). It presents description that embodies well-constructed themes/categories, development of context, and explanations of process or change over time. In-depth analysis is more likely to generate new knowledge and deeper understandings because it tends to go beyond what everyone already knows. Though these authors are biased toward taking the time to do a more in-depth analysis, they also recognize that researchers have different levels of motivation, training, direction, and resources to carry out their research projects and analyses. Many would-be qualitative researchers lack trained mentors to guide them. These researchers are often uncertain about how to proceed, lack confidence, or do not even know what constitutes "good" analysis. They may go from method book to method book, trying to figure out what to do, then finally do what they can best manage on their own, often settling for less than they intended or are capable of if they had proper guidance.

Then, too, some research projects do not demand a detailed analysis. There might be a few questions added to a quantitative study that necessitate some degree of qualitative analysis. In such projects, a summary of major themes may be sufficient. At the other extreme, it is possible to overdo analysis, making it so descriptive and detailed that reading the report becomes boring. Minutia are not what we aim for. The art of analysis comes in knowing what ideas to pursue, how far to develop an idea, when to let go, and how to keep a balance between conceptualization and description.

## Concepts Form the Basis of Analysis

Concepts/themes are the foundation for the analytic method described in this book. Blumer (1969) emphasizes the importance of concepts to research when he states:

> Throughout the act of scientific inquiry, concepts play a central role. They are significant elements in the prior scheme that the scholar has of the empirical world; they are likely to be the terms in which his problem is cast; they are the categories for which data are sought and in which data are grouped; they usually become the chief means for establishing relations between data; and they are the anchor points in interpretation of findings. (p. 26)

Concepts are derived from data. They represent an analyst's impressionistic understandings of what is being described in the experiences, spoken words, actions, interactions, problems, and issues expressed by participants. The use of concepts provides a way of grouping/organizing the data that a researcher is working with. If one thinks of a bird, plane, and a kite and asks what they all share in common, one can say "flight." The notion of "flight" enables the analyst to group these diverse objects together, then to explore each of these objects in greater depth, detailing their similarities as well as differences in terms of "flight." In doing so, the analyst discovers some interesting information about the concept of "flight" in general, as well as the peculiarities of flight as they apply to each group.

## Concepts Vary in Levels of Abstraction

Concepts vary in levels of abstraction. There are basic-level concepts and higher-level concepts that we call categories. Lower-level concepts point to, relate to, and provide the detail for the higher-level concepts. For example, in the example provided above, flight is a higher-level concept than is bird, kite, or plane. Flight explains what these objects have in common. Though there are flightless birds (e.g., flightless cormorants from Galapagos), many birds are able to fly long distances over continents and are therefore good examples of flight. A kite, by virtue of being a kite, should fly, otherwise it would be a dud. And if a plane did not fly, it would probably be called something else, such as a car with wings. So flight is integral to each of these objects. But if we want to understand and describe flight, we have to examine the individual properties and dimensions of flight as applied to each of these objects and by doing so develop an understanding of flight under various conditions.

By keeping lower-level concepts in any explanation of our higher-level concepts, we are never too far removed from the data and provide all of the detail that adds interest and variation to phenomena we are studying. The more one moves up the conceptual ladder, the broader and more explanatory the concepts become, yet as they move toward greater abstraction, concepts, while perhaps gaining in explanatory power, begin to lose some of their specificity. However, if the conceptual pyramid is carefully crafted, the higher-level concepts will rest on a solid foundation of lower-level concepts, which in turn go directly back to the data, bringing with them the detail and the power of description.

At first, analysis is open and free, much like brainstorming. The researcher identifies concepts, but early in analysis may not be certain if a concept is a lower-level concept or higher-level one, or what interpretive meaning must be given to events to group them and subsequently bring them to a higher level of abstraction. In the example of the bird, plane, and kite, it was easy to see what the three objects had in common. But when analyzing data, often what events, actions, interactions, emotions share in common is not so evident. With time and immersion in the data, a researcher gains insight and sensitivity. It is this insight that enables researchers to group events under a more conceptual label. Being open to all possible meanings in data, as well as potential relationships between concepts, is very important early in analysis. It prevents early foreclosure or jumping to conclusions, ones that might prove wrong later on as the analysis proceeds.

The open generative nature of early analysis is difficult for some persons, especially those steeped in the rigors of quantitative approaches. Novice qualitative researchers often worry that somehow they are "putting something" into the data if they brainstorm and list all possible meanings implied in certain events or actions. What novice researchers do not realize is how easy it is to jump to conclusions about the meaning of data. Taking the time to consider all possible meanings helps researchers to become more aware of their own assumptions and the interpretations they are placing on data.

## Aims of Research

There are different aims of qualitative research. The aim can vary from description, to conceptual ordering, to theorizing. Different researchers have different aims depending upon training, skill, type of qualitative method, and purpose. Since beginning researchers often have difficulty distinguishing between *description* and *theory,* we will take the opportunity to present a few words on these matters. Also, we will touch upon another mode of managing data that is often utilized in qualitative studies, a mode we call *conceptual ordering.* (For a similar but also somewhat different perspective on these same matters, see Wolcott, 1994.)

People commonly describe objects, people, scenes, events, actions, emotions, moods, and aspirations in their everyday conversations. Not only do ordinary people describe, so do, as part of their daily work, journalists, novelists, technical, travel, and other nonfiction writers. Description draws on ordinary vocabulary to convey ideas about things, people, and places. For example one might hear, "The streets were quiet early in the morning, and I looked forward to hitting the open road in my new convertible automobile." Description also makes use of similes and metaphors when ordinary words fail to make the point or more colorful mental pictures are called for (Lakoff & Johnson, 1981). Consider the following scene described by Márquez (1993), "It was a brilliant morning in early August. One of those exemplary postwar summer Sundays when the light was like a daily revelation, and the enormous ship inched along, with an invalid's labored breathing, through a transparent stillwater" (p. 117). The imagery is colorful, vivid, and it is easy for the reader to put him- or herself into the scene.

Persons literally could not communicate without the ability to describe. Description is needed to convey what was (or is) going on, what the setting looks like, what the people involved are doing, and so on. The use of descriptive language can make ordinary events seem extraordinary. Great writers, like Márquez and Flaubert, know this and strive to make their details so vivid that readers can actually see, taste, smell, and hear what is going on in a scene. Yet, even mere mortals, those of us with less well developed writing skills, use description to relate our adventures, thoughts, and feelings to others as we encounter new, and sometimes routine, situations.

Descriptions may seem objective, but they are not. Even basic description involves purpose (otherwise why describe?) and audiences (who will see or hear the description?) and the selective eye of the viewer (Wolcott, 1994). For example, police reports are focused on criminal or investigative issues. They are usually relatively straightforward and meant to be read by superiors and other interested parties, whereas a journalist's account of the same event is likely to be written more colorfully. The latter also tends to reflect some personal, political, or organizational stance, and is meant to inform and move newspaper readers.

In short, the descriptive details chosen by storytellers are usually consciously or unconsciously selective, based on what they saw or heard, or thought important. Though description is often meant to convey believability and to portray images, it is also designed to persuade, convince, express, or arouse passions. Descriptive words can carry overt and/or covert moral judgments. This can be true not merely of sentences, but entire books—as in exposés or in serious volumes that aim at reform. Even seemingly objective reports like those of police or journalists may reflect deep prejudice and moral judgments, without the individual being aware of those attitudes and feelings. Aesthetic judgments too are conveyed through descriptions: "The young soprano's voice was delicate, airy, though at the upper ranges she occasionally wobbled just the slightest, but generally conveyed the spirit of the character; she has a great future in opera." Sometimes the aesthetic and the moral are joined. Take for example the negative reaction of critics and audiences to early Impressionists paintings. Later these same paintings became the favorites of museum visitors and art collectors throughout the world, bringing to their present owners millions of dollars when they go up for auction.

It is important to understand that description is the basis for more abstract interpretations of data and theory development, though it may not necessarily lead to theory if that is not the researcher's goal. Descriptions already embody concepts, at least implicitly. Even at the highest levels of abstract science, there could be no scientific hypotheses and theoretical or laboratory activity without prior or accompanying descriptions. Though description is clearly not theory, description is basic to theorizing.

Description is also basic to what we call conceptual ordering. The latter refers to the organization of data into discrete categories (and sometimes ratings) according to their properties and dimensions, then the utilization of description to elucidate those categories. Most social science analyses consist of some variety—and there are many types—of conceptual ordering. Researchers attempt to make sense out of their data by organizing them according to a classificatory scheme, such as types or stages. In the process, items are identified from data and defined according to their various general properties and dimensions. Take restaurant ratings such as in the *Michelin Guide*. Restaurants are often rated dimensionally ranging from three stars to none based on properties such as quality, taste, presentation, ambience, value, and the complexity of the wine list. How each restaurant varies dimensionally across each property provides the basis for the more general rating. Ratings of restaurants are often biased toward reviewers' preferences, which do not necessarily reflect the taste of the general public. Yet to be given three stars, two or even one in the Michelin Guide is very prestigious and assures a restaurant's success. When presenting ratings, researchers are almost certain to include various amounts of descriptive material to explain their ratings. The chief reason to discuss conceptual ordering is because this type of analysis is a precursor to theorizing through its development of properties and dimensions.

Developing theory is a complex activity. What do we mean by theory? For us, theory denotes a set of well-developed categories (themes, concepts) that are systematically interrelated through statements of relationship to form a theoretical framework that explains some phenomenon (Hage, 1972, p. 34). The cohesiveness of the theory occurs through the use of an overarching explanatory concept, one that stands above the rest. And that, taken together with the other concepts, explains the what, how, when, where, and why of something.

Not everyone wants to develop theory. In fact, theory development these days seems to have fallen out of fashion, being replaced by descriptions of "lived experience" and "narrative stories." While we acknowledge that theory has its limitations, the relevance of theory development to the advancement of knowledge remained a constant throughout the life of Anselm Strauss (1995). This author agrees that not everything can or should be reduced to one clever theoretical explanatory scheme, as helpful as that scheme might be. However, theory development remains relevant as a research endeavor and should be recognized as such. A researcher considering developing theory should not be frightened off by recent antitheoretical trends. Trends have a way of coming and going. And even though any particular theory may become outdated as new knowledge comes to light, and even though theories do not represent "reality," and even though theories are reductionistic, they have over the years proven useful. A person has to wonder where the world would be if there were only "stories" and no "theories." We probably would never have been able to put a man on the moon, developed computers, or build houses out of glass. A researcher has to make choices and should choose the approach to, and aims for, research that are most suitable to the problem of study and most likely to make a professional contribution.

Theorizing is interpretive and entails not only condensing raw data into concepts but also arranging the concepts into a logical, systematic explanatory scheme. If a researcher is going to construct theory, then he or she should do it well and not settle for some poorly constructed, thin imitation of theory. The construction of theory necessitates that an idea be explored fully and considered from many different angles or perspectives. It is also important to follow through with the implications of a theory. The formulations and implications lead to "research activity" that entails making decisions about and acting in relationship to a multitude and variety of questions that enable the researcher to fully explore a topic. Decision making and subsequent action occur along the entire research course. At the heart of theorizing lies the interplay of making inductions (deriving concepts, their properties, and dimensions from data) and deductions (hypothesizing about the relationships between concepts; the relationships too are derived from data, but those data have been abstracted by the analyst to form concepts).

Theories may be substantive, middle range, or formal (Glaser & Strauss, 1967, pp. 32–34). A theory of how gays handle disclosure/nondisclosure (information management) of their sexual identity to physicians is an example of a theory derived from one substantive area. The notion of "information management" can also be used to study disclosure or nondisclosure of HIV status by gays to prospective partners. Studying "information management" by gays under varying conditions can lead to the development of a more middle-range theory of "information management" as applied to situations important to gay men's lives.

More formal theories are less specific to a group and/or place and apply to a wider range of disciplinary concerns and problems. To develop a formal theory of "information management," researchers could begin with a substantive or middle-range theory derived from previous studies, then use the theoretical formulations as a foundation for studying a wider range of related topics. For example, a researcher wishing to develop a general theory about "information management" derived from a study of sexually active adolescents disclosing about having a sexually transmitted disease could take the original framework and expand it, making it even more abstract by using it to study parents blocking access by their young children to certain Web sites; then secrets between governments; and finally "information management" in political campaigns. Because formal theories are usually derived from investigations of a concept under a variety of different related topics and conditions, they become much more abstract and have greater applicability than do substantive theories or middle-range theories.

## Delineating Context Is an Important Aspect of Analysis

When doing analysis, delineating the context or the conditions under which something happens, is said, done, and/or felt is just as important as coming up with the "right" concept. Context not only grounds concepts, but also minimizes the chances of distorting meaning and/or misrepresenting intent. A filmmaker and a novelist can take creative license with events, taking words or pictures out of context and inserting them into other contexts to make a political or social statement. Filmmakers or novelists can spin, twist, add, subtract, or embellish characters and events as they see fit to shape the story according their creative visions. But researchers are not working with the same creative license. That is what makes the difference between researchers and novelists. Both need a good eye, must be able to convey essence and emotion and feeling, but researchers cannot embellish the words or actions or feelings of participants to make a point. Researchers must locate the expressed emotions, feelings, experiences, and actions within the context in which they occurred so that meaning is clear and accurate. A researcher must stay close to the data during interpretation and present findings fairly even though the data may contradict the assumptions and expectations of the researcher (Sandelowski, 1994).

### Analysis Is a Process

Analysis is a process of generating, developing, and verifying concepts—a process that builds over time and with the acquisition of data. One derives concepts from the first pieces of data. These same concepts are compared for similarities and differences against the next set of data—either expanding concepts by adding new properties and dimensions, or, if there are new ideas in the data, adding new concepts to the lists of concepts. Or, there is still a third option of revising previous concepts if after looking at the new data it appears that another term would be more suitable. It is important to keep in mind that if a researcher knew all the relevant variables and relationships in data ahead of time, there would be no need to do a qualitative study.

### Analysis Begins With Collection of the First Pieces of Data

Novice researchers often ask, "When should I begin the analysis?" Ideally, the researcher begins the analysis after completing the first interview or observation (Glaser & Strauss, 1967; Strauss, 1987). This sequential approach to data collection and analysis allows a researcher to identify relevant concepts, follow through on subsequent questions, and listen and observe in more sensitive ways. The gathering of data based on concepts is termed *theoretical sampling* and more will be said about this later, in

Chapter 7. Though alternating data collection with analysis would be ideal, there is also the reality of sometimes having to collect data without being able to immediately begin the analysis. The danger here lies in the potential inability to follow through on relevant ideas, with the final result being that some themes or concepts may be better developed than others. But a researcher does what he or she has to do, and learns to work with what he or she has. Sometimes several interviews come all at once. Or, a researcher may go to another country or city to collect data and there is little or no time between interviews and observations. All we say is that if it is possible, analysis should begin after the first data have been collected.

In addition to allowing the researcher to follow up on, validate, and develop concepts, alternating data collection with analysis prevents the analyst from becoming overwhelmed by data. It is depressing to be faced with a pile of interviews or observations and have no sense of what to do with all that data. Being immersed in data analysis during data collection provides a sense of direction, promotes greater sensitivity to data, and enables the researcher to redirect and revise interview questions or observations as he or she proceeds.

## The Use of Analytic Tools in Analysis

**Analytic tools** are the mental strategies that researchers use when coding. Codes denote the words of participants or incidents as concepts derived from observation or video. Every analyst, whether conscious of it or not, uses some mental strategies during analysis. In this book we offer a variety of techniques—some our own, others borrowed from other published analysts. In fact, we have devoted a whole chapter to exploring these techniques (see Chapter 4). The choice of tools will depend upon the training, experience, and skill of the researcher. As an analyst becomes more comfortable working with data, he or she is likely to expand the use of tools, drawing from a repertoire developed over the years. Some of the analytic tools that we believe are most relevant to analysis are asking questions and making comparisons. In the end, analysts should use those strategies that they feel most comfortable with when matching the analytic tool to the analytic task at hand.

## Microanalysis and More General Analysis

In the 1998 edition of this book we talked about a form of analysis we termed "microanalysis," a form of open coding. One of the questions that students often have is, "How does microanalysis differ from other kinds of coding?" Microanalysis is not a different form coding. It is just a more detailed type of open coding. It is designed to break open the data to consider all possible meanings. One can say that coding varies in detail from the micro, meaning very detailed, to the more macro, or general, coding less for detail and more for the general essence. Microanalysis is most likely to be used at the beginning of a project, when the analyst is trying to break into the data, to make some sense out of the materials. Microanalysis is a very valuable tool. It is like using a high-powered microscope to examine each piece of data up close. We, the authors, often begin our research projects using microanalysis.

Why do we value microanalysis so highly? We value it because it enables us to think differently about things. Think about Einstein and Darwin. They were able to arrive at conclusions that went against the conventional wisdom of the time because they carefully observed, paid close attention to detail, and kept an open mind. Blumer (1969) says it a little differently but very well:

> How does one get close to the empirical social world and dig deeply into it? This is not a simple matter of just approaching a given area and looking at it. It is a tough job requiring a high order of careful and honest probing, creative yet disciplined imagination, resourcefulness and flexibility in study, pondering over what one is finding, and a constant readiness to test and recast one's views and images of the area. (p. 40)

Though most persons agree that microanalysis is a valuable analytic tool, especially once they see it in action, there remains the misconception that we advocate the use of microanalysis throughout a project or that we are making up data that are not there. The truth is that microanalysis is used selectively and usually at the beginning of a project. Its purpose is to generate ideas, to get the researcher deep into the data, and to focus in on pieces of data that seem relevant but whose meaning remains elusive. It also helps to prevent early foreclosure because it forces a researcher to think outside of his or her frame of reference. It is time consuming and takes some practice. But the payoff is considerable.

This reminds Corbin of a story often told by her husband. When he was working as an engineering manager, problems often arose in the product line and he would send his engineers to look for the problem. Often the engineers would speculate that the problem was due to "this or that" without closely studying the matter. Based on speculation rather than observation or testing, they wanted to make changes in a procedure, changes that could be very costly if their hypotheses were wrong. Whenever engineers behaved in this way, my husband would ask, "But how do you know this is the problem? Did you study the problem and gather all the data that would confirm or negate your assumptions?" Usually the answer was "no." My husband would send the engineers back to the "field" and tell them not to return until they microscopically studied the problem and had all the necessary detail in hand. Studying data closely in the beginning does take time, but it saves time later because the researcher has a solid foundation for progressing forward. In microanalysis we are generating possibilities and at the same time checking out those possibilities against data, discarding the irrelevant, and revising interpretations as needed.

Microanalysis complements and supplements a more general analysis. Whereas microanalysis looks at the detail, general analysis steps back and looks at the data from a broader perspective: "What are all these data telling us?" It is easier to do, especially for beginning analysts who might be unsure of themselves. Though microanalysis and general analyses are used together in our approach to analysis, some researchers prefer a more general analysis, especially if they are interested in gross identification of issues and problems, and not so interested in the details involved in concept development. However, even when doing a more general analysis, one must still challenge interpretations. In doing microanalysis, one is less likely to run away with any one interpretation. Rather, each possible interpretation is checked out against incoming data before arriving at any conclusions. With more general analysis, one could easily jump to conclusions because fewer possibilities are generated. Again, it gets down to balance, not over- or underdoing micro- or general analysis but knowing just when and how to use each.

Below is an example of a class session, probably conducted in the early 1990s, in which the class is doing microanalysis under the guidance of Dr. Strauss. (See also Strauss, 1987, pp. 82–108, for a longer example of open coding.)

---

**Class Session**

Before moving into the next chapter we'd like to provide a brief example of microanalysis taken from one of our class sessions. What is so interesting about the session is how many possibilities are generated from just one small piece of data and how words often take on different meanings depending upon how they are used or interpreted. Also note how the variety of interpretations guide deeper exploration of the data and give rise to comparative analysis.

Field note quotation:

> When I heard the diagnosis, it was scary. I panicked. Everything was doing well early in this pregnancy and I felt good, no morning sickness and I had a lot of energy. Then all of a sudden I was told I had diabetes. What a shock since this is my first baby. My main concern is for the baby. I worry about the baby. I want this baby so much. I am really scared 'cause I waited so long to have this baby and I don't want anything to go wrong.

## Class Discussion and Commentary

T = teacher
S = any student

T: Let's focus on the first word "when." What could "when" mean?

S: It represents time to me. A point in time. Some time, indeterminately, in the past.

T: Well, it could stand for some time in the future. Like, "When the telephone rings I will answer, because I anticipate he will be calling."

S: "When" also stands for a *condition,* something is happening that a question forced you to look at.

T: Suppose the word isn't "when," but "whenever." What then?

S: Then it means to me there's a repeated time. A pattern of something happening.

T: So, that's a different kind of *condition* for something that follows because of some event or events.

T: But suppose instead of "when" the speaker said "at the time"?

S: Oh, then it might mean telling a story with the "when" further back in time, maybe.

T: Okay, so far we have been minutely focused on that single word, and some variant alternatives. Now, what about possible *properties* of "when"?

S: It could be sudden. Or not sudden . . . Or unexpected (or not) . . . Or the accompanying events noticed only by you, and not by others; or noticed by others too . . . Or they might be unimportant—or very important.

T: We could dream up lots of properties of this "when" and its accompanying event(s). There's no end to them, and only some of them might be relevant to your investigation and in the data, though that has to be discovered. But, notice how my *question,* force you to look at *properties* and *dimensions.* Now, let's think about the phrase "I heard the diagnosis." What about that first word, "I"?

S: Could have been we who heard—or was told the diagnosis—or they, like parents. This would have made a difference.

T: And *under what conditions* maybe would it be told to a kinsmen, or parent, or to the patient? And what might be the different consequences of this? Now what about the verb, "heard"?

S: Oh, a diagnosis might be written. Or shown to the patient (also), like on an X-ray if she were diagnosed for TB or had a shattered hip.

T: Presumably there'd be different conditions in which each of those would occur, as well as perhaps different *consequences* of them. TB is interesting, because often the diagnosis is accompanied by the listener's skepticism, therefore the physician shows the X-ray. Of course the patient is unlikely to be able to interpret it, so he/she has to take the diagnosis on faith—or reject it if not trusting—so we are talking about the issue of legitimacy of the diagnosis. That gets us methodologically into the question of the possibly different relevant *properties* of diagnoses. What might some be?

S: A partial list of properties named by the students: "difficult to make, obscure versus well known, symbolic like cancer, or not particularly symbolic, important (to oneself, to others, to the physician, to all), expected or not, awful or actually reassuring when worst is expected or preceded by days of anxious waiting, easily believable."

T: Then there are some interesting *theoretical questions* about the announcements of diagnoses, and the *structural issues* behind the answer to each. *Who* (and why)? (Your well known family physician, a strange specialist, a resident in the hospital, or if you are a child then your mother?) *How* (and why this way)? (Think of the difference between a sudden and abrupt announcement in an emergency ward, by an attending resident, to a mother that "your child has died." As compared to how coroners pace their announcement of death after knocking on the door of a spouse. Another question might be *when?* Right away, after a judicious interval, etc. Or when other the father had arrived so that both could be told about their child's death? In hospitals, if someone dies at night, the nurse usually doesn't announce on the phone but is just likely to signal that things have gotten worse, and waits for the spouse or kin to arrive so that a physician can make the announcement. "When" here also includes a parent or spouse announcing the death to other kinsmen—later, sometimes hours later, and questions about how they do that, and whether face to face or on the telephone etc.

Can those kinds of questions also stimulate questions to be asked in interviews too? Yes, they certainly can stimulate descriptive questions.

T: Now, in the next phrase in that sentence notice "everything was going well." That could possibly turn out to be an in-vivo concept, a phrase used repeatedly by pregnant woman, and so representing events probably important to them—and so it should be to us as researchers. So we take note of it, just in case it should turn out to be relevant to our work . . . What could this phrase, as such, mean analytically?

S: Well, it strikes me as indicating temporality, a course of something . . . And the course is anticipated, there's a normal course (as well as ones that go off course) . . . Which means they are evaluating whether it's normal or not . . .

T: Yes, but that means there must be criteria (properties), which in fact she names later in the sentence . . . But note also that it's she who locates herself dimensionally on this course. Analytically we can ask why she (using commonsense criteria) and not the physician or a nurse is doing the locating? What we are talking about here is a locating process and the locating agents. If you think comparatively, you can quickly see that in other situations, for different structural reasons, there will be different locating agents. Like the economists will tell you that you are entering a recession—you might never recognize you were otherwise . . . Now a related phrase here is her "early in this pregnancy." Leaving aside the "this"—for here she is surely comparing it with other one or ones—think about "early in." How does she know this!

S: Every mother knows there's nine months in the course of a pregnancy, and so can locate herself. It is cultural, commonsense knowledge.

T: Again thinking *comparatively*—and to startle you a little with an extreme but analytically stimulating comparison—think of what happened in Germany when Hitler attained high office. People interpret this event in very different ways, though with hindsight we can see that Germany was by then deep into its evolution of Nazism. Who were the locating agents? How did they know where in the course Germany was? How did they achieve legitimacy for others—or not? What were the consequences for oneself (say you were a Jew) of correctly or incorrectly reading this evolutionary course? Such questions that are raised by these kinds of comparative cases (and even extreme ones are useful early in the research), can stimulate your thinking about the properties of women such as the interviewee who is thinking about and reacting to her pregnancy in the sense of applying the same questions about "locating" to her situation, (not the idea about Nazism) . . . Notice also that these kinds of comparisons, even when not as extreme as this one about Hitler, can stimulate you to ask questions about your own assumptions and interpretations of the pregnancy data. These kinds of questions jolt you out of your standard, taken-for-granted ideas about pregnancies and their nature, and force you to consider the implications of your assumptions in making the analysis.

S: It seems to me that there is a crisscrossing of two temporal courses. There's the mother's course of a hopefully successful pregnancy. And there's the baby's course, dependent biologically certainly on the mother's physiology, but involving a different set of concerns. (The rest of the quoted paragraph certainly suggests that.) Socially they involve different actions too, like preparing for the baby's entry into the family, and acting "right" during the pregnancy for the baby's foreseen welfare.

T: You are pointing to different phenomena and you could coin two different *concepts* to stand for these, also a concept to represent what you call "crisscrossing." I would call it "intersecting" or linking, as in axial coding. You are also pointing to sequence and phases of actions and events, another aspect of the temporality noted earlier. There is also *process* or movement through phases of action.

## Summary of Important Points

Analysis is the act of giving meaning to data. Our version of analysis involves taking data apart, conceptualizing it, and developing those concepts in terms of their properties and dimensions in order to determine what the parts tell us about the whole. In the beginning of a study, analysis is usually more detailed or "microscopic" because before arriving at any interpretation, the researcher wants to explore all possibilities. Later analysis tends to be more "general" in order to fully develop and validate interpretations. In our approach to qualitative analysis, concepts form the basis of analysis and are the foundation of research whether the aim is theory building, description, or case analysis.

In brief, describing is depicting, telling a story—sometimes a very graphic and detailed one—without a lot of interpretation or attempt to explain why certain events occur and not others. Conceptual ordering is classifying events and objects along various explicitly stated dimensions, without necessarily relating the classifications to each other or developing an overarching explanatory scheme. Theorizing is the act of constructing an explanatory scheme from data that systematically integrate concepts, their properties, and dimensions, through statements of relationship. Though findings are constructions, interpretations of data as seen through the eyes of the researcher, doing qualitative research remains a valuable endeavor. It is up to us as researchers to do the best that we can, taking our analyses apart and redoing them as necessary, and remain unwilling to settle for anything less than what "feels right."

## Activities for Thinking, Writing, and Group Discussion

1. Compare and contrast qualitative versus quantitative methods of analysis.
2. Think through and write a short paragraph describing what you think a researcher is trying to achieve analytically with qualitative analysis and how that differs from the aims of quantitative analysis.
3. From your professional research journals, choose three research articles, one that represents description, one that represents conceptual ordering, and one that represents theory development. Do not necessarily rely on how the article is presented by the author. Sometimes persons call their findings theory when in fact they are not. Bring the articles to group and explain why you think these articles represent examples of description, categorizing, or theory development.

# 4

# Strategies for Qualitative Data Analysis

*The purpose of an exploratory investigation is to move toward a clearer understanding of how one's problem is to be posed, to learn what are the appropriate data, to develop ideas of what are significant lines of relation and to evolve one's conceptual tools in the light of what one is learning about the area of life. (Blumer, 1969, p. 40)*

**Table 4.1** Definition of Terms

*Analytic Tools*: Thinking techniques used by analysts to facilitate the coding process.

*Asking of Questions*: An analytic device used to open up the line of inquiry and direct theoretical sampling (see Chapter 7).

*Coding*: Deriving and developing concepts from data.

*Constant Comparisons*: The analytic process of comparing different pieces of data for similarities and differences.

*In-Vivo Codes:* Concepts using the actual words of research participants rather than being named by the analyst.

*Theoretical Comparisons*: An analytic tool used to stimulate thinking about properties and dimensions of categories.

*Theoretical Sampling*: Sampling on the basis of concepts derived from data.

## Introduction

Analysis involves what is commonly termed **coding**, taking raw data and raising it to a conceptual level. Coding is the verb and codes are the names given to the concepts derived through coding. We want to emphasize very early in this discussion that coding is more than just a paraphrasing. It is more than just noting concepts in the margins of the field notes or making a list of codes as in a computer program. It involves interacting with data (analysis) using techniques such as asking questions about the data, making comparisons between data, and so on, and in doing so, deriving concepts to stand for those data, then developing those concepts in terms of their properties and dimensions. A researcher can think of coding as "mining" the data, digging beneath the surface to discover the hidden treasures contained within data. Here is how Miles and Huberman (1994) refer to coding and its relationship to analysis:

> To review a set of fieldnotes, transcribed or synthesized and to dissect them meaningfully while keeping the relations between the parts intact, is the stuff of analysis. This part of analysis involves how you differentiate and combine the data you have retrieved and the reflections you make about this information. (p. 56)

In interacting with data, analysts make use of thinking strategies. Each analyst has his or her own repertoire of strategies—useful techniques for making sense out of data. Howard Becker refers to the strategies that he uses as "Tricks of the Trade" (Becker, 1998). The naturalist Darwin, according to Blumer, also had his strategies for working with data and for good reason. Blumer (1969), paraphrasing Darwin, states:

> Darwin, who is acknowledged as one of the world's greatest naturalistic observers on record, has noted the ease with which observation becomes and remains imprisoned by images. He recommends two ways of helping to break such captivity. One is to ask oneself all kinds of questions about what he is studying, even seemingly ludicrous questions. The posing of such questions helps to sensitize the observer to different and new perspectives. The other recommended procedure is to record all observations that challenge one's working conceptions as well as any observation that is odd and interesting even though its relevance is not immediately clear. (pp. 41–42)

The purpose of this chapter is to provide readers with a number of analytic strategies that these authors have found useful for making sense out of data. We call these thinking strategies **analytic tools** because we use them as tools, strategically and purposefully. If used properly, these tools can stimulate the analytic process, and certainly help the novice analyst get a grasp on the mountains of data that might be in his or her possession. As with any tools, analytic tools are to be used with discretion and matched to the analytic problems that arise during coding.

## Summary of Purposes of Analytic Tools

Analytic tools help analysts to:

- Distance themselves from the technical literature and personal experience that might block the ability to see new possibilities in data
- Avoid standard ways of thinking about phenomena
- Stimulate the inductive process
- Not take anything for granted
- Allow for clarification or debunking of assumptions of researchers as well as those of participants
- Listen to what people are saying and doing
- Avoid rushing past "diamonds in the rough" when examining data
- Force the asking of questions that can break through conventional thinking
- Allow fruitful labeling of concepts and provisional identification of categories
- Identify properties and dimensions of categories

### Types of Analytic Tools

As stated, each analyst has his or her own repertoire of strategies for analyzing data and there is considerable variation. So much depends upon the type of qualitative research that the researcher is engaged in, for example whether the researcher is doing content analysis, or case analysis, and so on. Some researchers, such as Miles and Huberman (1994), begin with a list of codes derived from the literature, then they revise the codes as the researchers compare the codes against actual data. Glaser (1978) provides a list of eighteen coding families, the purpose of which is to sensitize researchers to possibilities in the data and to bring analysis up to a theoretical level. Schatzman (1991) has developed an analytic process that he refers to as "dimensional analysis." He states that research findings tell a story and that researchers need a perspective to select items from the data for the story, to create their relative salience, and to sequence them. Schatzman (1991) offers the following "Matrix" (similar to Strauss's [1987] notion of axial coding) as a means of framing the story in terms of its explanatory logic. His matrix looks something like this:

The Matrix for Explanatory Paradigm

(from) Perspective

(attributes) Dimensions–Properties

(in) Context (under) Conditions

Action/Process (with) Consequences

## Designations

Since Schatzman worked closely with Strauss, his emphasis on dimensions and their importance to analysis fits very nicely with our own approach to analysis. Other researchers use different types of analytic schemes to either organize or arrive at an understanding of the data. For example, Lofland, Snow, Anderson, and Lofland (2006, p. 119) suggest the use of "focusing" as a prelude to analysis. The purpose of focusing is to do just that, get the researcher focused in on the research process. Focusing includes strategies such as examining the data for possible topics on which to concentrate, arriving at an understanding of those topics by asking questions of them, and treating them in a manner that will arouse interest. When it comes time to do the actual analysis, Lofland et al. offer the following group of sense-making strategies. These include social science framing, normalizing and managing anxiety, coding, memoing, diagramming, and thinking flexibly. (For a more in-depth discussion of these strategies see Lofland et al. [2006], pp. 119–219.) Another qualitative researcher, Dey (1993), proposes strategies such as: "using check-lists," "transposition" (what-if questions), "making free association," and "thinking by shifting sequence" to get at the essence of data.

We have developed our own strategies for probing data. These are "tried and true" strategies that Strauss has used over the years and that have proven their usefulness for analyzing data. Many of the strategies used by Strauss are similar to heuristic devices proposed by Wicker (1985) to stimulate new insights on familiar research problems. Among those heuristic devices suggested by Wicker are: (a) playing with data by applying metaphors, imagining extremes, making diagrams, and looking at process; (b) considering context by placing problems within larger domains and making comparisons outside the problem domain; (c) probing assumptions and making opposite assumptions; and (d) scrutinizing key concepts (p. 1094).

While there are many analytic strategies, two stand out. These are asking questions and making comparisons. These two strategies are the mainstay of analysis and are used by us and many other qualitative researchers. The other strategies we discuss in this chapter are also important but are used less often and for the purpose of solving analytic problems. In previous editions of this book, there was a separate chapter devoted to "Basic Operations," which includes asking questions and making comparisons. In this 3rd edition, the chapter on basic operations has been combined with the chapter on analytic tools to form one comprehensive chapter. The analytic tools we will be discussing in this chapter include:

- The use of questioning
- Making comparisons
- Thinking about the various meanings of a word
- Using the flip-flop technique
- Drawing upon personal experience
- Waving the red flag
- Looking at language
- Looking at emotions that are expressed and the situations that aroused them
- Looking for words that indicate time
- Thinking in terms of metaphors and similes
- Looking for the negative case
- "So what?" and "what if?"
- Looking at the structure of the narrative and how it is organized in terms of time or some other variable

## The Use of Questioning

The first analytic tool we will discuss is the use of questioning. It is, as Blumer (1969) emphasizes and Darwin tells us, fundamental to analysis. Every researcher wants to ask good questions, ones that will enhance the discovery of new knowledge. **Asking questions** enables the researcher to:

- Probe
- Develop provisional answers
- Think outside the box
- Become acquainted with the data

The asking of questions is a tool that is useful at every stage of analysis, from the beginning to the final writing. It helps researchers when they are blocked and having difficulty getting started with their analysis. In a book by Ann Lamott titled *Bird by Bird* (1994), the use of questions is suggested as a way getting a writing project off the ground. Lamott believes that asking questions helps a writer get past that initial block of not knowing where to start. Though Lamott is talking about writing and not data analysis, the two have a lot in common. Both have the potential to create a situation of being "blocked" and having difficulty "getting off the ground."

The questions that are asked of data in the beginning of analysis need not be earth shattering or clever. They just have to start the analyst thinking about the data. Suppose we are studying spousal caregivers and the first paragraph of our first field note says something like:

> It was a very difficult decision to put my husband in a nursing home but I couldn't physically or emotionally care for him anymore. I am 85 and it was just getting to be too much. But he died only six months after I put him there. Now I wish I had kept him home.

When we look at a piece of data, such as this one, and are just brainstorming, we can ask questions that are very exploratory. What does "getting to be too much" mean? What is this woman trying to tell us about herself, about her spouse, their relationships, and about nursing home "placement" in the context of that relationship? What is the meaning of putting her husband in the nursing home? What if she had kept him home? Then what? Would the outcome have been different? How does age of the caregiver affect placement? If the wife were younger, would she have been able to continue to care for her husband? How long did she care for him? Did she have help? All of these questions are directed at getting us to think bout what it is like for a woman to be eighty-five, in a long-term marriage, and having to place her husband in a nursing home.

Asking questions and thinking about the range of possible answers helps us to take the role of the other so that we can better understand the problem from the participant's perspective. Any answers to the questions are only provisional, but they start us thinking about what ideas we need to be looking for in the data, both from this participant as well as future ones.

To give another illustration of questioning the data, let us refer to a concept that has to do with illegal drug use among teens. While interviewing a young woman about teen drug use, the participant comments, "Getting drugs is easy for teens. There is an obliging supply network." The concept we chose to work with is "obliging supply network." To get us thinking about this topic and give us ideas of where to probe next, we might ask the data questions such as the following: Who is doing the supplying? What does the word "obliging" indicate? Where is this obliging supply network most likely to be operating—at parties, during school breaks on campus, when students go off campus for lunch, around the campus after school, at local teen hangouts? This question helps us to think about "site" and provides places to go to sample theoretically and gather more information about the concept of "supplying." What drugs are being supplied, and to whom? Now we are crosscutting "type of drug" with "supplier."

We turn now to "how." How does one go about tapping into this supply network, or how does one go about letting others know that one is in the business of "supplying"? Are there verbal or nonverbal codes that kids use to indicate the desire to purchase or sell drugs? Is there a testing-out process to determine if one is a legitimate user or seller and not a cop? What about visibility of exchange, drugs for money—how is that done to keep it hidden? What happens to kids if they can't pay for their drugs or if they get caught selling or using? If drugs are supposedly available everywhere, why is it that not everyone knows that or that not everyone uses? Still another question is "how much." How much of a supply is there, of what kind of drugs. Is the supply unlimited; that is, one can buy any drug, any time of day? Are there enough drugs at parties for everyone to get stoned, or is the purpose more to create group cohesiveness, so just a puff or two for every person is sufficient?

We are not saying that analysts must ask all of these types of questions of every single bit of data on a page. Analyzing every bit of data in this manner is not practical. It would take forever to get through one set of field notes. Being an analyst means using common sense and making choices about when and what bits of data to ask questions about. Analysts have to follow their instincts about what seems important in data and take off from there. There is no right or wrong about analysis. Nor is there a set of rules or procedures that must be followed. Analysis is, for a large part, intuitive and requires trusting the self to make the right decisions.

Asking questions of data does not take a lot of time. One can do it while driving down the street. Its value is that once one starts asking questions about data, more questions come to mind, enabling the analyst to probe deeper into the data. Deeper analysis is necessary to avoid shallow and uninteresting findings. What becomes obvious when we ask questions of the data is how much we do not know about a concept such as "obliging supply network" and how much more we need to find out. When we probe and develop a concept it becomes not just a "label" for a piece of data, but a whole new set of ideas about a phenomenon.

In addition to the questions of who, what, when, where, how, and with what consequences, there are other types of questions that are useful to think about when doing analysis. For example, it is useful to ask temporal questions, such as frequency, duration, rate, and timing. Another type of questions are spatial ones: how much space, where, circumscribed or not, open or closed. Questions of this nature give us even more insight, like where kids who sell and those who buy drugs carry out their transactions, or do they keep them hidden while on campus? Where do they sell them, how often, how long does a sale take, and is it visible or invisible to others? One could ask technological questions such as whether special equipment is needed to sell or use drugs. If so, where does this equipment come from? Or, one can ask informational questions such as who knows who is using, selling, where to buy, and so forth. Also, one could ask questions about rules, cultural values or morals, and standards (purity, in the case of drugs). All of these questions stimulate thinking about teens and drugs, and make us more sensitive to what to look for in these and future data.

For emphasis, we have categorized below the types of questions that can be asked of data, but questioning is not limited to these.

- First there are *sensitizing questions.* These tune the researcher in to what the data might be indicating. Questions of this type might look something like this: What is going on here; that is, issues, problems, concerns? Who are the actors involved? How do they define the situation? Or, what is its meaning to them? What are the various actors doing? Are their definitions and meanings the same or different? When, how, and with what consequences are they acting, and how are these the same or different for various actors, and various situations?

- Second, there are *theoretical questions.* These are questions that help the researcher to see process, variation, and so on, and to make connections between concepts. They might look as follows: What is the relationship of one concept to another; that is, how do they compare and relate at the property and dimensional level? (See section below in the making of theoretical comparisons.) What would happen if . . . ? How do events and actions change over time? What are the larger structural issues here and how do these events play into or affect what I am seeing or hearing?

- Third, there are the questions that are of a more *practical* nature. They are the questions that provide direction for theoretical sampling and that help with development of the structure of theory (if theory development is one's goal). These questions include, among many others, the following: Which concepts are well developed and which are not? Where, when, and how do I go next to gather the data for my evolving theory? What kinds of permission do I need? How long will it take? Is my developing theory logical, and if not, where are the breaks in logic? Have I reached the saturation point?

- Fourth, there are the *guiding questions.* These are the questions that guide our interviews, observations, document gathering, and analyses of these.

The questions we ask over the course of a research project will change over time. Questions are based on the evolving analysis and are specific to the particular research. Usually at the beginning of the research, questions are open ended, then tend to become more focused and refined as the research moves along. A question at the beginning of a series of interviews might look like this: Have you ever taken drugs and if so, what was the experience like for you? In later interviews, the same general question will still be relevant; however, the researcher will also want to ask questions that give further information about specific concepts, their properties, and dimensions. Later questions might resemble the following one that puts the two concepts "easily available" and "drug using" together. How does the fact that drugs are "easily available" influence the frequency, amount, and type of "drug using" that you engage in?

## Making Comparisons

Doing comparative analysis is another one of those staple features of social science research, and it is for us also. Usually, it is built into a project's design, whether explicitly or implicitly. For instance, when a sociologist compares gender behavior with respect to sexual activity; a criminologist compares the rates of homicide between ethnic groups; or an anthropologist comments on the differences between ritual, or other cultural behavior, he or she is making comparisons. Such comparative studies are often very valuable. As analytic tools, we offer two different types of comparison making. One is the making of constant comparisons and the other is the making of theoretical comparisons. We explain each below.

**Constant Comparisons.** Comparing incident with incident (as in Glaser & Strauss, 1967) in order to classify data is not difficult to comprehend. As the researcher moves along with analysis, each incident in the data is compared with other incidents for similarities and differences. Incidents found to be conceptually similar are grouped together under a higher-level descriptive concept such as "flight." This type of comparison is essential to all analysis because it allows the researcher to differentiate one category/theme from another and to identify properties and dimensions specific to that category/theme. Let's return to our study of spouse caregivers for an example. In the next paragraph our eighty-five-year-old woman caregiver goes on to say:

> Since my husband's death my life has seemed so empty. You know we were married for 65 years. That's a long time to be with somebody. Even though he was ill and in the nursing home, at least I knew he was there. Now I'm alone. I know it was time for him to die but I don't know if I'll ever get over the loneliness.

In comparing this passage with the earlier passage by the same elderly woman, we can see that each section of her interview is addressing a different phenomenon. In the first quotation, the woman is dealing with the issue of "placement" and her feelings about this. In the second, she is not only mourning her husband's death but also dealing with "loss" and having to live alone after sixty-five years of marriage. "Placement" and "loss," though related concepts in this case, are different conceptually. The analyst, in comparing the two quotes, uncovers those differences and puts them under separate codes. In subsequent interviews, incidents that are coded as "placement" will be compared with other incidents labeled as "placement" for similarities and differences within the same code. The purpose of this within-code comparison is to uncover the different properties and dimensions of the code. Each incident has the potential to bring out different aspects of the same phenomenon.

**Theoretical Comparisons.** There are times during coding when analysts come across an incident and are at loss to identify its significance or meaning. In other words, the analyst is unsure of how to classify it or is unable to define the incident in terms of its properties and dimensions. At these times, we turn to what we call theoretical comparisons. The making of theoretical comparisons requires further explanation. People are constantly thinking comparatively and make use of metaphors and similes (which is a kind of comparison making by allowing one object stand for another) when they speak. We use these techniques to clarify and to increase understanding. For instance, we might say, "Yesterday, work was like a zoo because our deadline for finishing the project was upon us. Everyone wanted something from me at once and people were running around in all directions." When we speak in this manner, "work was like a zoo," it is not the specifics of zoos that we are trying to convey but a mood or tone. It is the properties of the situation that convey these, and the properties transcend the specific situation. Words such as "demanding," "hectic," and "tense" are all properties of the situation and convey something about the tone and experiences of the day. We are not saying that we were at a zoo, but that some of the properties that we might think of as pertaining to daily life at a zoo also apply to our day.

We'll give another, more specific, example. While out grocery shopping, we come across two bins of oranges, each priced differently. In order to comprehend why they are priced differently, we might compare the two oranges, one from each bin, along certain properties such as color, size, shape, perhaps smell, firmness, juiciness, sweetness (if samples are provided), and so on. We hope that, by examining the two groups of oranges according to these dimensions or specific properties, we will come to understand why there is a price difference, then choose the oranges that are the better deal, which may not necessarily be determined by price alone. If the cheaper ones are dry, small, and tasteless, they may not be much of a bargain after all. However, commonsense comparisons are not always as systematic as those that are used in research, nor do they address theoretical issues, such as how the two bins of oranges relate to each other, or how they came to be of different sizes, shapes, and degrees of sweetness. This in turns gets us into issues such as care, soils, and temperatures, then into lobbying, price controls, and so on. The first is a way of making a classification, the second approach, which requires examining concepts in terms of their properties of dimensions, leads us to rich thick description, concept analysis, and theory development.

To summarize briefly, comparisons at the property and dimensional level provide persons with a way of knowing or understanding the world around them. People do not invent the world anew each day. Rather, they draw upon what they know to try to understand what they do not know. And, in this way, they discover what is similar and different about each object and thus define them. For example, take a bed and a sofa. We know that a bed can be used as a sofa and vice versa, but at the same time each object has its own functions and characteristics that make them unique objects, and when we study concepts we want to know what that uniqueness is.

We use theoretical comparisons in analysis for the same purposes as we do in everyday life. When we are confused or stuck about the meaning of an incident or event in our data, or when we want to think about an event or object in different ways (a range of possible meanings), we turn to theoretical comparisons. Using comparisons brings out properties, which in turn can be used to examine the incident or object in the data. The specific incidents, objects, and actions that we use when making our theoretical comparisons can be derived from the literature and experience. Take note that it is not that we use experience or literature as data, but that we use the properties and dimensions derived from the comparative incident to examine the data in front of us. Just as we do not reinvent the world around us each day, in analysis we draw upon what we know to help us understand what we don't know. Theoretical comparisons are tools designed to assist the analyst with arriving at a definition or understanding of some phenomenon by looking at the property and dimensional level. We come to know something through its properties and dimensions. If the properties are evident within the data, then we do not need to make theoretical comparisons in order to flush these out. However, because details are not always evident to the "naked" eye, and because humans are fallible in their understandings, persons sometimes do not recognize what is staring them in the face.

The mechanics of making theoretical comparisons is quite simple. We take an experience from our own life or the literature that might be similar to a phenomenon that we are studying and start thinking about it in terms of its properties and dimensions. We can do this because it is not the specifics of an experience that are relevant but the concepts and understanding that we derive from them. For example, suppose a researcher is interviewing a nurse and hears the following: "When working alone at night, I prefer to work with another experienced nurse. When I work with an inexperienced nurse I end up carrying most of the workload." To gain some understanding of what she means by "experience," we can do a theoretical comparison using an incident in our life or the literature to which "experience" and "inexperience" are relevant.

Since it is the concepts "experience" and "inexperienced" that are the focus of interest here, we can draw our comparisons from any area or situation in which experience and inexperience might make a difference, such as with driving or house painting. In making theoretical comparisons, we are still comparing incident with incident but here we are using incidents from our own experience. We are looking at the properties and dimensions of experience and inexperience in order to determine if any of those properties or dimensions that are applicable to drivers or house painters might offer some insight into what our nurse is telling us. The properties and dimensions that are derived from the "outside" situation will not be applied to the data, rather they give us ideas of what to look for in the data, making us sensitive to things we might have overlooked before. An inexperienced house painter or driver might have the properties of being cautious, apprehensive, frequently seeking direction, afraid to deviate from the pattern, prone to making errors, unsure or him- or herself, afraid to act in a crisis, and so on. Now, with some idea of what the properties of being "inexperienced" might be, we can look to the data to see if any of these are in the data, and thus understand more specifically what the nurse meant when she made her remark.

What is significant about this tool is that it forces the analyst to think at the property and dimensional level and not just at the specifics or raw data level. If we were continuing on with our study of novice nurses, we could ask questions or make observations that would give us more specific and defining information about the notions of experience and inexperience. We might want to triangulate our study by adding observation. Here we could watch for similarities and differences in how experienced and inexperienced nurses function and handle problems under various conditions, such as during routine tasks, during crisis situations, or when overworked—thereby doing **theoretical sampling**. See Chapter 7 for more information on theoretical sampling.

Sometimes when making theoretical comparisons, we use similar types of situations as the basis for comparison. Other times we use what we call "far-out comparisons." In doing so, we are following the example of the sociologist

E. C. Hughes (1971), who enjoyed making such striking and sometimes shocking comparisons as between the work of psychiatrists and prostitutes. Both belong to professions, have clients, get paid for their work, and as he states, "Take care not to become too personally involved with clients who come to them with their intimate problems" (p. 316).

The making of theoretical comparisons has a function of moving the researcher more quickly away from describing the specifics of a case, such as this particular garden is very pretty, to thinking more abstractly about what properties various gardens share in common and what is different about them. One of the difficulties that plagues beginning qualitative analysts is that they become focused on pinning down the "exact" facts or following the procedures "exactly." In doing so, they expend a great deal of energy needlessly when they should be thinking about data in the abstract.

To use another example, when we go out to buy a horse, the issue is not how many teeth a particular horse has. Rather, what is important is what the teeth—their properties (number, size, shape, if there are any sores, the pinkness of its gums, etc.) tell us about this horse's state of health and how its state of health compares to that of other horses that will be running against it in the next race. Of course, if having the winning bet is our aim, there is a lot more we would want to know about a horse, such as how fast it runs, again as compared with other horses in its class.

Both constant and theoretical comparisons can be used throughout the analysis whenever the need for such techniques arises. As the analysis proceeds, the basis for making comparisons tends to become more sophisticated and more abstract.

### *Summary of the Use of Comparisons*

- Helps analysts obtain a grasp on the meaning of events that might otherwise seem obscure
- Helps sensitize researchers to possible properties and dimensions that are in the data but remain obscure due to a lack of sensitivity on the part of the researcher
- Suggests further interview questions or observations based on evolving theoretical analysis
- Helps analysts move more quickly from the level of description to one of abstraction
- Counters the tendency to focus on a single case by immediately bringing analysis up to a more abstract level
- Forces researchers to examine their own basic assumptions, their biases, perspectives, and those of participants
- Forces examination of findings, sometimes resulting in the qualification or altering of the initial interpretations

- Makes it more likely that analysts will discover variation as well as general patterns
- Ensures the likelihood of a more fluid and creative stance toward data analysis
- Facilitates the linking and densification of categories

## Various Meanings of a Word

A third analytic strategy is to think about the various meanings of a word. During the course of an interview, we often think that we know what respondents are indicating when they say something. Then, when we get home and take a closer look at the interview we discover that perhaps we didn't really understand what was being said. And perhaps even our participant didn't know at a conscious level all that was implied in a statement.

There are various levels of meaning and various meanings that can be contained in a word or statement, especially if the meaning is left vague by the speaker. There is also the additional problem of accepting one's own interpretation of what is being said; that is, assigning meaning without careful exploration of all possible meanings.

When we talk about exploring the meaning of a word or a phrase, we do not mean that analysts should use this strategy on every word in a document. The researcher has to be selective about the choice of which words to spend time on, and explore only those that will further the analysis. Sometimes the meaning of a word is obvious from the context. Sometimes it is not so obvious. Or, there may be the concern that our taken-for-granted interpretation is not the only meaning that can be assigned to that word or phrase and that there is something deeper there but we can't get our hands around it. When this happens it is time to sit down and consider all the possible meanings that might be contained in a word or phrase. In other words, it calls for brainstorming about meaning, and including even the most far-fetched possibilities, then discarding those meanings that are improbable and irrelevant to the data.

Technically, doing analysis of a word, phrase, or sentence consists of scanning the document, or at least a couple of pages of it, then returning to focus on a word, or phrase, that strikes the analyst as being significant and analytically interesting. Then, the analyst begins to list all of the possible meanings of the word that come to mind. With this list in mind, the analyst can turn to the document and look for incidents, or words, that will point to meaning. For instance, take a phrase mentioned by a teen when talking about taking drugs, namely that teens use drugs as a "challenge to the adult stance." The word "challenge" can have many different meanings. Since our interviewee did not specify what she meant, we must speculate about what she intended. "Challenge" could mean not agreeing with the parent's understanding, asking the parent to rethink or reconsider his or her stance, making a statement of independent thinking that is judging for the self, a way of rebelling, a way of learning something about oneself or about drug use, a way of escaping from parental authority, and a way of defining who one is. All of these are possible interpretations. It is up to the analyst to discern which interpretation is most accurate by looking to the data for cues, for instance, following through with the data and searching it for cues as to which of these possible meanings makes sense within the framework of the rest of the interview. We might find that none of the possible meanings we came up with during brainstorming hold up to scrutiny when examined in light of the data. But, at least we have some ideas to rule out.

## Using the Flip-Flop Technique

Flip-flopping consists of turning a concept "inside out" or "upside down" to obtain a different perspective on a phrase or word. In other words, one looks at an opposite or extreme range of a concept to bring out its significant properties. To use another concept pertaining to teens and drug use, let us look at the word "access," which is described by our respondent as being "easy." In order to better understand what is implied by "easy" access, we can ask the opposite: What would happen to teen drug use if access was "difficult"; that is, if one had to travel a long distance to obtain drugs, ask around a lot, or pass a certain test before obtaining a drug? Would "difficult access" make a difference in the amount or type of teen drug use? Once we think through what "difficult access" might mean, we then can return to our interview with more questions to ask about what "easy" access means in terms of amount, type, and frequency of drug use. To continue with this example, if one thinks about "difficult access," one might conclude that that there might be fewer places to buy the drugs, that they might be less available at parties, or that the drugs might be more expensive. Returning to the concept "easy access," one might look for properties such as degree to which they are accessible, how much they cost, and places of purchase.

This raises other important questions: If "easy access" makes it easier for teens to use drugs, why is it that not all teens use drugs? What makes some teens take advantage of the easy accessibility, while others do not? Are some teens more adventurous, more rebellious, more curious, or more vulnerable to peer pressure? These questions then lead to further sampling along conceptual lines during data collection. Another approach would be to turn teen drug use around and look at teen "nondrug use" to see what insights that might provide. The researcher could then interview teens who have not used drugs and compare their interviews to those who do, always, of course, thinking not about specific interviews per se, but in terms of incidents of concepts, their properties, and dimensions.

## Drawing Upon Personal Experience

When we share a common culture with our research participants, and sometimes even if we don't share the same culture, we, as researchers, often have life experiences that are similar to those of our participants. It makes sense, then, to draw upon those experiences to obtain insight into what our participants are describing. It is not that our experience is identical to that of our participants, but that certain elements of our experience may be similar to theirs. For example, if we are studying elderly people and want to know how they adapt physical space to meet their functional needs, each one of us probably has an elderly parent or aunt who comes to mind when thinking about this problem. Since it is impossible to completely void our minds of our parent's or relative's experience, why not put that knowledge to good use? We can use the experience of Mom or Aunt Julia, not as data per se, but as a comparative case to stimulate thinking about various properties and dimensions of concepts.

As the authors of this book, we can hear our critics saying bias—bias at the suggestion of using personal data. We are not suggesting that a researcher impose his or her or our experience upon the data. Rather, we want to use our experiences to bring up other possibilities of meaning. Our experience may even offer a negative case, or something new to think about that will make us confront our assumptions about specific data. And, if we stay at the conceptual level when making comparisons, looking at them in terms of their properties and dimensions, it might get us to start to think more closely about what properties might be the data.

## Waving the Red Flag

Analysts, as well as research participants, bring to the investigation biases, beliefs, and assumptions. This is not necessarily a negative happening; after all, persons are the products of their cultures, the times they live in, their genders, experiences, and training. The important thing is to recognize when either our own or the respondents' biases, assumptions, or beliefs are intruding into the analysis. Recognizing this intrusion is often difficult because meanings are often taken for granted. Sometimes researchers become so engrossed in their investigations that they don't even realize that they have come to accept the assumptions or beliefs of their respondents. The researcher must walk a fine line between getting into the hearts and minds of respondents, while at the same time keeping enough distance to be able to think clearly and analytically about what is being said or done—a good reason for the researcher to keep a journal of his or her responses and feelings.

Whenever researchers hear terms such as "always" or "never," these should wave a red flag in their minds. So should terms such as, "it couldn't possibly be that way," or "everyone knows that this is the way it is." Remember, as analysts we are thinking in dimensional ranges, and words such as always, never, everyone, and no way, represent only one point along a continuum. We want to also know the "sometimes" and the conditions that are likely to lead to those dimensional variations. For example, a student in one of our seminars was studying the use of interpreters in clinics treating Asian women. The student explained that a male interpreter is called upon to interpret for a female client when no female is available. The use of men in these cases is problematic because some problems, like those involving sexual or gynecological problems, are considered too sensitive to be discussed in mixed gender company.

From an analytic standpoint, the concepts of "taboo" and "never" stand out, immediately waving a red flag in our mind. It would be very easy for persons familiar with Asian cultures to accept this stance and not raise any further questions about the matter. Yet, the concept of "taboo" brings up some very interesting questions. What happens in life-threatening situations, when a woman's life is immediately at stake? Would the woman and/or the interpreter let her die because no one is willing to talk about what is happening? Or, are there subtle ways of getting around taboos by making inferences, by providing subtle clues, or using nonverbal communication? Would a sensitive clinician who is familiar with this population pick up on what is not being said and follow up on it? Would the woman find an excuse to come back at another time? To simply accept what we are told and never question or explore issues more completely forecloses on opportunities to develop more encompassing and varied interpretations.

The analytic moral is not to take situations or sayings for granted. It is important to question everything, especially those situations where we find ourselves or our respondents "going native" or accepting the common viewpoint or perspective. Also, when we hear a term such as "sometimes," we want to explore the conditions that bring about "sometimes," and determine if there are other situations that also produce "never" or an "always." We would want to look for contradictory or opposite cases so that we might find examples of how concepts vary when conditions change. And, even if "never" is the situation, we want to know why this is, and what conditions enable this to be so. We should remember that people are very resourceful. Over the years, they seem to find strategies for managing or getting around many different types of situations. Finding these variations adds depth and gives concepts greater explanatory power.

Certain words, such as never and always, are signals to take a closer look at the data. We must ask: What is going on here? What is meant by never or always? It is never, but under what conditions? How is a state of never maintained? Or, by what strategies are persons able to circumvent a happening? What happens if a state of never is not maintained? That is, if some unaware person breaks the rules or taboos of society? Finally, we need to ask under what conditions are the rules likely to be broken and kept, and what happens then?

## Looking at Language

People often use language in interesting ways. Examining how respondents use language can tell us a lot about a situation. Take the passage we quoted earlier about the elderly woman who put her husband in a nursing home. The woman says "it was a difficult decision when I put my husband in a nursing home." Notice that she is using the first person language of I and not "we," which tells us that she views putting him in the nursing home as her decision. Does it mean that she had to somewhat "distance" herself emotionally from her husband in order to be able to place him in a nursing home? Did her husband have any input into the decision? Did she have children, and were they involved in her decision? If not, why? Does the fact that she alone made the decision potentially increase her sense of guilt?

Language is also interesting in the sense that persons often conceptualize events for us. Often the terms that they use to express something are so conceptually expressive that we can use them as a code. When someone says, "I guess I'll just have to come to terms with my disabilities," they are giving us the concept "coming to terms." When we think about it, this term is very descriptive of what happens and it would be difficult for the analyst to find a better term. When we use the words of respondents as a code we call this an **in-vivo code**, indicating that the term came out of the data. The interesting thing is that as analysts we immediately respond to such terms and most likely so will our audience. Language is often rich and very descriptive and worth paying attention to, because it can provide considerable insight into the people we are studying and where they are coming from.

## Looking at Emotions That Are Expressed

Situations or events that are significant enough to be mentioned in an interview may provoke a range of emotions in participants and in the researcher. When doing analysis it is important not to overlook expressed emotions and feelings, because they are part of context and often follow and/or are associated with action or inaction. Emotions and feelings cue the analysts as to the meaning of events to persons. Take the following lines from a field note regarding a spouse's reaction to the discovery of a lump in his wife's breast:

> When we first discovered the lump in her breast we probably reacted like most people do. At first we thought it was probably nothing, but it should be checked. I think secretly we were both very upset and scared. She did get it checked and then it became apparent that it was probably suspicious and that she would probably have to have surgery. Then we became very frightened because we had both been educated that cancer was a very life-threatening thing. And you have to act quickly to do something about it and that is what we did. (Excerpt from field notes)

What is obvious when looking at the above field note excerpt is the meaning of cancer to this couple. When the lump was "first discovered" they were upset and scared because a breast lump is often associated with cancer (though not always). Then, when the couple found out she would have to have surgery, they became "very frightened" because surgery meant cancer and cancer to them was synonymous with a "life threat." The implication of this fear was that it led the couple to a quick decision for the wife to have surgery on the affected breast. It was not the surgery that aroused such strong emotions, rather it was the word "cancer" and the fear of death.

## Looking for Words That Indicate Time

The use of "time" related words often denote a change or a shift in perceptions, in thoughts, events, or interpretations of events. Time words are words such as when, after, since, before, in case, and if. Time words help us frame events, indicate conditions, and are important when we are trying to identify context and process. Re-examine the quote from the field notes above where the husband is describing events surrounding his wife's surgery for breast cancer. The word "when" makes us take notice. It frames the events that followed and marks entry into the cancer experience. The word "then" that follows several lines below denotes a shift in the experience from it "might be cancer" and we are secretly afraid, to it is probably cancer, we need surgery and we are "very frightened" because cancer is life threatening.

## Thinking in Terms of Metaphors and Similes

We frequently use metaphors in our everyday lives to explain things to others and ourselves. When we call someone "a fox" we are implying that he or she is sly and cunning, perhaps intelligent and purposeful. If we say that someone is "like a turtle" we mean that a person is slow but persistent. Our research participants often use metaphors and similes to describe events and convey emotions. (See Lakoff & Johnson [1981] for a description of how persons use metaphors to talk about things.) Researchers can use metaphors to help them understand or explain events. For example, a person might describe undergoing cancer treatment as "going through hell" or "fighting a battle." The use of even a few words like these can carry with them a lot of meaning and paint a vivid picture. Immediately, the mind grabs on to the scenario and understands, perhaps not all but some of, what it must be like.

## Looking for the Negative Case

The negative case is a case that does not fit the pattern. It is the exception to the action/interaction/emotional response of others being studied. Though a researcher might not find a negative case, searching for that case is useful because it enables a researcher to offer an alternative explanation. Looking for the negative case provides for a fuller exploration of the dimensions of a concept. It adds richness to explanation and points out that life is not exact and that there are always exceptions to points of view.

## Other Analytic Tools

Another analytic tool is asking "so what?" to get at meaning. A researcher could ask of the couple above, "Why is making this discovery significant?" So what if there is a lump? What does it mean to this couple now and to their future? Answers to these questions can better help the researcher to understand why the couple felt that they needed to take immediate action.

Still another technique involves looking at the structure of the narrative; that is, looking at how it is organized in terms of time, or at what point in the life story the narrative starts, how it proceeds, and ends. Are there gaps in the story? Is context brought into the narrative? Looking at how participants structure the story gives the analyst some sense of how the participants locate events in their lives and the salience of these events.

Then, too, the analyst could play the "what if?" game with data. "What if" the couple described above ignored the lump; how would the scenario be different? Or "what if" the lump was discovered on a mammogram or routine physical exam rather than by the couple? Or, if the lump was discovered when the couple was on vacation rather than home? Letting the mind drift and thinking about other scenarios helps analysts return to the data with a "fresh eye." It helps them to let go of assumptions and challenges respondents' assumptions. For example, in the quotation above about the breast cancer, the respondent implies that the most natural thing in the world for a woman to do when she finds a breast lump is to run to the doctor. But is it? There are women who ignore such things and who dismiss a lump as unimportant, or are too fearful to follow through. One might ask, what is the background, education, and experiences of this couple that they became suspicious and went to a doctor? Asking questions such as these enable researchers to get at contextual factors. Why did this couple become suspicious when others might not have?

It is important to ask of data: What are the assumptions, cultural beliefs, and knowledge levels of our participants? In addition, one should also ask what the data tell us about broader societal beliefs about cancer and the American health care system. By asking these types of questions, the analyst is exploring for possible relationships between contextual factors and any action/interaction/emotional response that is noted. Asking a variety of questions enables analysts to develop categories in terms of all their ramifications.

## Summary of Important Points

In this chapter we've presented a set of analytic tools. We expect that analysts will use the tools like any good craftsman, flexibly and as extensions of their own abilities. As analysts we want to generate findings that have substance and contribute to knowledge development in our professional fields. To generate new knowledge requires sensitivity to the multilayers of meaning that are embedded in data. Analytic tools are heuristic devices that promote interaction between the analyst and the data, and that assist the analyst to understand possible meaning. Analytic tools are used to probe the data, stimulate conceptual thinking, increase sensitivity, provoke alternative interpretations of data, and generate the free flow of ideas.

In addition, the thoughtful use of analytic tools fosters awareness of how bias and assumptions influence the direction of analysis. Though some analysts claim to be able to "bracket" their beliefs and perspectives when analyzing data, we have found this impossible. Bias and assumptions are often so deeply ingrained and cultural in nature that analysts often are unaware of their influence during analysis. We find it more helpful to acknowledge our biases and experiences and consciously use experience to enhance the analytic process. In addition to the use of analytic tools described here, we suggest that analysts keep a personal journal during the data gathering and analytic processes. Journal keeping provides a record of the thoughts, actions, and feelings that are aroused during the research. An important part of doing analysis is reflecting back on who we are and how we are shaped and changed by the research. Inevitably we are shaped by, as well as shapers of, our research.

## Activities for Thinking, Writing, and Group Discussion

1. As a class or alone, apply some of the techniques described above to analyze passages from the field notes that accompany this chapter in Appendix A. These field notes are taken from a biographical study exploring the meaning of life-threatening events to persons and how they incorporate these events into their lives. The event in these field notes is chest pain. If you prefer, you may use some of your own or a group member's field notes. What is important is that you take the time to practice these activities and make them part of your own thinking process. Without practice, the use of these tools becomes forced rather than skillful. Share the results of your analysis with the group and explain how you think their use enhanced the analysis—what did you think about that you might not have if you had not used them?

2. Think about other analytic techniques that you as an analyst might add to the list of analytic techniques. Discuss these with your group.

# 5

# Introduction to Context, Process, and Theoretical Integration

*The relation of the event to its preceding conditions at once sets up a history, and the uniqueness of the event makes that history relative to that event. . . . All of the past is in the present as the conditioning nature of passage, and all the future arises out of the present as the unique events that transpire. (Mead, 1959, p. 32)*

**Table 5.1** Definition of Terms

*Conditional/Consequential Matrix*: An analytic strategy useful for helping analysts to consider the wide range of possible conditions and consequences that can enter into context.

*Context*: Structural conditions that shape the nature of situations, circumstances, or problems to which individuals respond by means of action/interaction/emotions. Contextual conditions range from the most macro to the micro.

*Integration:* Linking categories around a central or core category and refining the resulting theoretical formulation.

*Paradigm*: An analytic strategy for integrating structure with process.

*Process*: The flow of action/interaction/emotions that occurs in response to events, situations, or problems. A change in structural conditions may call for adjustments in activities, interactions, and emotional responses. Actions/interactions/emotions may be strategic, routine, random, novel, automatic, and/or thoughtful.

## Introduction

In the previous chapter, we presented a series of analytic tools or strategies that we have found useful for analyzing data. In this chapter, we will take up where that chapter left off, presenting strategies for analyzing data for context, process, and theoretical integration. We'll begin the chapter with a discussion of context, move on to process, and finish with a section on theoretical integration. This is a rather lengthy chapter, so we won't bother with a long introduction. However, we would like to add that once readers of this text have read the chapters demonstrating analysis—that is, Chapters 8 through 12—they may want to return and study this chapter more deeply.

## Context

**Context** doesn't determine experience or set the course of action, but it does identify the sets of conditions in which problems and/or situations arise and to which persons respond through some form of action/interaction and emotion (process), and in doing so it brings about consequences that in turn might go back to impact upon conditions. Readers might recognize in the above statement our philosophic beliefs about the nature of events and human response to these as arising out of Symbolic Interactionism and Pragmatism as discussed in Chapter 1. That is, persons play an active role in shaping their lives by the way they handle or fail to handle the events or problems they encounter, and their action/interactions/emotional responses based, of course, on their perceptions of those events.

Analyzing data for context is not much different from analyzing data for concepts or categories. The researcher will continue to ask questions and make comparisons. In fact, when a researcher is doing initial coding early in the analysis, it is likely that some of the concepts delineated from data will eventually be identified as pertaining to context. However, once a researcher has identified a concept as pertaining to context and wants to open up or elaborate upon contextual concept(s), additional strategies might be necessary. While more experienced researchers may intuitively be able to identify the wide range of conditions that enter into situations and define problems, novice researchers may need more direction on how and where to look for context. This is where the Paradigm and the Conditional/Consequential Matrix become useful as analytic tools.

## The Paradigm

The data that qualitative researchers work with are complex. They consist of multiple concepts existing in complex relationships that are often difficult to tease out of the data. Having a way to think about those relationships can be helpful. One tool for helping the researcher to identify contextual factors and then to link them with process is what we call the **paradigm.** The paradigm is a perspective, a set of questions that can be applied to data to help the analyst draw out the contextual factors and identify relationships between context and process.

The terminology used in the paradigm is borrowed from standard scientific terms and provides a familiar language facilitating discussion among scientists. In addition, the basic terms used in the paradigm follow the logic expressed by persons in their everyday descriptions of things. The basic components of the paradigm are as follows:

1. *There are conditions.* These allow a conceptual way of grouping answers to the questions about why, where, how, and what happens. For example, if one hears something like, "When I first heard that she said I lied about the relationship, I became so angry that I walked out of the room." The word "when" here is an analytic cue. It focuses the researcher's attention on what follows. The individual is revealing the circumstances or conditions that lead him or her to make a particular response.

2. *There are inter/actions and emotions.* These are the responses made by individuals or groups to situations, problems, happenings, and events. In the example above, "hearing that she said I lied" was the condition or reason given for the respondent becoming angry and walking out of the room.

3. *There are consequences.* These are outcomes of inter/actions or of emotional responses to events. Consequences answer the questions about what happened as a result of those inter/actions or emotional responses. Take the following description, "After I walked out of the room, I realized how foolish I appeared and went back to apologize."

Answers to questions denoting conditions or reasons such as why, when, where, and what may be implicit or explicit in field notes. That is, sometimes persons use words that cue analysts that they are about to explain or give a reason for behavior, such as "since," "due to," "when," and "because" followed by the event or action. In reading a memoir about the Vietnam War, a researcher might read something like, "When the enemy started putting more mines along the trails we normally follow through the jungle, our casualty rates increased" (a change in conditions). "To minimize casualties we had to be very watchful of where and how we walked down the trails, always looking for cues like fine wires" (action/interaction/emotional strategies for handling a problematic situation). "If we missed one, we had our legs or worse blown off" (consequence). This descriptive incident tells us that something changed in the conditions. As a result, American soldiers had to be more careful as they carried out their patrols. If they missed the cues, then they were likely to suffer the consequences.

As a researcher codes data, the conceptual names that are placed on categories do not necessarily point to whether a category denotes a condition, inter/action, emotional response, or consequence. The analyst has to make these distinctions. An important point to remember is that the paradigm is only a tool and not a set of directives. The analyst is not coding for conditions or consequences per se, but rather uses the tool to obtain an understanding of the circumstances that surround events and therefore enrich the analysis. A common mistake among beginning analysts is that they fixate on the specifics of the paradigm rather than thinking about the logic behind its use and what this use of paradigm is designed to do. Being overly concerned about identifying "conditions" or "strategies" or "consequences" in data rigidifies the analytic process. The final results may be technically correct but there is something missing, and that something is the creativity and feeling that give qualitative research its soul.

## The Conditional/Consequential Matrix

The paradigm provides cues for how to identify and relate structure to process. It suggests looking for key words that signal a line of action or an explanation for something, then following that thought through in the data. While the paradigm is helpful for thinking about context, in of itself it is incomplete. What the paradigm doesn't do is: (a) address the many possible theoretical sampling (see Chapter 7) choices that an analyst must make during the research process or where to look for contextual factors; (b) explain the varied, dynamic, and complex ways in which conditions, inter/actions, and consequences can coexist and impact upon each other; (c) account for the different perceptions, constructions, and standpoints of the various actors; (d) put all the various pieces together to present an overall picture of what is going on; and (e) most of all, it emphasizes that both micro and macro conditions are important to the analysis. In this next section we introduce the **Conditional/Consequential Matrix**, henceforth referred to as the

Matrix. The Matrix enriches analysis by helping the analyst sort through the range of conditions/consequences in which events are located and responded to. We do not believe that every possible condition must be brought into the research. What is important is that research findings don't oversimplify phenomena, but rather capture some of the complexity of life. It is up to the researcher to determine just how complex he or she wants the final findings to be. For researchers who want to bring complexity into their findings, we suggest thinking in terms of the Matrix. (For an excellent discussion of the Matrix and how it can be used in research see Hildebrand, 2007.)

## Ideas Contained in the Matrix

The ideas contained within the Matrix are as follows:

1. *Conditions/consequences do not exist in a vacuum.* They are always connected through action/interaction/emotional responses. Since one event often leads to another, and to another, like links in a chain, the relationships between events are often complex and difficult to sort through. Furthermore, the relationships between conditions and subsequent inter/action and consequences rarely follow a linear path. Conditions and subsequent action are more likely to bounce off one another like billiard balls, leading to consequences that one cannot always predict in advance. This point is brought out very clearly in the following quote taken from a book written by McMaster (1997) about the Vietnam War:

> The Americanization of the Vietnam War between 1963 and 1965 was the product of an unusual interaction of personalities and circumstances. The escalation of U.S. military intervention grew out of a complicated chain of events and a complex web of decisions that slowly transformed the conflict in Vietnam into an American war. (p. 323)

2. *The distinction between micro and macro is an artificial one.* Most situations are a combination of micro conditions (those closer to the individual) and macro conditions (those more distant from the individual, i.e., historical, social, political, etc., conditions). When appropriate, the analyst should trace the relationships between micro and macro to problems, situations, and events. As analysts, we are interested in the interplay between micro and macro conditions, the nature of their influence on each other and subsequent inter/action, and the full scope of consequences that result, then how those consequences feed back into conditions that become part of the situation and subsequent inter/action or emotional responses. For example, some American soldiers serving in Vietnam during the war made the connection that the war

was escalating by noting that an increased number of American troops were coming into Vietnam and that there was an increase in the number of casualties. The soldiers may not have been aware of the governmental policies that brought about that change, but they felt the results in the front lines. Analysts can add complexity to their research by picking up on an incident such as "the noting of an increase in the number of casualties" and tracing it back through the different levels of the Matrix to determine why this might be happening.

Consider the following quotation, for it demonstrates how the micro and the macro conditions fuse to create certain events or situations. Here, McMaster (1997) is explaining his thoughts about how the United States got into the Vietnam War. One can see him putting the micro (Lyndon Baines Johnson's personal issues and agenda) and the macro (wider social issues) in this one paragraph:

> Between November 1963 and July 1965, L.B.J. made the critical decisions that took the United States into war almost without realizing it. The decisions, and the way in which he made them, profoundly affected the way the United States fought in Vietnam. Although impersonal forces, such as the ideological imperative of containing Communism, the bureaucratic structure, and institutional priorities, influenced the president's Vietnam decisions, those decisions depended primarily on his character, his motivations, and his relationships with his principal advisers. (p. 324)

3. *The full range of possible interrelationships between micro/macro conditions are not always visible to individual research participants.* Each comes to the situation from his or her standpoint or perspective and rarely has a grasp of the whole situation. It takes listening to many voices to gain understanding of the whole.

4. *Conditions and consequences usually exist in clusters and can associate or covary in many different ways, both to each other and to the related inter/action* Furthermore, with time, and the advent of contingencies, the clusters of conditions and consequences can either change or rearrange themselves so that the nature of relationships or associations that exist between them and the inter/action also changes.

5. *Action/Interaction and emotional responses to events are not confined to individuals.* They can be carried out by representatives acting in behalf of nations, organizations, and social worlds. Furthermore, inter/action and emotional responses can be directed at individual or groups representative of nations, organizations, social worlds, and so on. This point is illustrated in an event reported by Lt. Alvarez, a navy pilot shot down early in the Vietnam War and taken prisoner by the North Vietnamese. One day

a group of prisoners, including Alvarez, was taken from their prison cells and paraded through the streets of Hanoi. As the prisoners made their way through the streets, the Vietnamese people who were lining the streets began to verbally and physically abuse the prisoners. Though these men were prisoners and no longer dropping bombs, they remained symbols of America and the war, thus were viewed as legitimate targets of the North Vietnamese people's anger (Alvarez & Pitch, 1989, pp. 144–149).

## Diverse Patterns of Connectivity

The analytic picture presented in the discussion above is one of multiple and diverse patterns of connectivity with discernible shifting patterns of inter/action over time. Though experienced researchers often have their own devices for keeping track of these complex sets of relationships, a researcher new to qualitative analysis may feel overwhelmed. It is important to remember that not every path a researcher follows will lead to discovery of an analytic gold mine. Nor is it ever possible to discern all the possible connections between conditions, action/interaction, and consequences. Every analyst has to accept that there are limitations to what can be discovered based on access to data, degree of analytic experience, and amount of personal reserves. We acknowledge that doing this complex analytic work is not easy and that persons reading this book will pick and choose where to go with their research efforts.

## Description of the Matrix

All this time we have been talking about the Matrix as a set of ideas. The problem lies in translating abstract ideas into an easily understood diagram. The diagram we have devised does not capture the complexity of the set of ideas we have presented above.

The Matrix consists of a series of concentric and interconnected circles with arrows going both toward and away from the center. The arrows represent the intersection of conditions/consequences and the resulting chain of events. Conditions move toward and surround the inter/action to create a conditional context. Other arrows move away from inter/action, representing how the consequences of any inter/action move from inter/action to change or add to conditions in often diverse and unanticipated ways. One of the limitations of the diagram is that the flow appears linear. A more likely metaphor is billiard balls each striking the other at different angles, setting off a chain reaction that ends with knocking the appropriate ball(s) into the pockets.

The Matrix is meant only to be a conceptual guide and not a definitive procedure. The Matrix can be modified to fit each study and data. To maximize

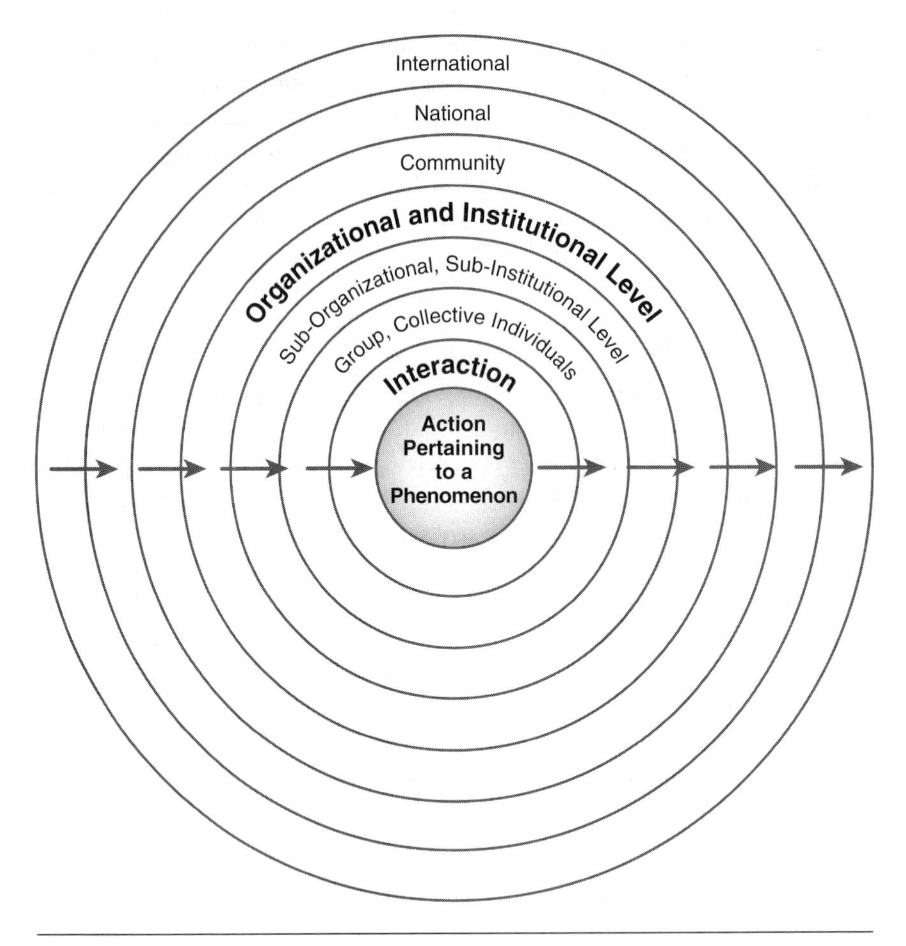


**Figure 5.1** The Conditional/Consequential Matrix

use of the Matrix as an analytic tool, each area is presented in its most abstract form. Items (sources of condition/consequences) to be included in each area will emerge from the study, thus depend upon the type and scope of the phenomenon being studied. Usually researchers using the Matrix have altered the classification scheme to suit their own purposes or, based on their critiques, developed alternative approaches (see Clarke, 2005; Dey, 1999; Guessing, 1995).

Beginning at the outer edges of the circle we have placed the most macro area represented by the term "international" or "global." This area includes, but is not limited to, such items as: international politics, governmental regulations, agreements or differences among governments, culture, values, philosophies, economics, history, and international problems and issues, such as "global environmental warming." The next area we have designated as the "national" or "regional" area. Included in this area are potential conditions such as: national/regional politics, governmental regulations, institutions, history, values, and national attitudes toward gender relationships and behaviors. Next come conditions from what has been designated as the "community" area. Included in this area are all of the above items but as they pertain to a particular community, giving it singularity among all other communities. The next circles represent the "organizational" and "institutional" areas. Each organization or institution has its own purposes, structure, rules, problems, histories, relationships, spatial features, and so forth, which provide sources of conditions. (Some institutions, such as religious ones, might be international in scope, but how rules are interpreted and practiced are often individualized to communities or even to individuals.) Still another circle represents the "sub-organizational" and "sub-institutional" areas. Most important, in the center, action/interaction/emotional responses are located.[1]

A researcher could study any substantive topic within any area of the Matrix. For example, he or she might study health care at a national level, focusing on recent legislation, policies, and emerging organizations and trends. Or a researcher could move down several levels of the Matrix and study the management of chronic illness by families. Regardless of area of focus, it is important to keep in mind that conditions in the outer levels such as health care policies of a nation (e.g., national health care or the lack of it) will affect individual and family management of illness and, conversely, problems that arise in individual or family illness management can have an impact on future legislation affecting health care policy. Other substantive areas that might be studied include, but are not limited to: identity, decision making, social movements, arenas, conflict and consensus, awareness, social change, work, information flow, and moral dilemmas. Each of these can be studied within any area. Time, history, biography, space, economics, gender, power, politics, and so on, are all potential conditions that *can be* relevant to any substantive area studied in any area outlined in the Matrix. The important thing is that no item (be it gender, age, power, etc.) should be stated as being relevant to the evolving story unless there are data to support it.

## Process

**Process** is an elusive term. It is as difficult to explain as it is to capture in data. Perhaps the best way to begin our discussion is to present two scenarios unrelated to research but that nevertheless will help students to understand.

When listening to a piece of music (well, some music anyway), one can't help but be struck by all the variations in tone and sound. We know that music, whether it be jazz, popular, or classical, is composed of a series of notes, some played fast, some slow, some loud, others soft, sometimes in one key, sometimes in another, often with movement back and forth between keys. Even the pauses have purpose and are part of the sound. It is the playing of these notes with all their variations and in coordinated sequences that gives music its sense of movement, rhythm, fluidity, and continuity.

To us, process is like a piece of music. It represents the rhythm, changing and repetitive forms, pauses, interruptions, and varying movements, that are part of sequences of inter/action and that give rise to emotional responses to events. The next scenario is perhaps an even more graphic illustration of our understanding of process. Recently, one of the authors was seated in the waiting room of a small airport. Having nothing to do but wait, she began to take an interest in what was going on in the coffee shop nearby. It was a modest shop, of a type that can be found in any small town in the United States. There were between twenty and twenty-five persons seated around the room at tables and at the counter. There was one waitress and one cook. The waitress moved from table to table, taking orders, bringing the orders to the cook, who after preparing the food, gave it to the waitress to be delivered to the waiting customers. The same waitress also received the money from customers and rang it up in the cash register. From time to time, the waitress stopped to talk to customers, poured more coffee, cleared the tables, and generally kept moving, her eyes ever watchful for signs of customer needs. Though each situation was a little different, and her inter/actions differed in form and content over the time she was observed, they were all part of a series of acts pertaining to an overall process that might be called "food service work." While the waitress was doing her work, the patrons were eating, talking, and watching the small private planes arrive and depart.

The scene was not a very dramatic one. In fact, it was quite routine and surely repeated day after day in much the same way in coffee shops all over the country. Though routine, there was continuous flow of activity, with one sequence of actions flowing into another. There were interruptions to the flow of work and small problems to be solved, but these tended to be resolved as part of the ongoing flow of action. Watching the scene made the observer realize, "Ah, now that is process!"

## What Is Process?

Process is ongoing action/interaction/emotion taken in response to situations, or problems, often with the purpose of reaching a goal or handling a problem. The actions/interactions/emotions occur over time, involve sequences of different activities and interactions and emotional responses (though not always obvious), and have a sense of purpose and continuity. Structure (context) and process are related because persons act in response to something, the something being the issues, problems, situations, goals, and events occurring in their lives. The relationship between structure and process is very complex, leading to infinite variation in the intensity, type, and timing of action/interaction/emotional responses. As contextual conditions change, adjustments are made in action/interaction/emotions. Of course, any action/interaction/emotion response to goal accomplishment, situation, event, or set of circumstances depends upon how the individual or group perceives or defines it, and the meanings that they give those situations. That is why one sees so much variation in action/interaction/emotion in similar situations; persons are likely to define situations differently or give them different meanings.

This means that if one or more persons are acting together to reach a goal or manage a problem, they must bring their actions/emotions into alignment or the flow and continuity will be disrupted. Take the restaurant described above. The waitress and the cook must align their actions in order to serve the customers within a reasonable period of time. The cook prepares the food, and the waiter or waitress serves the food. Customers also have to align their actions with those of the cook and waitress or waiter, meaning that they have to wait patiently or impatiently until the food is served. Now imagine what would happen to the action or the "flow of work" if we varied the conditions. What if several large groups of persons come in at the same time, with still only one cook and one waitress to wait on them? Imagine how this would change the pacing of the work—the cook's ability to take orders, prepare the food, or the waiter/waitress's ability to talk with customers, to pour that extra cup of coffee, and serve the food before it cooled down. Customers might have to wait longer for their food and their emotions might flare as a response. What if the cook suddenly became ill and the waitress had to cook as well as serve the food? Or if the cook asked everyone to leave and eat someplace else? What if there were five waitresses but only twenty customers? What would all the different waitresses be doing to pass the time away? Suppose that a waitress was inexperienced and slow and the customers got tired of waiting for their food. Would the pleasant friendly inter/action taking place between customers and waitress turn into one of impatience, frustration, and even anger? Each of these different scenarios could potentially alter or shift the nature of the inter/action/emotional response. Since structure over time tends to change (think of change in the number and type of customers coming and going from the restaurant), there must be adjustments in inter/action and emotion to stay aligned with the flow of customers and the circumstances.

## Variable Nature of Process

One could say that, at best, process is like a coordinated ballet or symphony, each movement graceful, aligned, purposeful, sometimes thoughtful, other times routine, with one action flowing into another. At its worst, it might resemble a soccer riot, the acts misaligned, disrupted, random, uncontrolled, nondirected, and sometimes hurtful. Most human inter/action and emotional responses probably lie somewhere in between. The sequences of action/interaction/emotion do not proceed as gracefully as in a ballet, nor are they as chaotic as in a riot. In fact, much of what we see and hear as analysts can be dull and routine. Process demonstrates an individual's, organization's, and group's ability to give meaning to and respond to problems and/or shape the situations that they find themselves to be in through sequences of action/interaction, taking into account their readings of the situations and emotional responses to them. In addition, process illustrates how groups can align or misalign their inter/actions/emotional responses and in doing so maintain social order, put on a play, have a party, do work, create chaos, or fight a war. As researchers, when we analyze data for process, we are trying to capture the dynamic quality of inter/action and emotions.

## Conceptualizing Process

Process in data is represented by sequences of action/interaction/emotions changing in response to sets of circumstances, events, or situations. How one conceptualizes or describes process is determined by the content of the data and a researcher's interpretation of these.

Process is often described in developmental terms such as phases or stages, implying a linear or progressive nature to it. However, not all process is developmental or progressive. It can be chaotic. It can move upward for a while, then turn downward, or it may proceed circularly. Think of the example of the restaurant above. There was nothing developmental or progressive about the waitress, cook, or customers' action/interactions/emotional responses. However, to say that a waitress "takes the order," "communicates the order," "picks up the order," and "delivers the order" is a rather dull description and requires one to stretch the imagination to think of the action/interaction as developmental. Such description does not capture the dynamic quality of the busy scene. In the case of the restaurant, the action/interaction is much more circular, beginning and ending with the customer.

Another way of describing process is as sequences or a series of actions/interactions/emotions taken in response to situations or problems, or for the

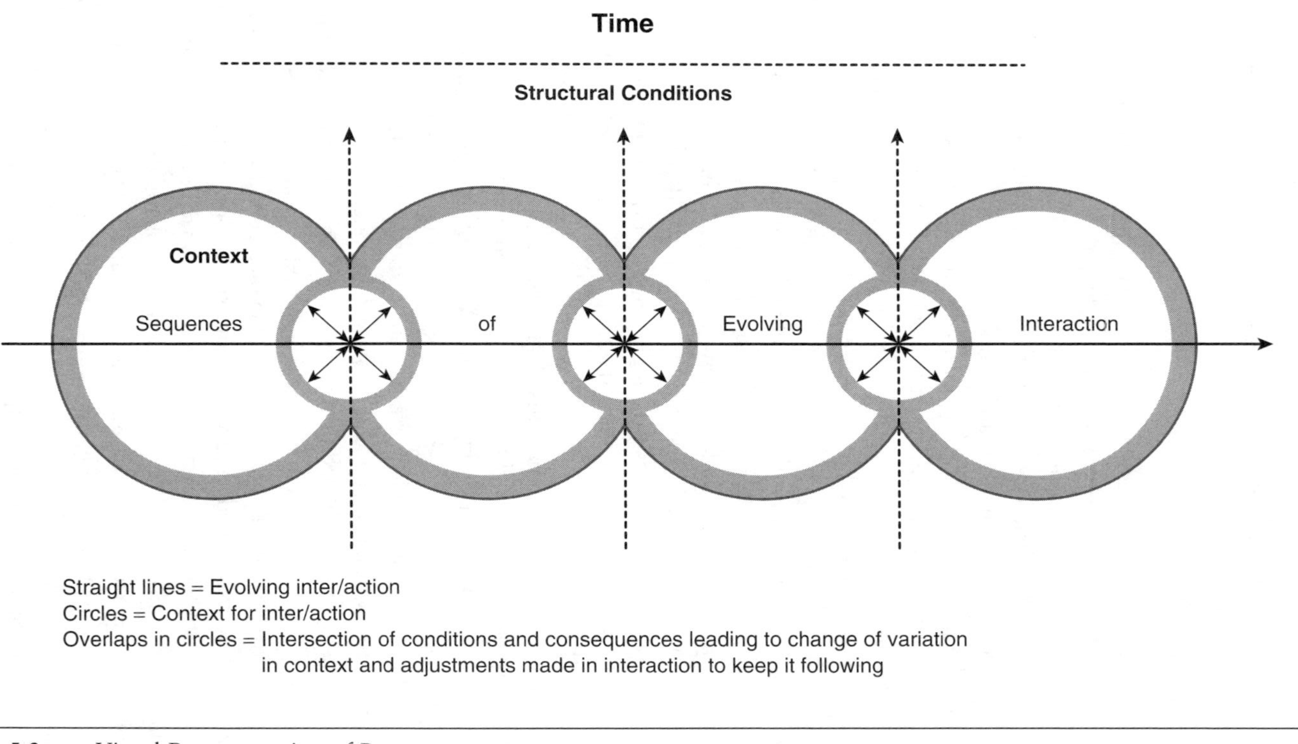

**Figure 5.2** Visual Representation of Process

purpose of reaching a goal as persons attempt to carry out tasks, solve certain problems, or manage events in their lives. Processes can be psychosocial. They can also be educational, legal, managerial, political, military, and so on.

## Analyzing Data for Process

Analyzing the data for process has certain advantages. In addition to giving findings a sense of "life" or movement, analyzing data for process encourages the incorporation of variation into the findings. Along with variation, process can lead to the identification of patterns as one looks for similarities in the way persons define situations and handle them. And, if one's final goal is theory building, analyzing data for process is an essential step along the way. Finally, in relating process to structure, one is in fact linking categories.

Just because an inter/action is routine does not mean that it is not important. Studying routine has broad implications for knowledge development. It enables researchers to identify the patterns of inter/action/emotional response that make it possible to "establish" and "maintain" social and personal stability in the face of contingencies (possible but uncertain or unpredicted happenings), thereby expanding our understanding of everyday life. For example, when Corbin and Strauss were studying work in hospitals, they discovered that each hospital unit had established routines in the form of tradition, policies, procedures, and rules that enabled the units to function on an ongoing basis and deliver patient care. There were even policies established to handle unplanned events, such as when patients have a cardiac arrest.

Some questions that one might ask of data when analyzing for process include the following: What is going on here? What are the problems or situations as defined by participants? What are the structural conditions that gave rise to those situations? How are persons responding to these through inter/action and emotional responses? How are these changing over time? Are inter/actions/emotions aligned or misaligned? What conditions/activities connect one sequence of events to another? What happens to the form, flow, continuity, and rhythm of inter/action/emotions when conditions change; that is, do they become misaligned, or are they interrupted, or disrupted because of contingency (unplanned or unexpected changes in conditions)? How is action/interaction/emotion taken in response to problems or contingencies similar or different from inter/action that is routine? How do the consequences of one set of inter/actions/emotions play into the next sequence of inter/actions?

The latter question is extremely important because it enables researchers to see how inter/actions/emotions have consequences, and these often become part of the conditional context in which the next set of inter/action/emotional responses occurs. Consequences that feed back into the original context or situation may alter inter/action, to maintain the status quo or disrupt it. For instance, being overtaken by fear in the heat of battle may cause a soldier to freeze in the face of enemy fire with a consequence of his being killed or injured. Whereas controlling that fear and channeling it into productive action may help that soldier to survive.

There is another major point to be made about process before moving on. Sometimes persons will ask, what is the difference between a phenomenon and a process? This is rather confusing at times. To us, phenomenon stands for the topic, the event, the happening, the goal, or the major idea (category or theme) contained in a set of data. Process stands for the means of getting there. For example, "survival" is a phenomenon. Combatants who went to Vietnam wanted to survive the war experience. Process represents the strategies or the means by which combatants attempted to handle the problems that stood in the way of their surviving.

## Sub-Processes

Process can be broken down into sub-processes. Sub-processes are also concepts; they explain in more detail how the larger process is expressed. For example Corbin, in her study of pregnant women with chronic illness, defined the major process to be "protective governing." Protective governing represented the means (the process) by which the pregnant woman, her partner, and the health team worked together to minimize the risks associated with such a pregnancy and maximize the chances of a positive outcome for mother and baby. Protective governing was broken down into the sub-processes of "assessing," "balancing," and "controlling" the risks. Each sub-process in turn consisted of certain risk-managing tactics or strategies that tended to change as definitions of the level of risks changed over the course of the pregnancy. However, the major process of "protective governing" and its sub-processes of "assessing," "balancing," and "controlling" remained consistent throughout the course of the pregnancy. It was the specifics of action/interaction/emotion (the management strategies) that changed in response to changes in the status of the pregnancy and illness conditions, and that in turn brought about a reduction in those risks.

## Process Analysis at Both Micro and Macro Levels

Process can be studied at any level of the Conditional/Consequential Matrix as presented previously. For example, one could analyze data at a national level in order to determine how the United States came to be fighting in Vietnam, looking at the historical and political processes that led us there. Or, one could look at how social activism, beginning at a local grassroots level, then progressing on to a national level (even international level if one includes the role of the North Vietnamese), shaped the outcome of the war. Or even further, one could look at the peace movement as a collective movement, with the idea of studying how such movements begin, are maintained, and finally fold. One could also study the interrelationship of biography (a more inner level of the Matrix) to war by examining how President Johnson's character, fears, policies, dreams, failings, and the like, shaped the war and its outcome. And in turn, one could examine how the outcome of the war impacted Johnson's biography. Additionally, one could study how a particular village or community in Vietnam managed to keep going despite war going on within and around it. What we are trying to point out is that one can study process at any level of the Matrix and that any study of process often crosses levels of the Matrix. In the pages of the book that follow, Corbin studies front-line soldiers' experience because she came upon an interview with a Vietnam veteran and got hooked on the topic. But an equally interesting study might be the history of Vietnam and the interplay between history, culture, and politics in generating and maintaining the war.

## Analyzing Data for Process at a Formal Theory Level

What happens when the focus of the research is building "formal," rather than generating "substantive," theory? Is the analysis much different? When building general theory it is not so much the questions that are different, it is the approach to data collection that is different. General theory building is concept driven. One begins with a concept such as "awareness" or "stigma" and samples theoretically, but in this case compares and contrasts data across research contexts. The concept chosen for study usually is derived from and builds upon a researcher's previous line of research, though it need not be. The idea is to raise the concept of the study up to a more abstract level where it can have broader applicability but at the same time remain grounded in data. For example, Strauss (1978) wanted to formulate a formal theory of "negotiations." He began with a concept of negotiation, and examined "negotiations" in a variety of contexts. He looked at negotiations between representatives of nations, judges in law courts, political machines, clans and ethnic groups, and also insurance companies and their clients. By comparing and contrasting these various groups for similarities and differences, he was able to identify the components of the negotiation process that transcended all groups, providing a more abstract and broader understanding of negotiations. At the same time, he was able to identify those aspects of negotiation that were particular to each group, showing the wide range of variation between groups. On the other hand, a researcher interested in concept development at a substantive level of negotiation would confine data collection to one main group, for instance negotiations between buyer and seller in a housing transaction.

## Techniques for Achieving Theoretical Integration

As stated throughout this book, not every qualitative researcher is interested in building theory. But for those who do want to proceed to theoretical **integration** we offer the following suggestions.

An umbrella has many spokes. The spokes provide structure and give the umbrella form or shape. But it is not until the spokes are covered with some kind of material that the object becomes an umbrella and useful for keeping rain off of the person. In other words, spokes alone don't make an umbrella. The same is true for theory. Concepts alone do not make theory. Concepts must be linked and filled in with detail to construct theory out of data. Admittedly, integration is not easy for novice researchers. As stated by Paul Atkinson, a coauthor of an excellent textbook on field research (Hammersley & Atkinson, 1983) in a personal communication:

> This aspect—making it all come together—is one of the most difficult things of all, isn't it? Quite apart from actually achieving it, it is hard to inject the right mix of (a) faith that it can and will be achieved; (b) recognition that it has to be worked at, and isn't based on romantic inspiration; (c) that it isn't like a solution to a puzzle or math problem, but has to be created; (d) that you can't always pack everything into one version, and that any one project could yield several different ways of bringing it together.

This section presents several analytic techniques designed to help researchers achieve integration, the final analytic step for those interested in theory building. The techniques are especially useful when analysts are perplexed, that is sensing that the data are beginning to "gel," but not quite sure how to explicate those intuitive feelings on paper. This section will also discuss procedures for refining the theory, once an analyst has committed to a theoretical scheme.

There are several important ideas to keep in mind while reading this section:

1. As stated earlier in the book, concepts that reach the status of a category are abstractions. They represent the stories of many persons or groups reduced into and depicted by several highly conceptual terms. Any theoretical formulation that is generated based on these concepts should have general applicability to all the cases in a study. It is the details included under each category and/or subcategory, through the specification of properties and dimensions, that bring out the differences and variations in each case.

2. If theory building is indeed the research goal, then findings should be presented as a set of interrelated concepts, not just a listing of themes. It is the overall unifying explanatory scheme that raises findings to the level of theory. The subconcepts with all their properties and dimensions provide the detail. Concepts are related through statements that denote the nature of the relationship. These statements, like concepts, are derived through analysis of data. Just as with concepts, relational statements represent the analyst's interpretation of what is going on in the data. Rarely are the concepts or relational statements the exact words of one respondent or case (though they can be as in "in vivo" codes). Usually, the upper-level concepts and their descriptors and the relational statements linking those concepts are derived from, and apply to, all participants in a study.

3. There is more than one way of expressing relational statements. In our publications, they tend not to be presented as explicit hypotheses or propositions. Rather, when we write, we tend to weave the relationships into the narrative. How one expresses relationships is a stylistic matter, largely the result of training and the discipline(s) for which the researcher is writing. The essential element of theory is that categories are interrelated into a larger theoretical scheme

### *The Central or Core Category*

The first step in integration is deciding upon a central category. The central, or as it is sometimes called, the "core category," represents the main theme of the research. It is the concept that all the other concepts will be related to. To identify the central category, the researcher must choose from among the many categories developed over the course of a study: the category that appears to have to greatest explanatory relevance and highest potential for linking all of the other categories together.

A central category has analytic power. What gives it that power is the category's ability to explain or convey "theoretically" what the research is all about. For example, in the demonstration study on Vietnam veterans that will begin in Chapter 8, there is a final question driving the quest to continue probing the data after context and process are delineated. That question is, what is that special something that ties together all of different categories to create a coherent story about survival of Vietnam combatants? What is being searched for in the memos and data is a coherent overarching story, something larger than the sum of all its individual parts (see Chapter 12).

A central category may evolve out of the list of existing categories. Or, a researcher may study the categories and determine that though each category

**Table 5.2** Criteria for Choosing a Central Category

1. It must be abstract; that is, all other major categories can be related to it and placed under it.
2. It must appear frequently in the data. This means that within all, or almost all, cases there are indicators pointing to that concept.
3. It must be logical and consistent with the data. There should be no forcing of data.
4. It should be sufficiently abstract so that it can be used to do research in other substantive areas, leading to the development of a more general theory.
5. It should grow in depth and explanatory power as each of the other categories is related to it through statements of relationship.

tells part of the story, none capture it completely. Therefore, another more abstract term or phrase is needed, a conceptual idea under which all the other categories can be subsumed. Strauss (1987, p. 36) provides a list of criteria that can be applied to a category to determine if it qualifies.

## Choosing Between Two or More Possibilities

Sometimes analysts are confronted by two or more possible core categories. Our suggestion, especially for beginning analysts, is to select one idea as the central category. The notion of theory development implies that all ideas are incorporated into one theoretical scheme. Having two central categories means developing two different theories. If a researcher has the time and expertise, then he or she should feel free to do so. But usually if a researcher looks hard enough at the data, he or she can come up with one unifying idea.

## Difficulty Deciding Upon a Central Category

Sometimes researchers have difficulty moving beyond the level of description with their research. They want to develop theory and are reaching for that level of analysis but cannot seem to make the final analytic leap. Other students have difficulty committing to a central unifying concept. To these students, every idea in the data is of equal importance.

One of the reasons researchers have difficulty formulating theory is that they fail to write long thoughtful memos throughout the research process. A researcher cannot expect to understand the analytic story behind the data, if at the end of the research the only thing an analyst has to work with is a list of concepts or codes and some quotes from the raw data pertaining to each code, but no real memos. Theory building is a process of going from raw data, thinking about that raw data, delineating concepts to stand for raw data, then making statements of relationship about those concepts linking them all together into a theoretical whole, and at every step along the way recording that analysis in memos.

Another reason some researchers have difficulty with final integration is that they do not quite understand the difference between description and theory. In theory building it is the word "explanatory" that makes the difference. Description describes something. For example, "This is a red ball. It bounces, fits into the palm of the hand, and retains its shape even when bounced repeatedly." But description does not explain why a ball bounces or what it is made from, or why it retains its shape or what happens if you change its size, shape, or the material that it is made out of. Theory, on the other hand, gives you those explanations. For example, "a ball is able to bounce because it is round and is constructed out of rubber. If a person gives a ball a different shape or makes a ball out of a different material, then it may not bounce as well or maybe better. It will have different abilities, forms, and functions, like a football or volleyball."

A third reason why a researcher may have difficulty identifying a central category is a lack of trust in his or her analytic ability. Students are often concerned about the possibility of reading something into the data that is not there. A researcher has to trust that by the time this point in a study is reached, the integrative story is in him or her. It just needs to be drawn out. Some students may need the help of an outside person to encourage them to take that conceptual leap. The "outside person" can ask a series of directed questions forcing the analyst to reply with abstract rather than descriptive renditions of the story. Simply being able to tell the analytic story to another person often helps an analyst gain perspective and confidence.

## Techniques to Aid Integration

There are several techniques that can be used to facilitate identification of the central category and the integration of concepts. Among these are writing the story line, making use of diagrams, and reviewing and sorting of memos either by hand or by computer program (if a program is being used).

### *Writing the Story Line*

By the time the researcher begins thinking about integration, he or she has been immersed in the data for some time and usually has a "gut" sense of what the research is all about even though he or she may have difficulty articulating what that "something" is. One way to move beyond this impasse is to sit down and write, in a few descriptive sentences, about "what seems to be going on here." It may take two, three, or even more starts to be able to articulate thoughts about the data concisely. Eventually a story emerges. Often, returning to the raw data and rereading several interviews or observations helps to stimulate thinking. This tends to work if the researcher does not read the interviews or observations for detail but for the general sense, standing back and asking, "What is the main issue or problem that these people seem to be grappling with? What keeps striking me over and over when I read these interviews or observations, or watch the videos? What comes through in the data though it may not be said directly?" (See Chapter 12 for an example of a descriptive story.)

### *Moving From the Descriptive Story to the Theoretical Explanation*

Once an analyst has written a few descriptive sentences about what the research is all about, he or she is ready to move on to integrating the main categories or themes into a unified theoretical explanation. Integrating means choosing a core category, then retelling the story around that core category using the other categories and concepts derived during the research. Integrative Memo 2 in Chapter 12 is an example of a theoretical memo; it utilizes concepts derived from a study of Vietnam veterans to explain how combatants were able to "survive" the war experience. Readers can jump ahead to Chapter 12 and read that memo for an example of a theoretical memo. Not every concept (only the major ones) that evolved from the analysis of Vietnam veterans' experiences in Vietnam is included in that memo, as the purpose of this book is methodological. It is not a research report on Vietnam veterans, which would certainly merit a book of its own. Notice that linkages that are made between concepts are not written in a cause-and-effect fashion as hypotheses. However, a researcher wishing to write such statements could do so if the researcher chose to.

### *The Use of Integrative Diagrams*

There are times, when either through preference or because an analyst is more of a visual person, that diagrams are more useful than storytelling for sorting out the relationships between categories. Diagrams can be valuable tools to integration because integrative diagrams are abstract but visual representations of data. Constructing diagrams is helpful because it enables analysts to gain distance from the data, forcing them to work with concepts at the category level rather than the details contained in the many memos.

Diagramming also demands that analysts think very carefully about the logic of relationships, because if the relationships are not clear, then the diagrams will come across as muddled and confused. If analysts have made use of diagrams throughout the research process (and some analysts are very visual thinkers and good with diagrams), the succession of operational diagrams should lead up to the integrative story. However, if there are few diagrams or if after reviewing previous diagrams the researcher is still unclear about the nature of relationships between concepts, sitting down with a teacher, consultant, or colleague and explaining to them diagrammatically "what the research is all about" can facilitate the integrative process. The listener can ask directed questions or request that the researcher present a few representative cases diagrammatically. This should stimulate the analyst to think about relationships. Usually there are several attempts at putting the concepts together in a diagram before the conceptualization feels right.

A diagram need not contain every concept that emerged during the research process, but should focus on those that reach the status of major categories. Diagrams should flow, the logic apparent without a lot of explanation. Also, integrative diagrams should not be too complicated. Diagrams with too many words, lines, and arrows make it difficult for the reviewer to know what the major point is. The details should be left to the writing.

### *Reviewing and Sorting Through Memos*

Memos are the running logs of analytic thinking. They are the storehouses of ideas generated through interaction with the data (see Chapter 6). Generally, memos start off quite simple (dealing with mainly one concept) and descriptive, and as the research progresses memos generally become more summary-like, abstract, and integrative (exploring relationships between two or more concepts). This means that later memos often contain the clues to integration, especially summary memos. Sometimes, as readers will see in Chapter 12, the main concept emerges early in the research; it is just that the researcher doesn't recognize its significance until much later.

Memos are usually sorted by categories. However, sorting can become more and more difficult as memos begin to link two or more concepts. This is where the retrieval function of computer programs can be most useful. They allow the researcher to sort and resort until a logical theoretical structure is constructed. It is our experience that students are able to discern patterns and processes, but even with all the memos in front of them they have difficulty making that last analytic leap. Confronted by many different concepts and categories, they become confused and uncertain. This is to be expected. Rereading summary memos can be very helpful, especially if the researcher listens to his or her own words and looks for recurrent themes. Sooner or later the "aha!" experience will come.

Ideas for a unifying concept sometimes come out of nowhere, or so it appears because of that "aha" experience. Sometimes merely thinking about a metaphor that describes the situation is the stimulus that is needed. Or, some researchers turn to the literature to look for a unifying concept. Though this is not our usual practice, sometimes the literature does get the analytic juices flowing. We are all for using any device that stimulates thinking. Furthermore, examining the literature can help a researcher to start thinking about how to place his or his findings within the larger body of professional literature when writing.

## Refining the Theory

Once a researcher has outlined the overarching theoretical scheme, it is time to refine the theory. Refining the theory consists of: (a) reviewing the scheme for internal consistency and for gaps in logic, (b) filling in poorly developed categories and trimming excess, and (c) validating the scheme.

### Reviewing the Scheme for Internal Consistency and Logic

A theoretical scheme should flow in a logical manner and should not have inconsistencies. If the story line, memos, and diagrams are clear, consistency

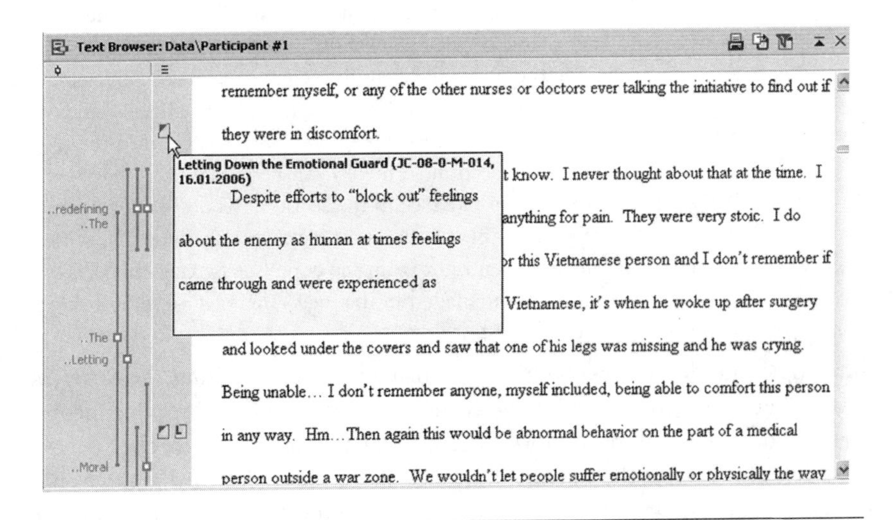


**Screenshot 2a & 2b** *(Continued)*

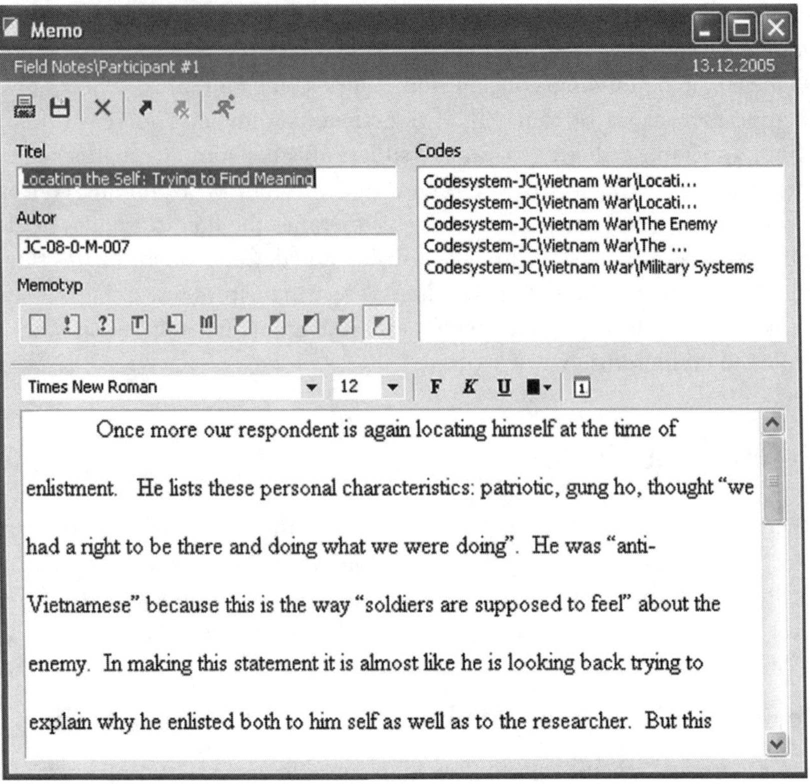

Memo
Field Notes\Participant #1 13.12.2005
Titel
Locating the Self: Trying to Find Meaning
Autor
JC-08-0-M-007
Memotyp
Codes
Codesystem-JC\Vietnam War\Locati...
Codesystem-JC\Vietnam War\Locati...
Codesystem-JC\Vietnam War\The Enemy
Codesystem-JC\Vietnam War\The ...
Codesystem-JC\Vietnam War\Military Systems
Times New Roman 12 F K U

Once more our respondent is again locating himself at the time of enlistment. He lists these personal characteristics: patriotic, gung ho, thought "we had a right to be there and doing what we were doing". He was "anti-Vietnamese" because this is the way "soldiers are supposed to feel" about the enemy. In making this statement it is almost like he is looking back trying to explain why he enlisted both to him self as well as to the researcher. But this

**Screenshot 2a & 2b** The big screenshot shows a MAXQDA memo sheet. A memo can be attached at any place in a text in the Text Browser (see Screenshot 2b). Clicking in the margin beside the text will open up a memo form (see Screenshot 2a). The researcher is free to define eleven different memo types. Up to twenty-six pages may be written into the memo text section. A memo can be linked to as many codes as you like (see in the section "Codes"). Thus you are always able to access all memos concerning a specific category even right from the code (see Screenshot 3). All memos are managed in the memo manager (see Screenshot 6)

and logic should follow. Sometimes during the final writing, however, a researcher senses that "something" is not quite right. One or more concepts or the final ideas still need work. When this happens, the researcher should go back and review the memos and once more make use of diagrams. But unless the analyst knows what he or she is looking for, or what is missing, diagramming will not help.

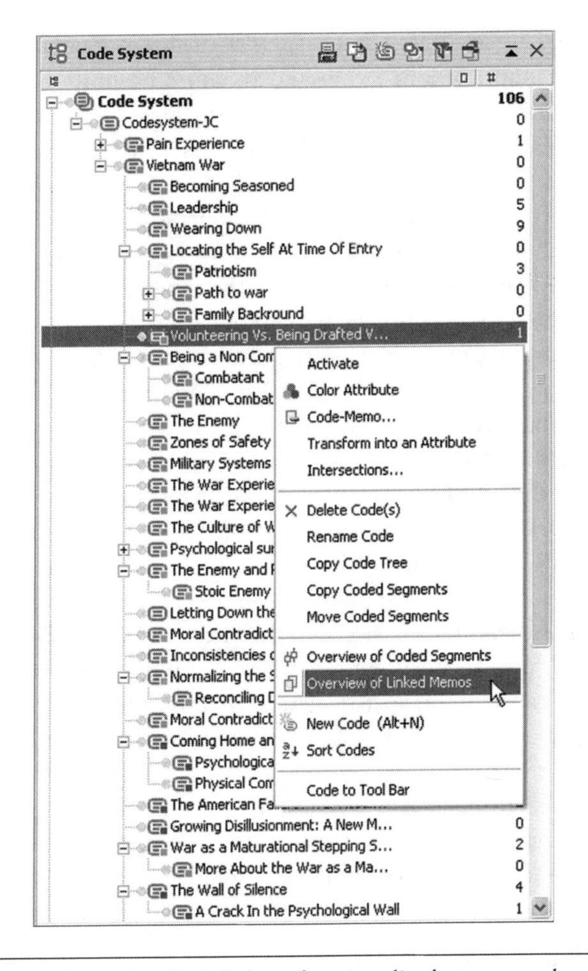


**Screenshot 3** The MAXQDA linking functionality between codes and memos makes it very easy to keep track of all memos written concerning a single code: right clicking a code allows for opening up the context menu, offering a range of management options for the code system. One click will show you, at one glance, all memos you have linked to that specific code.

A place to begin is with the central category itself. A central category, like any category, must be defined in terms of its properties and dimensions. The definition should come out of the data. Even if the central category was not named in earlier memos, when the analyst reviews the memos, he or she should find references in the data to the idea, along with properties and dimensions.

To check for internal consistency and logical development, the analyst can stand back and ask him- or herself (because by now the analyst is so immersed in the data) what he or she thinks the properties are, then go back to see how much of this has been built into the scheme. If it is still not clear, or there are areas that seem to be missing, then the analyst should go back to memos and sort through them for clues. Sometimes the analyst is almost there, but without realizing it has taken the wrong stance toward the data. That is, it is easy to look at the data from the perspective of the analyst and not tune into the respondents even though one thinks one is. For example, as stated earlier, when Corbin was writing her dissertation, a research project that looked at management by women of pregnancies complicated by chronic illness, something seemed awry with the logic, in that some women's protective governing strategies did not match the level of risks. Finally, it dawned on Corbin that though she thought she was being impartial, in reality she was assigning risk levels to the women based on her (Corbin's) perspectives of what the risks were. Women often viewed the risks quite differently from health professionals, and acted more on the basis of their definitions than on the definitions of health professionals—that was why women's actions sometimes appeared to be inconsistent with the risks. It was not Corbin's assignation of risks or those of the doctors that were important. Rather, it was what the women perceived the risks to be. In other words, it is easy to be blinded by the researcher's own perspective without even being aware of it. Often, it is not until the analytic scheme doesn't quite seem right that the role of professional bias comes to light. With that realization, a researcher can return to the data with a clearer view of what the respondents are saying.

## Filling In Poorly Developed Categories

In theory building, the analyst is aiming for density as well as abstraction. By density, we mean that all (within reason) the salient properties and dimensions of a category have been identified and variation is built in. Density and variation are what give a category precision and increase its explanatory power. Poorly developed categories are evident when making diagrams and sorting memos.

Filling in can be done through review of either memos or raw data, looking for data that might have been overlooked. Or, an analyst can go back into the field and selectively gather additional data about that category through theoretical sampling (see Chapter 7). Filling-in often continues into the final writing phase. Analysts always find gaps when they begin to write. The problem is deciding when to let go of the research. Not every detail can be well developed or spelled out. In the research example of Vietnam War combatants that follows in Chapters 8 through12, not much was done with the concept "Homecoming." Though I (Corbin) realized that this was an important concept, because of time constraints I was not able to gather the data necessary to elaborate on this concept.

The ultimate criterion for determining whether or not to end the data gathering processes remains "theoretical saturation." This term is perhaps the most widely misunderstood and incorrectly used concept in grounded theory and other qualitative research. It is often used as an excuse to legitimate discontinuing data gathering after five to ten interviews. But theoretical saturation is not that simple. It means taking each category and spelling out in considerable detail its properties and dimensions, including variation. It requires lots of memo writing and a very conscious effort to fill in gaps in data.

## Trimming the Theory

Sometimes the problem is not insufficient data, but an excess of data. That is, some ideas don't seem to fit the theory. These are usually extraneous concepts, nice ideas, but ones that were never developed, most probably because they did not appear much in data or seemed to trail off into nowhere. Our advice is to drop them for this study, especially if the researcher wants to graduate within a reasonable amount of time. If the concepts are interesting, the analyst can pursue them at a later date, but there is no reason to clutter a theory with concepts that lead nowhere or contribute little to its understanding.

## Validating the Theoretical Scheme

When we speak of validating, we are not talking about testing in the quantitative sense of the word (more will be said about this topic in Chapter 14). What we mean is the following. A theory is constructed from data, yes, but by the time of integration, it represents an abstract rendition of that raw data. Therefore, it is important to determine how well that abstraction fits with the raw data and to also determine if anything salient was omitted from the theoretical scheme. There are several ways of validating the scheme. One way is to go back and compare the scheme against the raw data, doing a kind of high-level comparative analysis. The theoretical scheme should be able to explain most of the cases. Another way to validate is to actually tell the story to respondents or ask them to read it, and request that they comment upon how well it seems to fit their case. Naturally, it won't fit every aspect of each case because the theory is a reduction of data and built on a compilation of cases, but in the larger sense, participants should be able to recognize themselves in the story that is being told.

## What If a Case Doesn't Fit?

It is not unusual to find outlying cases, those that fall at either extreme in the dimensional range of a concept, or that seem quite contrary to what is going on. For the most part, these outliers represent variations of the theory or present alternative explanations.

## Building in Variation

One of the problems with some theoretical schemes is that they fail to account for variation. They present process in a developmental fashion without accounting for variation in that developmental process. This is problematic because it makes the theory appear artificial, like every person or organization falls into these neat and distinct types or steps in a process. We know that life does not fit into neat little boxes. There are always variations of every process, some persons move slower, some faster, some drop out, and some follow a different passage. This means that even within patterns and categories there is variability with different people, organizations, and groups falling at different dimensional points along some properties.

## Summary of Important Points

Analyzing data for context and process are essential aspects of any analysis. To do anything less is to misrepresent or distort the situations that are studied and to present only a partial explanation of what is happening and why. In other words, persons or collectives do not live or act within a vacuum, but rather exist and act within a larger framework of structural conditions. Structural conditions do not determine action/interaction/emotional responses. Rather they lead to certain events circumstances, situations, and/or problems that individuals and collectives respond to through some form of strategic action, interaction, or emotional response (process). Thus, context and process are necessarily linked and should be part of an explanation of any phenomenon

For researchers who are interested in theory construction, this chapter presented some strategies to facilitate integration. It was explained that integration occurs around a central explanatory concept. Integration occurs over time, beginning with the first analysis and often does not end until the final writing. Once a commitment is made to a central idea, major categories are related to it through explanatory statements of relationships. Several techniques can be used to facilitate the integration process. These include telling or writing a storyline, the use of diagrams, and the sorting and reviewing of memos.

Once the theoretical scheme is outlined, the analyst is ready to refine the theory, trimming off excess and filling-in poorly developed categories. Poorly developed categories are saturated through further theoretical sampling (see Chapter 7). Finally, a theory is validated by comparing it to raw data or by presenting it to respondents for their reactions. A theory that is grounded in data should be recognizable to participants and the larger concepts should apply to each case even if some of the details specific to their case are missing or don't seem to fit.

## Exercises for Thinking, Writing, and Group Discussion

1. Take a situation in your life, one that you don't mind sharing with the group. Think about the circumstances or conditions from the most macro to the micro that frame that situation.
2. Think about the strategies that you used for handling the situation and how you responded when and if the situation changed, bringing process into the analysis.
3. Write a detailed memo about the above and bring the memo to class to discuss it with the group.
4. As a group, discuss what other contextual factors and strategies might be involved in the individual's memo that the individual might have missed.
5. What does theory mean to you and how might you bring theory into your study?
6. Discuss those features of the research process that you think are important for theory building.

# Note

1. I want to thank Adele Clarke for picking up on an error made in the second edition of this text. The diagram of the Matrix in that text was one borrowed from Guessing. The reason for choosing that diagram was that it showed a more fluid relationship between conditions and consequences. However, it was never meant to omit action/interaction from the center. Action/interaction was and remains the center of the Matrix.

# 6

# Memos and Diagrams

*To exercise maximum control over his experiences, the researcher requires an efficient system for recording them. Novices may think of note-taking and recording principally as devices that help with remembering and with the storage and retrieval of information. They are correct, but only on a rather mechanical level. . . . What our researcher requires are recording tactics that will provide him with an ongoing, developmental dialogue between his roles as discoverer and as social analyst. (Schatzman & Strauss, 1973, p. 9)*

**Table 6.1** Definition of Terms

| |
|---|
| *Diagrams*: Visual devices that depict relationships between analytic concepts. |
| *Memos*: Written records of analysis. |
| *Theoretical Sampling*: Data gathering based on evolving concepts. The idea is to look for situations that would bring out the varying properties and dimensions of a concept. |

## Introduction

The purpose of this chapter is to introduce the reader to memos and diagrams. **Memos** are a specialized type of written records—those that contain the products of our analyses. **Diagrams** also arise from analysis. They are visual devices that portray possible relationships between concepts. But memos and diagrams are more than just repositories of thought. They are working and living documents. When an analyst actually sits down to write a memo or do a diagram, a certain degree of analysis occurs. The very act of writing memos and doing diagrams forces the analyst to think about the data. And it is in thinking that analysis occurs. Strauss (1987) states, "Even when a researcher is working alone on a project, he or she is engaged in continual internal dialogue—for that is, after all, what thinking is" (p. 110).

In the 2nd edition of *Basics,* we broke memos down into several types—code notes, theoretical notes, and operational notes. In this 3rd edition, we want to get away from thinking about memos in a structured manner. The reason is that novice researchers often become so concerned with "getting it right" that they lose the generative fluid aspect of memoing. It is not the form of memos that is important, but the actual doing of them. However, one of the early reviewers of this text came up with an organizational scheme for the memo examples, used later in this chapter, that is quite descriptive and helpful in explaining the many types of memos that researchers can write. We present that scheme below. There are memos for:

- Open data exploration
- Identifying/developing the properties and dimensions concepts/categories
- Making comparisons and asking questions
- Elaborating the paradigm: the relationships between conditions, actions/interactions, and consequences
- Developing a story line

—courtesy of anonymous reviewer

The important thing for the reader is that he or she not be concerned with writing memos according to each type. More important is to just get into the habit of writing memos.

Writing memos should begin with the first analytic session and continue throughout the analytic process. Doing diagrams is more periodic but nevertheless very important. Doing memos and diagrams should never be viewed as chores, or as tasks to be agonized over. They are also not to be confused with finished papers ready for publication. Rather, memos and diagrams begin as rather rudimentary representations of thought and grow in complexity, density, clarity, and accuracy as the research progresses. One of the complaints we often hear from students is that writing memos and doing diagrams is just too time consuming. They say that they would rather make a few notes in the margins of their field notes. We puzzle over those remarks. Writing memos and doing diagrams is part of the analysis, part of doing qualitative work. They move the analysis forward and as such are just as important to the research process as data gathering itself.

Qualitative analysis involves complex and cumulative thinking that would be very difficult to keep track of without the use of memos. Furthermore, most research projects go on for several months at a minimum. Some extend for years. How could researchers remember what they were thinking months earlier unless those thoughts are written down someplace? Then, too, many studies are conducted by teams of two or more persons and researchers need a way to store and share their individual as well as mutual analytic sessions. Without memos and diagrams, it would be difficult to keep the lines of communication open between researchers or to retrace the process by which the researchers arrived at their final findings.

## General Features of Memos and Diagrams

There are some general features of memos and diagrams that would-be analysts should be familiar with. We turn to these next.

- *Memos and diagrams vary in content, degree of conceptualization, and length, depending upon the research phase, intent, and the materials one is coding.* In the beginning stages of analysis, memos and diagrams appear awkward and simple. This is of no concern. Remember, no one but the analyst (and possibly committee members) has access to the memos and diagrams

- *Though analysts can write on the actual interview or field notes, this is not practical, except perhaps in the earliest phases of open coding.* We say this for several reasons: (a) It is difficult to write memos of any length or to do diagrams on field notes because usually there is insufficient space to develop ideas; (b) some of the original concepts may be revised as the analysis proceeds and these might be misleading and confusing when analysts return to a document to recode and are confronted by the old codes written in the margins; (c) it is difficult to retrieve information, in other words to combine or sort memos, if the margin of a field note or interview transcript is the only place where information has been stored; and (d) there are many computer programs available to assist with memo writing and diagramming, making it unnecessary to write in the margins of a document. Some of the texts and papers that are useful for introducing analysts to the pros and cons of computers for data analysis are Bong (2002); Fielding and Lee (1991, 1998); Kelle (1995, 1997); Lonkila (1995); Pfaffenberger (1988); Roberts and Wilson (2002); and Weitzman and Miles (1995).

- *Each analyst develops his or her own style for doing memos and diagrams.* Some analysts use computer programs, others use color coded cards, while still others prefer putting written memos into binders, folders, or notebooks. The method that the analyst uses for recording and managing memos is not as important as just doing them. However, we might add that computer programs now facilitate this process greatly.

- *While the contents of memos and diagrams are crucial to keeping a record of analyses, they have functions in addition to storing information.* Among the most important of these is that they force the analyst to work with concepts rather than raw data. Also, they enable analysts to use creativity and imagination, often stimulating new insights into data.

- *Another function of memos and diagrams is that they are reflections of analytic thought.* A lack of logic and coherence of thought quickly manifests itself when analysts are forced to put ideas down on paper.

- *Memos and diagrams provide a storehouse of analytic ideas that can be sorted, ordered and reordered, and retrieved according to the evolving analytic scheme.* This ability becomes useful when it comes time to write about a topic, or when analysts want to cross-reference categories or evaluate their analytic progress. Studying diagrams and reviewing memos can also reveal which concepts are in need of further development and refinement.

- *Analysts should code after every analytic session.* In fact, the writing of memos often is the analytic session, especially for analysts working alone. However, it is not always necessary to do long memos or diagrams. When stimulated by an idea, an analyst should stop whatever he or she is doing and capture that thought on paper. A few generative ideas or sentences would suffice. When an analyst has more time, he or she can write a lengthier memo.

- *Summary memos can be written that synthesize the content of several memos.* As the analysis moves along it is important that the analyst takes the time sit down and write a summary of where he or she thinks the analysis is at this point. Doing so really helps later with integration. Integrative diagrams can be used to display those ideas visually.

## Specific Features of Memos and Diagrams

In addition to the general features of memos, we offer some suggestions to make memos and diagrams more useful. There is nothing more frustrating than attempting to retrieve a memo that you recall writing but can't find because it lacks identifying information. Some ideas for making memos and diagrams more useful include:

- *Date memos and diagrams.* It is also helpful to include a reference to the document and raw data from which the memo was derived. (Including an excerpt from the raw data is facilitated by the use of computer programs.) The reference can include the code number of the interview or observation; document; the date on which the data were collected; the page (and line number for those using computer programs); and any other means of identification that might prove useful later when retrieving the data.

- *Create a heading for each memo and diagram.* This makes the contents more readily accessible. One can cross-reference memos or diagrams that relate two or more categories to each other.

- *Include short quotes or phrases of raw data in the memo.* (Plus include date, page number, and all other identifying information for easy retrieval.)

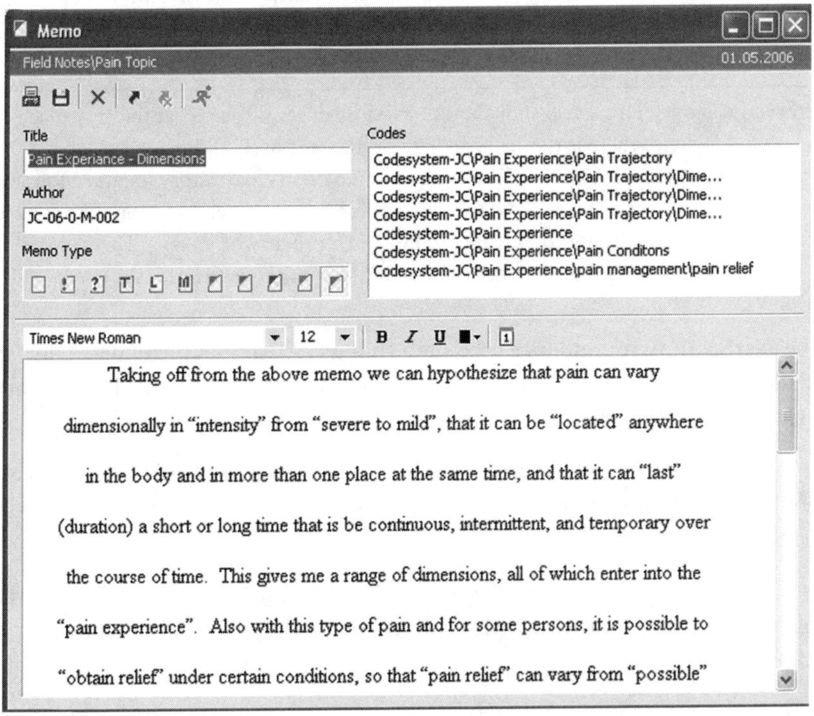

Memo
Field Notes\Pain Topic 01.05.2006
Title
Pain Experiance - Dimensions
Author
JC-06-0-M-002
Memo Type
Codes
Codesystem-JC\Pain Experience\Pain Trajectory
Codesystem-JC\Pain Experience\Pain Trajectory\Dime...
Codesystem-JC\Pain Experience\Pain Trajectory\Dime...
Codesystem-JC\Pain Experience\Pain Trajectory\Dime...
Codesystem-JC\Pain Experience
Codesystem-JC\Pain Experience\Pain Conditons
Codesystem-JC\Pain Experience\pain management\pain relief
Times New Roman 12

Taking off from the above memo we can hypothesize that pain can vary dimensionally in "intensity" from "severe to mild", that it can be "located" anywhere in the body and in more than one place at the same time, and that it can "last" (duration) a short or long time that is be continuous, intermittent, and temporary over the course of time. This gives me a range of dimensions, all of which enter into the "pain experience". Also with this type of pain and for some persons, it is possible to "obtain relief" under certain conditions, so that "pain relief" can vary from "possible"

**Screenshot 4** A memo is automatically "stamped" with the name of the author. In this example we also used the field to identify each memo within this book in the corresponding MAXQDA project. So, you can easily work simultaneously with the book data and the project data. Moreover, each memo can be given a title and can be linked to any of your codes (see Chapter 5). A memo can be exported or printed out. A very useful feature is the option to freely define different memo types.

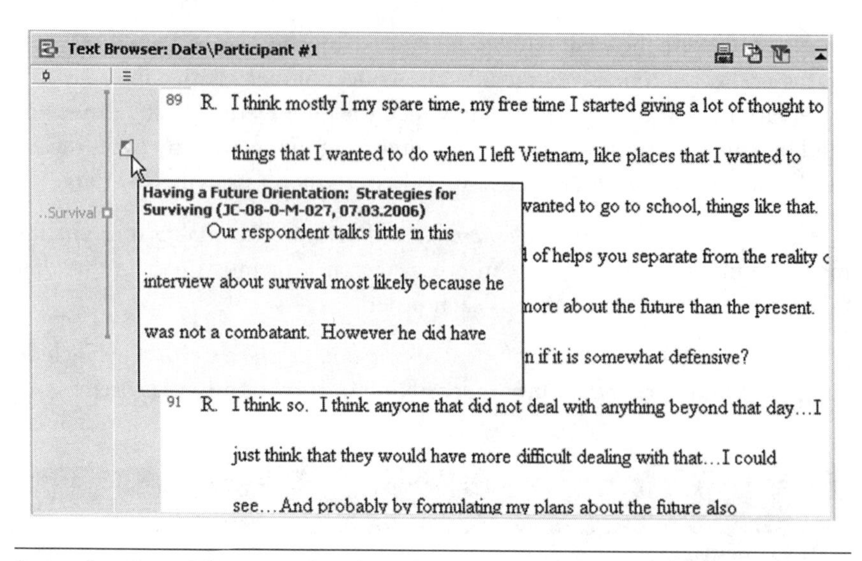


**Screenshot 5** This screenshot shows how a memo is displayed at the very place where it is attached: between the text itself and the code visualisation. This is an example of an exploratory memo. Mousing over a memo will show an information screen indicating the memo title, author, creation date, and a preview of its text. The information fades out as soon as you move the mouse away.

These are handy reminders of the data that gave rise to a particular concept or idea (this happens automatically with computer programs). Later, when writing, the actual data can be used to illustrate that concept.

- *Regularly update memos and diagrams.* As the analysis progresses, new data lead to increased insight, therefore evolving into memos of more depth and complexity.

- *Keep a list of concepts and subconcepts available for reference* (again, this happens automatically with computer programs). This helps prevent duplication and oversight.

- *Take notice if several memos on different codes begin to sound alike.* The analyst can recompare the concepts for similarities and differences, perhaps combining them or making notations of how they are different.

- *Keep multiple copies of memos or a computer backup of one's work.* A researcher who has lost important data due to computer failure knows how frustrating it can be not to have backup of the work.

- *Memos indicate when a category appears saturated—meaning when a category is well developed in terms of its properties and dimensions.*

Data collection can then be directed toward other categories still in need of development.

- *If, as an analyst, you come up with two or more exciting ideas at the same time, you should jot down a few notes about each immediately.* This way neither idea is lost and later you can write a memo on each.

- *Be flexible and relaxed when doing memos and diagrams.* Worrying about correctness can stifle creativity and freeze thought.

- *Be conceptual rather than descriptive when writing memos.* Memos are not so much about specific incidents or events, but about the conceptual ideas derived from these. It is the denoting of concepts and their relationships that moves the research from raw data to findings.

- *Develop your own style and techniques for writing memos and doing diagrams.*

- *Use a notebook or running log, separate from memos, to write up impressions of participant and researcher's reactions during interviews or observations.* A diary works well as a means of keeping an account of self-reflections during the entire research process.

## Memos and Field Notes

One of the reviewers of this text suggested that field notes are "in a way a form of memo." Patton (2002) says, "Recording and tracking analytical insights that occur during data collection are part of fieldwork and the beginning of qualitative analysis" (p. 436). Whenever observations of events are made, the observations are filtered through the eyes of the researcher who can't help but start thinking about and classifying the information. It just kind of happens spontaneously because persons tend to think consciously, or not in terms of concepts. And there is no reason not to jot down analytic ideas while in the field, for as Patton (2002) goes on to say, "Repressing analytical insights may mean losing them forever, for there's no guarantee they'll return" (p. 406).

The point to be made is that if a researcher is out in the field collecting data, theoretical ideas will be stimulated by data and it is very appropriate to jot those theoretical ideas down before the researcher forgets them. In fact, it is almost impossible to be purely descriptive when writing about incidents out in the field because we naturally name and categorize what we see (Wolcott, 2001). However, I (Corbin) want to make a distinction between field notes and memos in order not to confuse novice researchers about the nature or importance of each. Field notes are data that may contain some conceptualization and analytic remarks. Memos, on the other hand, are lengthier and more in-depth thoughts about an event, usually written in conceptual form after leaving the field. And as such, they are much more complex and analytical than any remarks that I might make on my field notes. For persons who are interested in ways of keeping memos and field notes separate in or out of the field, Schatzman and Strauss (1973) offer the following scheme. They suggest writing observational notes (ONs) that describe the actual events, writing theoretical notes (TNs) denoting the researcher's thoughts about those events. And finally they suggest writing methodological notes (MNs) or reminders about some procedural aspect of the research (pp. 99–101). Complex note writing while out in the field might be difficult as the researcher might become so engrossed that he or she loses sight of what is going on.

Lofland, Snow, Anderson, and Lofland (2006) help clarify "observational notes" as being reports of events or interactions observed in the "field." Such notes might also include a description of the setting and perhaps some informal interviewing.

Corbin follows a similar process. When she is out in the field, she writes observational notes. Then when doing the analysis at home, she writes memos. For example, in the study of the articulation of patient care by head nurses, Corbin and Strauss wrote many memos based on Corbin's observations. During each fieldwork session, Corbin followed a head nurse, as a shadow, writing down to the best of her ability each thing that each head nurse did and said along with descriptions of the setting (there was never any problem with writing in notebooks as action occurred except for during psychotherapy sessions). At the end of the observational session, Corbin reviewed the day's notes with the head nurse, going over the incidents and obtaining the head nurse's explanations for his or her actions/interactions, serving as a kind of informal interview and verification session.

Within the next day or so, Corbin would meet with Strauss and they would take each incident, analyze it, and write memos of their discussions, using the same approach to analysis as they did with interviews. It was during the analytic sessions that interpretations and impressions of those incidents were derived. Our suggestion for field researchers would be to write observational notes documenting each incident, including as much description as possible, then to write memos from the observational notes—incident by incident—in a manner similar to interview data, always keeping in mind there is perhaps some conscious and unconscious analysis that occurs when gathering data.

## Diagrams

Everyone is familiar with diagrams. They are conceptual visualizations of data, and because they are conceptual, diagrams help to raise the researcher's thinking out of the level of facts. Diagrams enable researchers to organize their data, keep a record of their concepts and the relationships between them, and to integrate their ideas. Diagrams help researchers to explain their findings to colleagues and others in very systematic and organized ways. Most of all, doing diagrams force a researcher to think about the data in "lean ways"; that is, in a manner that reduces the data to their essence. And if an analyst can do that, he or she has it all together. One can do qualitative analysis without doing diagrams, but as is so often said, "a picture is worth a thousand words." Miles and Huberman (1994) are two researchers who use diagrams extensively for organizing data and illustrating conceptual relationships. They have this to say about diagrams:

> Conceptual frameworks are best done graphically, rather than in text. Having to get the entire framework on a single page obliges you to specify the bins that hold the discrete phenomena, to map likely relationships, to divide the variables that are conceptually or functionally distinct, and to work with all of the information at once. (p. 22)

Early diagrams are not elaborate. Like early memos, they are quite simple and hint at, rather than describe, relationships. Here are some diagrams from previous studies. Notice that they are very simple and help the researcher think about possible relationships.

Complex diagrams showing multiple relationships often pertain more to theory building than description, though there can be descriptive diagrams

| | *Homogeneous Patients* | *Heterogeneous Patients* |
|---|---|---|
| Easy work | | |
| Difficult work | | |

**Figure 6.1** Homogeneous/Heterogeneous Patients: Easy/Difficult Work

| *Phases of Illness* | *Number of Machines* | | *Frequency* | | | *Duration* | |
|---|---|---|---|---|---|---|---|
| | *Few* | *Many* | *Few* | *Intermittent* | *Often* | *Short* | *Forever* |
| Early | | | | | | | |
| Middle | | | | | | | |
| Late | | | | | | | |

**Figure 6.2** Illness Course: Machine-Time Dimension

| | Consequences For | | | | | | |
|---|---|---|---|---|---|---|---|
| *Pain Tasks* | *Illness Trajectory* | *Life and Death* | *Carrying On* | *Interaction* | *Ward Work* | *Sentimental Order* | *Personal Identity* |
| Diagnosing | | | | | | | |
| Preventing | | | | | | | |
| Minimizing | | | | | | | |
| Inflicting | | | | | | | |
| Relieving | | | | | | | |
| Enduring | | | | | | | |
| Expressing | | | | | | | |

**Figure 6.3** A Balancing Matrix

as well. When constructing theory, even though concepts can be put together in different ways, the relationships proposed by the researcher are based on data and therefore can be said to have some grounding in the data. With continued comparison of concepts against actual data, proposed relationships become substantiated in that they continue to make sense and offer one possible explanation. With time, diagrams become more integrative and complex. Notice the diagrams from previous studies below. Though still relatively simple, many revisions took place in these diagrams before the authors arrived at the final versions.

For other examples of the changes that take place in integrative diagrams over time, see Strauss (1987, pp. 174–178).

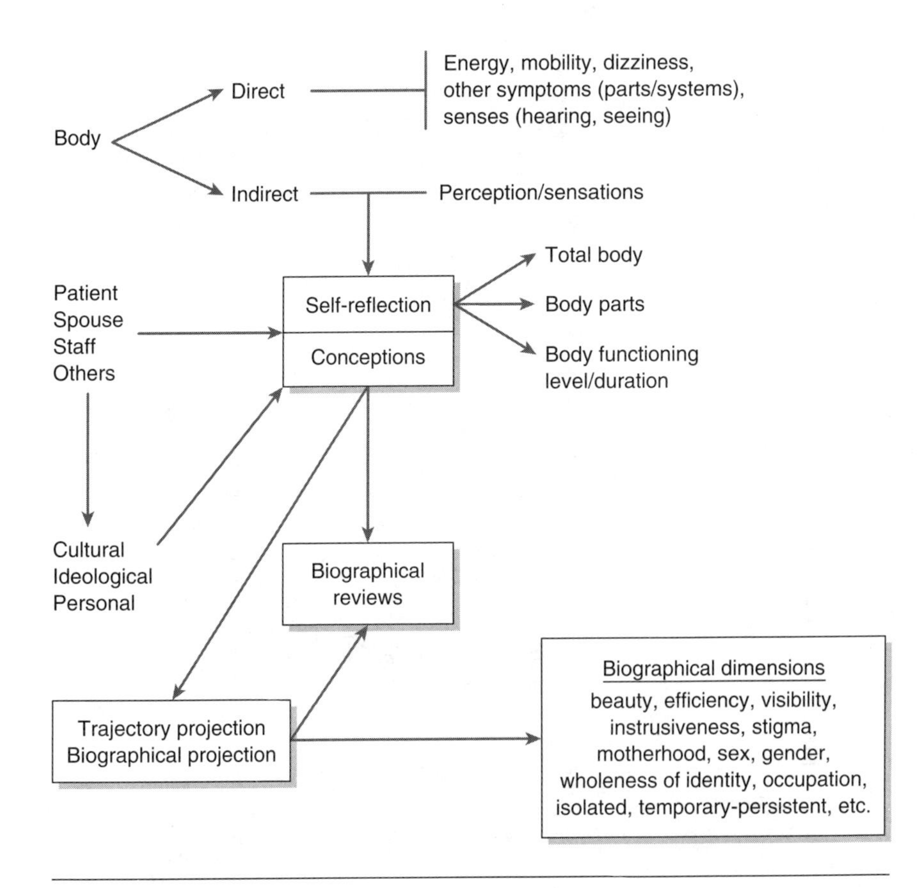


**Figure 6.4** Body, Biography, and Trajectory

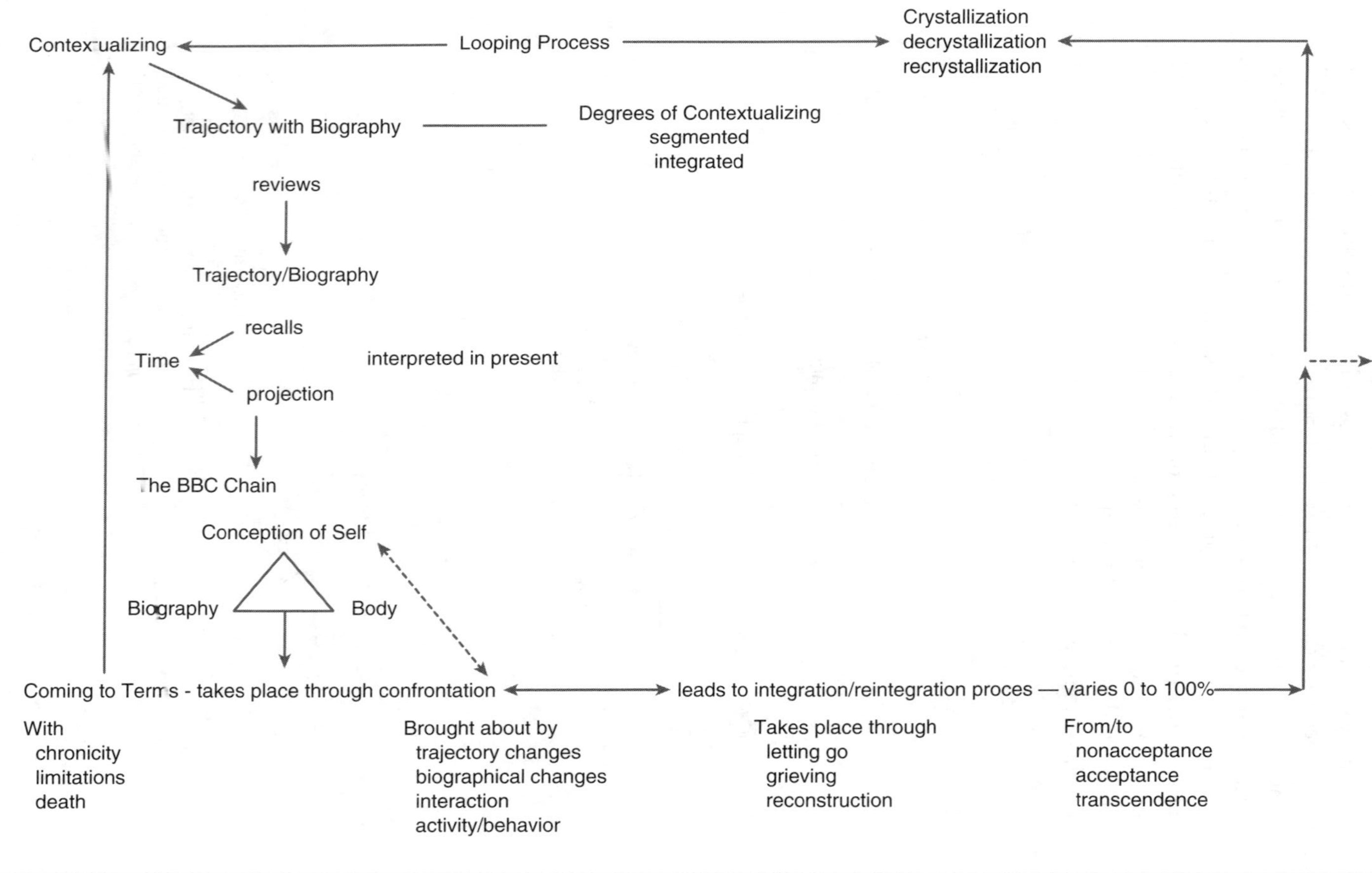


**Figure 6.5** Time Reflection Process

## Writing Memos and Doing Diagrams

Now that we've had this rather lengthy discussion about memos and diagrams and their importance to the analytic process, we would like to provide some examples. The illustrations should reassure readers that there is nothing magical about writing memos or doing diagrams. The memos and diagrams presented pertain to analysis of "the pain experience." As readers will notice, all a researcher needs to get the analytic process going is a small bit of data.

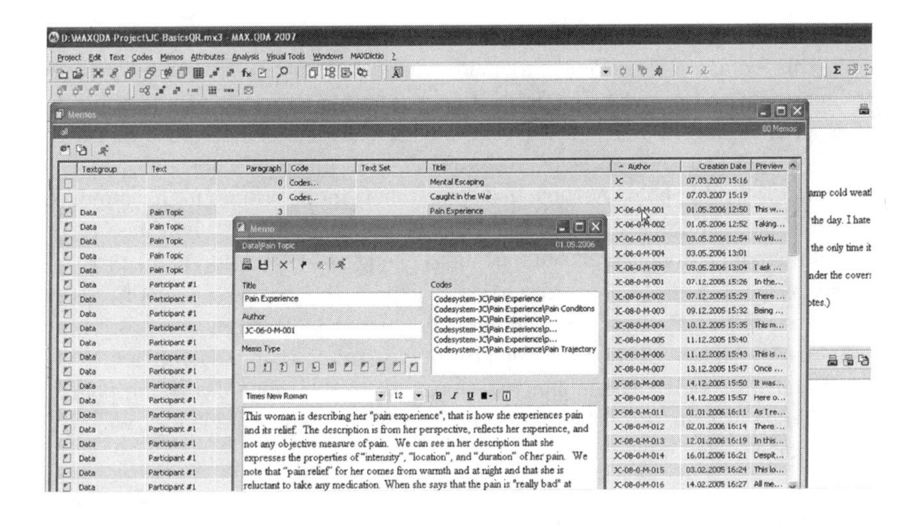


**Screenshot 6** The picture shows the MAXQDA Memo Manager, which is accessible from the upper menu bar of the main screen (see "Memos"). It allows for having all memos of your project at a glance. Double clicking a line will open the memo itself, and—at the same time—the text to which it is attached will be opened up in the MAXQDA Text Browser window. The column "Author" shows who created the memo. In this project the author column is used to identify the memo. Each memo in this book has its ID in brackets and thus allows one to work easily with the book data parallel to the data prepared in the project "JC-BasicsQR.mx3." You may download the project from www.maxqda.com/Corbin-BasicsQR or www.sagepub.com/corbinstudysite. Each column of the Memo Manager can become a sort criterion for the table by just clicking on the column header. Thus, you may, for example, sort memos according to the type of memo, the title, and so on.

## Memo 1[1] Example of a Memo Showing Open Data Exploration

### April 4, 1997

### Properties and Dimensions of the Pain Experience

(Field note excerpt)

*The pain in my hands from arthritis is really bad in damp cold weather. I wake up with it in the morning and it lasts throughout the day. I hate to take medication because of all the possible side effects so the only time it seems to get better is at night when I am warm in bed and under the covers. (Quote from interview with subject #1, p.1 of field notes.)*

In my memo I write:

This woman is describing her "pain experience," that is, how she experiences pain and its relief. The description is from her perspective, reflects her experience, and is not any objective measure of pain. We can see in her description that she expresses the properties of "intensity," "location," and "duration" of her pain. We note that "pain relief" for her comes from warmth and at night and that she is reluctant to take any medication. When she says that the pain is "really bad" at times she is giving us a dimension of the property of "pain intensity." The "location" of the pain, she tells us, is in her hands and it is of "long" duration lasting throughout the day. "Pain relief" is possible "under conditions of warmth." The possible use of medication is dismissed at least at this point in her "pain trajectory." All of the properties of the pain including its management and relief are part of her "pain experience," which of course is very individualized.

### Commentary on Memo and Diagram

The above memo and the diagram based on the memo (Figure 6.6) are not complicated or extensive. But they do get us started on our analysis. We have a concept, "the pain experience," and we've identified some of the properties and dimensions of this particular woman's "pain experience" (intensity, duration, location). We also have the concepts of "pain management" and "pain relief." Now we could go back to the data to look for further descriptions of her "pain experience," in the same interview, or we could turn to the interview with another research participant and compare the two descriptions. Or, before doing the latter, we could write additional memos off of the field note above to help us think about what we want to look for when we go back to the field to do another interview. Here we want to make a very important point. Every analyst must choose the depth of analysis he or she is reaching for and how much time he or she is willing to put into a study. We don't want to oversimplify the analytic process but

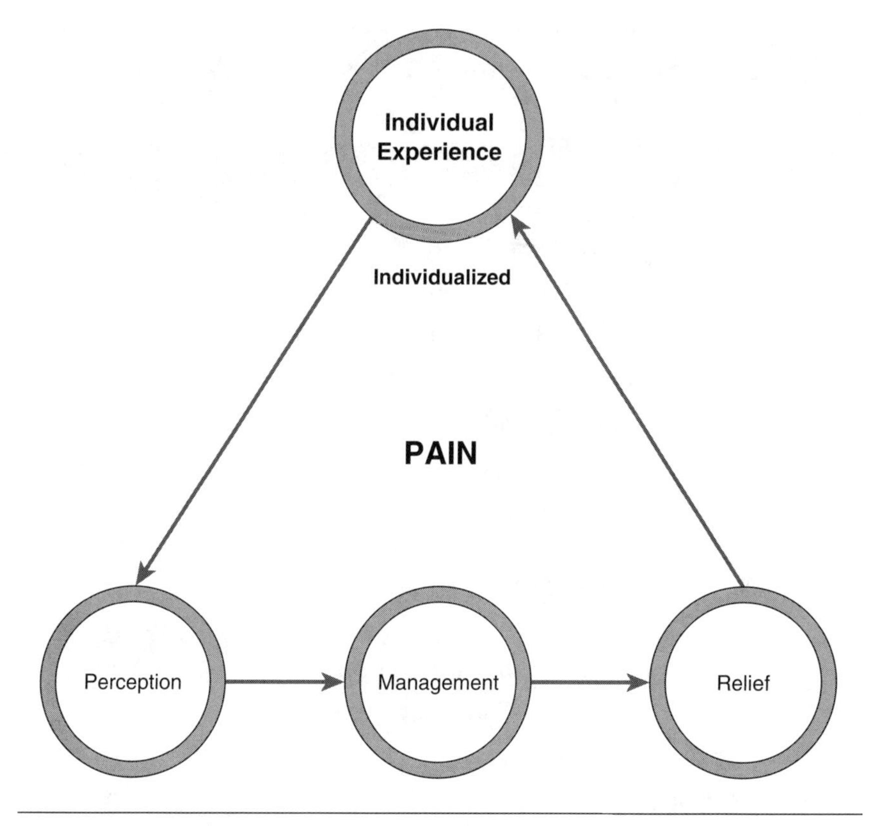


**Figure 6.6** Early Diagram on the Pain Experience

neither do we want to overwhelm our readers with memo writing. We, the authors, continue to write memos off previous memos until we run out of ideas because that's how we work. But every analyst has his or her own approach, style, and work rhythm. In this book we want to provide information that is of benefit to experienced as well as novice researchers. We would be doing our readers a disservice if we did not provide a range of both complex and less complex memos. We leave it to our readers to select the level of complexity of analysis they are reaching for and how much time they are willing to invest in memo writing.

After writing the above memo, we continued with our analysis, writing another memo bouncing off of the first.

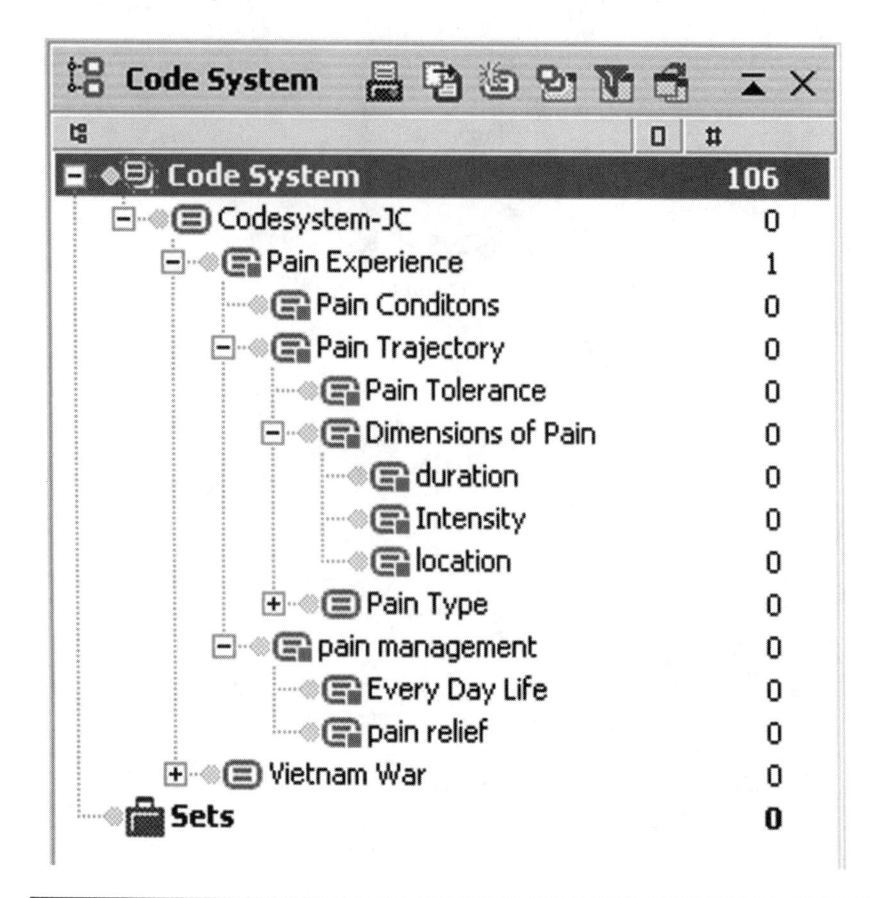


**Screenshot 7** The hierarchy of the Code System can be used to expand the dimensions of your categories. Furthermore, you may map your codes in MAXMaps, the graphics tool in MAXQDA. The connection between your data and its graphic representation in a MAXMap is vivid; that means you may open a code in MAXMaps and look at and import the coded segments as well as the memos, and so on. Moreover, you may also use the mapping tool independently of your data and work with free objects to display your ideas and conceptions, like the MAXMap displayed in Figure 6.6.

## Memo 2 Example of a Memo That Identifies/Develops the Properties and Dimensions of Concepts/Categories

### April 4, 1997

### The Pain Experience

Taking off from the above memo we can hypothesize that pain can vary dimensionally in "intensity" from "severe to mild," that it can be "located" anywhere in the body and in more than one place at the same time, and that it can "last" (duration) a short or long time that is be continuous, intermittent, and temporary over the course of time. This gives me a range of dimensions, all of which enter into the "pain experience." Also with this type of pain and for some persons, it is possible to "obtain relief" under certain conditions, so that "pain relief" can vary from "possible" to "impossible," be "temporary" or "permanent" depending upon the person, the type or cause of pain, and a person's response to it. To make it more complicated, it seems that "perception" of pain or the "pain experience" can vary depending upon many factors or conditions such as "location" of pain in the body—some areas being more sensitive than others, "degree of activity" one engages in, "time of day," and even "odd things" like weather. Finally there is the property of "duration" of the pain. Duration can vary dimensionally as "continuous," "intermittent," or "temporary." In the above case, one might say that the pain is "intermittent." But how do all the various dimensions or variations along the properties of pain enter into the "pain experience"? Also, I have another question: what is the meaning of pain to this person? Hmm. The "subjective experience" of pain incorporates many factors and it is up to me to tease all of these out of the data. Other factors that might influence the pain experience but that are not brought out in this particular field note are "pain history" and both present and previous experiences with "pain relief," and also if one believes that relief is possible in the future. Oh my! Pain relief and treatment are big areas that I've not yet explored but will have to before this study is over. As I continue to collect data and analyze the interviews I'll be looking for data about these areas.

### Commentary on Memo and Diagram

In this memo and the diagram based on the memo (Figure 6.7), we are laying out possible dimensions of pain and how they relate to the "pain experience." Notice that memos and the diagram reflect a systematic thinking about the topic and that both come from the actual data. The analyst is using the data to stimulate thinking. Another point about memos is that they force the analyst to ask questions of the data and the questions direct theoretical sampling, as indicated in the memo. When the analyst returns to collecting data he or she will be increasingly sensitive to those areas brought out in the analysis, listening

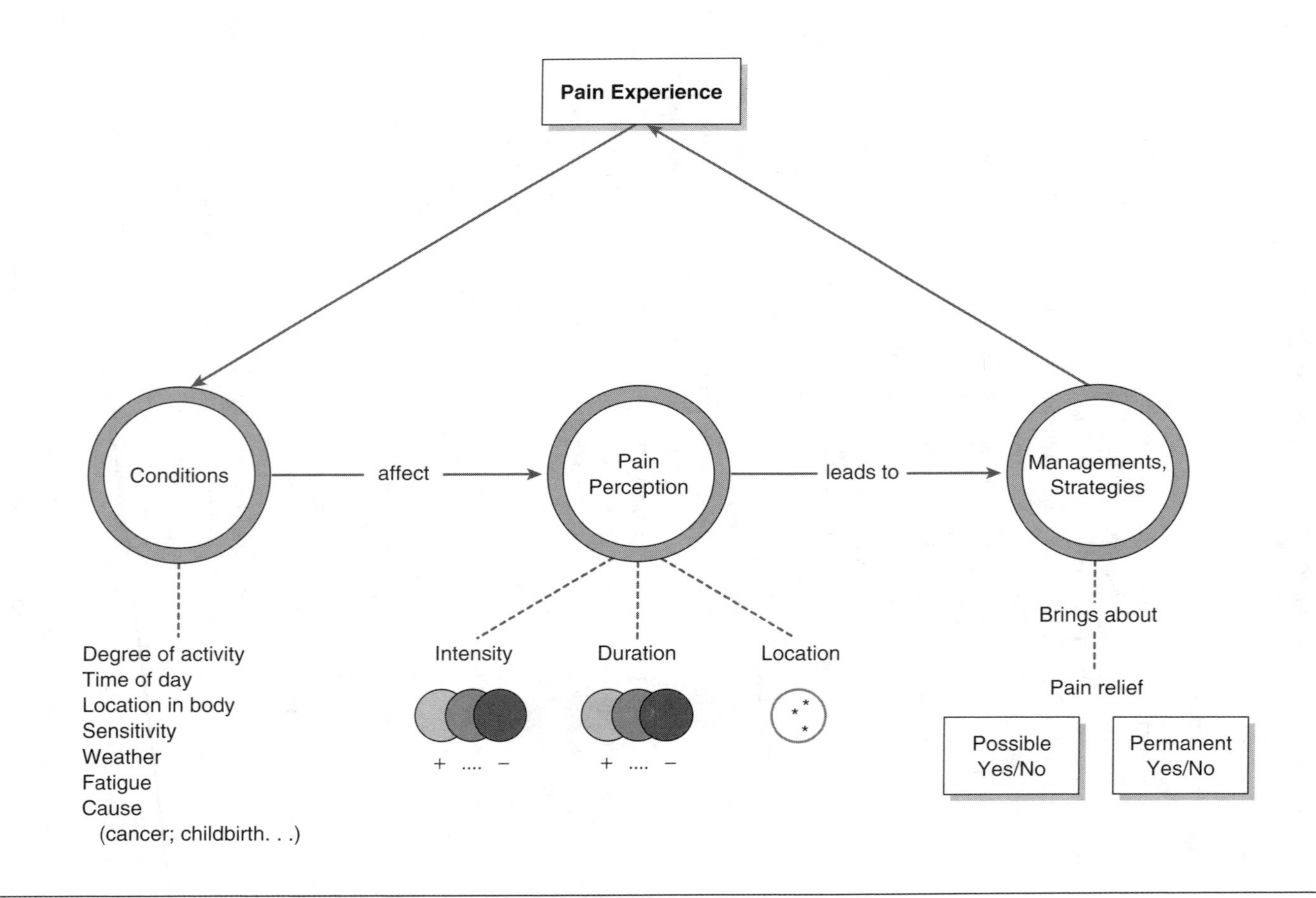


**Figure 6.7** Further Development of the Pain Experience

carefully to how persons describe their pain experience, including the pain history, experiences with pain relief, and treatment. These ideas need further exploration through data collection.

Here is another memo written off the same field notes. This memo is more speculative than previous memos that came out of the data and is meant to give direction for theoretical sampling and to help the analyst break out of the analytic ruts that block thinking. The ideas that arise in a brainstorming memo are not incorporated into the study.

---

## Memo 3 Example of a Comparison and Question Asking Memo

**April 4, 1997**

**Brainstorming Memo About "The Pain Experience" and Its Properties and Dimensions**

Working from my personal experience, professional training, and the literature, I know that arthritis is certainly not the only cause of pain. One can also have pain from an injury, say a pulled muscle or a mild burn. Pain then can vary in "type" from "burning" to "sharp," "dull," and even "throbbing." It can be described as "horrible," "overwhelming," "disruptive" or just an "irritant." Pain is "perceptual." That means no two persons experience pain in the same way because of who they are and what they bring with them to the pain experience. Some persons after surgery need a lot of pain medication. Other persons need less. It's because each person has a different "pain threshold" and different "reactions" to pain. Another point, the "pain experience" has a "trajectory" or course. The experience of pain does not begin with this pain but reaches back into the past, has a present, and enters into the next pain experience. Also this particular pain experience can vary being more or less intense over time. Thus, I now have some ideas for theoretical sampling such as looking for situations of temporary vs. chronic pain, intense pain versus more mild pain, and a pain history that includes relief from pain vs. a history unsatisfactory pain management. As I think about it, the pain experience is influenced by a combination of many factors such as intensity, duration, and whether or not it can be relieved partially, entirely, for good, or temporarily, as well as history. I recall a woman who had post-herpetic pain that never went away. She eventually died not from the pain per se but probably because she was worn down by it. Her history of looking for relief was a long one. In the end she just had to learn to live with the pain and decided that fighting it every day just was no longer feasible, getting at the meaning of pain and implications for daily life or biography. I can see that I have a lot of work to do to discover the relationships between pain and its properties, pain relief, and the pain experience. I also have a few questions that I would like to ask when I do that sampling. I should look at individuals with "chronic pain." It is the property of chronicity that is driving me to collect data on persons, such as those with rheumatoid arthritics, herpetic pain, sickle cell anemia, and cancer. I should also look at those with "temporary" pain. Here it is the property of pain being "temporary" that is driving the data collection. I should go to persons with pain related to childbirth, surgery, or an injury to find temporary pain. Burns and amputations are both interesting areas because the pain may be "temporary" or "chronic" depending upon complications. Still another question is, what are the various patterns of the pain experience? Are there various patterns of experiencing pain that cross cut these various properties? How is the meaning of pain derived? Does whether or not pain is expected or not expected make a difference? Does the ability or probability to obtain relief make a difference? I mean, if one expects or believes that relief will come with treatment vs. the belief that there will be no relief despite treatment, does this make a difference in the experience? If pain is expected, what are the steps that are taken to prevent or lessen it? How do persons control their lives or activities to minimize pain? How do factors such as culture, age, gender, how long the pain has been going on, intensity, and efforts at relief affect the pain experience?

### Commentary on Memo and Diagram

The long memo above and the diagram based on the memo (Figure 6.8) are not complete, but they do demonstrate how a researcher thinks comparatively "chronic" and "temporary" to extend her or his thinking about properties and dimensions before going back into the field. The idea is to collect data on these two extremes, then compare those data to see how properties and dimensions vary. Understanding a phenomenon like the "pain experience," which is very complex and personal in nature, takes a lot of thought and data collection from a variety of areas. In the above memo and diagram the reader can also see how categories and subcategories become linked around a phenomenon such as the "pain experience" as he or she works with the data. But before anything hypothesized about in a memo or diagram is built into findings the researcher needs to collect data and make the necessary comparisons.

---

## Memo 4 Example of a Memo That Elaborates the Paradigm: The Relationships Among Conditions, Actions/Interactions, and Consequences

**June 18, 1998**

### A More Advanced Memo on the Pain Experience

After months of collecting data and immersing myself in the pain stories of others, what is the overall pain story to be told? I think the story is somewhat

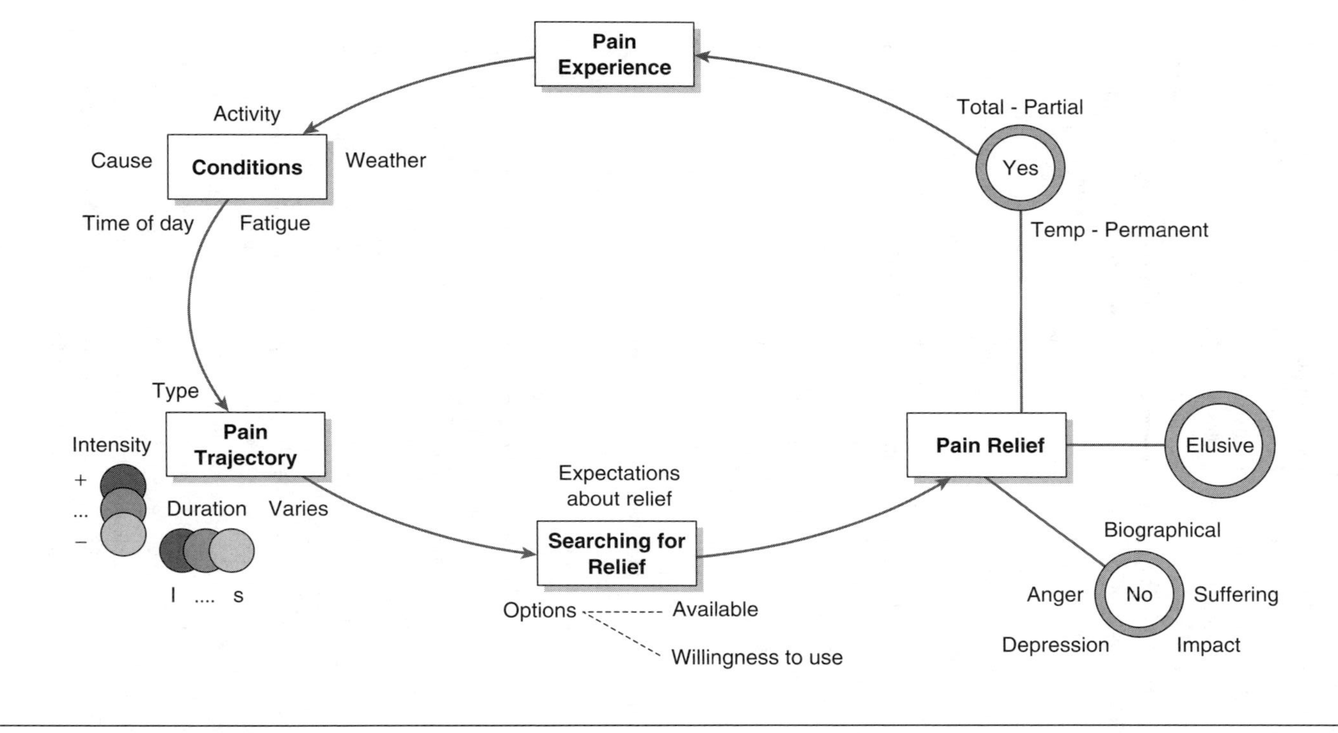


Figure 6.8 Diagram for Pain Experience

as follows. Pain is a difficult experience unless the pain is "very mild" and of very "short duration." Every time I did an interview, I could feel the "intensity" of "suffering" of people who were experiencing "severe pain," as they spoke about their experience. These people are driven to find "relief" but relief is often "elusive" (dimension of the ability to relieve of pain which can vary from relief being "obtainable" to being "elusive"). The search for relief often takes them down dead end paths with emotions ranging from "anger" to "depression" for many reasons (property of emotional response to pain). Not the least of which the "lack of control" over their lives and the "suffering" that they "endure." Though there are many "treatment options" out there, finding one that works is not easy. There are a lot of "trial and error strategies" involved. It seems to me that "pain tolerance," which is an interesting concept, diminishes when the pain is of "long duration" and people are "fatigued" or worn down by its "constancy." "Searching for relief" can be compared to being lost in a dark forest at night, as one is trying to find a way through, an escape, but the escape path is blocked and difficult to locate in the dark. People sometime become "desperate" wondering if the suffering will ever end and sometimes wishing for death as an "escape" from their pain. Pain can "take over" a life. Sometimes life "revolves" around periods of pain and relief. "Every day life" is a very important concept here because of the potential impact of pain on every day life. Every day life has the potential to be disrupted "very little" or "greatly." People seek relief not only to "get away" from the suffering but also so that they can "get on" with their lives. Pain is such a personal experience that it is difficult to explain and acute temporary pain is very different from chronic severe pain. I do see some patterns emerging. There are those persons who experience "acute temporary pain." Their pain experience may have been intense but for the moment that intensity is forgotten when the situation has passed. Their pain experience is defined by how the pain was handled and the treatments made available for controlling it. Some persons within this group describe the pain experience as horrible, or poorly managed. Others describe it as not so bad. Whatever the pain experience, it becomes incorporated into a persons "pain history" coming back into play in future episodes of pain whether acute or chronic. There are those who suffer from "chronic pain." They have developed management strategies for controlling its intensity and impact on daily life. They often describe their pain as bearable though they would rather be without it and continue to search for a cure. Their pain experience is modulated by the support and recognition that they receive from others and the hope that the situation is temporary. There are still avenues open to them. Then there is the group in which "every day" is a "pain experience," the "constant pain sufferers." Suffering defines their lives. Every day activities are severely limited. Depression is moderate to severe, as one would expect. There is little hope that the situation will improve. Their stories are touching.

## Memo 5 Example of Memo Developing the Story Line

### June 20, 1998

### Exploring Story Line Options

I ask myself, what is the main concept or storyline that integrates these various groups? I am left perplexed. I know there is "Searching for Relief," but that seems such a logical and common explanation. It is a process that goes on but doesn't explain or do justice to these varied experiences. There has to be an even better explanation. Hmm! I want to focus on the pain experience itself, what it is like to have pain, to suffer whether that pain is temporary or permanent. I keep coming back to the imagery of a forest at night and the darkness, which so reminds me of living with pain, being in darkness that is suffering both physically and often psychologically, the fear, the stumbling, the fatigue, and the discouragement. There is "Wandering in the Darkness of Pain," or "Pain, A Story of Suffering" but neither of these ideas seem to quite capture it. I can't yet put the feeling into words. I'll have to keep thinking about the problem and hopefully the right conceptualization will emerge.

### Commentary on Memo

As the reader can see, memos do wander as the analyst tries to think things out. And sometimes the analyst isn't ready to do an integrative diagram on the major theme because he or she hasn't arrived at an overarching scheme yet. Memos express the analyst's own emotions and frustrations at having an inner sense of what is going on but being unable to articulate it at this time. And the inability to complete a diagram tells the analyst that he or she still has more thinking to do. Writing memos and doing diagrams force the analyst to keep searching for the "right" conceptualization. That is why memos and diagrams are such powerful analytic tools. Notice how the later memos in this chapter demonstrate expanded thinking about the pain experience in comparison to the ones developed earlier in the study.

### Sorting of Memos

The image that comes to our mind when we think about sorting memos is of an inexperienced researcher standing with stacks of memos in his or her hands; then, dropping them one by one on the floor, letting them fall where they will. The piles that result represent a fortuitous sorting of the concepts. There are times when we all feel this way, especially when we are inundated with conceptual ideas but can't quite understand how they come together.

Yet those of us with experience know that the research does eventually come together. After months of gathering data, studying the data, writing memos, and doing diagrams there is that inner sense, or "gut feeling" of what these data are all about. It's difficult to explain, but the story of our participants becomes part of us. It's not that we have a chronic illness, or are drug addicts, gamblers, or new mothers. Rather it is that we've listened to their words, observed their actions, felt their emotions, taken on their burdens, and so understand what it is like for them. The final story may not be easy to synthesize into just a few words but it's there in our minds. From our general reading of the memos we can write a descriptive story. Then using the categories/themes we've developed over time, we can translate our descriptive story into an analytic one. Yes, the story or theory (should theory be our aim) is a construction, but a construction grounded in data.

With construction of a storyline the researcher can try out the scheme on research subjects, colleagues, committee members, friends, spouses, and companions. The final grouping of memos into specific topics as well as a whole enables researchers to write on each topic in detail as well as to present an integrated story/theory.

## Summary of Important Points

When analysts sit down to analyze those first field notes, they often feel overwhelmed by the task in front of them. It is difficult for novices to know where to start, what to look for, or how to recognize "it" when they see it. The words on a page may appear as an undifferentiated mass with little or no meaning other than what is most obvious. It happens to all of us, so do not be concerned if it happens to you. The idea is just to take that first piece of data and sit down and write a memo about it. Don't be concerned about how those first memos or diagrams look. It is not unusual for this early confusion and uncertainty to be reflected in memos and diagrams. Just remember that whatever a researcher writes in a memo or puts into an early diagram is less important than getting started with the analysis.

Early analysis is about gaining insight and generating initial concepts. In order to make sense out of data one must first "chew" on it, "digest" it, and "feel" it. The researcher has to take the role of the other and try to understand the world from the perspective of participants. This can be done in memos. Memos and diagrams are essential aspects of analysis whether the research aim is description or theory. As explained, memos and diagrams are more than just repositories of codes. They stimulate and document the analytic thought processes and provide direction for theoretical sampling.

Furthermore, without memos and diagrams there is no accurate way of keeping track of the cumulative and complex ideas that evolve as the research progresses. Diagrams are visual representations of the relationships between concepts. The purpose of diagrams is to facilitate, not hinder, the analytic process. They too evolve and become more complex as the research progresses. Some persons are more adept at doing diagrams than other persons.

There is no need for concern if one has difficulty doing diagrams. Some persons are just not visual.

As a final note, there are no rules governing the writing of memos or the doing of diagrams. Each analyst develops his or her own style that carries him or her through the research process. Both memos and diagrams are useful later, when writing for publication and giving talks about the research.

## Activities for Thinking, Writing, and Group Discussion

1. Turn ahead to the field notes located in Appendix A. Now begin to analyze the data, putting your thoughts down in memos. Try to write different kinds of memos using the memos provided in this chapter as patterns. Naturally, with these limited data, you'll not be able to write an integrative memo but you should be able to write a couple of memos about concepts you've identified and about some of the properties and dimensions of those concepts.
2. Make a couple of diagrams reflecting your analysis to this point.
3. Now bring your memos and diagrams to the group meeting and discuss them with the other group members.
4. Each group member can present a memo for discussion.
5. Based on the group discussion, write a group memo summarizing all of the ideas.
6. Do a group diagram putting all of your thoughts together.

# Note

1. The identification code in brackets after each memo allows the user to find easily the memo in the prepared MAXQDA project that you find at www.maxqda.com/Corbin-BasicsQR or at www.sagepub.com/corbinstudysite. See also Screenshot 6.

# 7

# Theoretical Sampling

*The most productive scientists have not been satisfied with clearing up the immediate question but, having obtained some new knowledge, they make use of it to uncover something further and often of greater importance. (Beveridge, 1963, p. 144)*

**Table 7.1** Definition of Terms

*Saturation*: Saturation is usually explained in terms of "when no new data are emerging." But saturation is more than a matter of no new data. It also denotes the development of categories in terms of their properties and dimensions, including variation, and if theory building, the delineating of relationships between concepts.

*Theoretical Sampling*: A method of data collection based on concepts/themes derived from data. The purpose of theoretical sampling is to collect data from places, people, and events that will maximize opportunities to develop concepts in terms of their properties and dimensions, uncover variations, and identify relationships between concepts.

## Introduction

Sometimes a writer is not quite certain where to locate material in a book. And so it is with this chapter. The notion of theoretical sampling may be easier to comprehend once a reader sees how it works during the research process, yet it is a procedural technique and some understanding of it is necessary before beginning the chapters on analysis. Perhaps the best approach is to suggest that students read the chapter twice—once before reading Chapters 8 through 12, then again after reading those chapters. We leave to readers and to teachers' discretion when and how to use this chapter.

Moving on with the chapter, one of the major issues to confront analysts is knowing what data to collect, when, where, and how. This chapter explores the notion of **theoretical sampling**, a method of data collection based on concepts derived from data. What makes theoretical sampling different from conventional methods of sampling is that it is responsive to the data rather than established before the research begins. This responsive approach makes sampling open and flexible. Concepts are derived from data during analysis and questions about those concepts drive the next round of data collection. The research process feeds on itself. It simply keeps moving forward, driven by its own power. Also, rather than being used to verify or test hypotheses about concepts, theoretical sampling is about discovering relevant concepts and their properties and dimensions. In this form of research, the researcher is like a detective. He or she follows the leads of the concepts, never quite certain where they will lead, but always open to what might be uncovered.

Questions to be addressed in this chapter include: What is theoretical sampling? What advantage does theoretical sampling have over other forms of sampling? How does one proceed? How does one keep the sampling systematic and consistent without rigidifying the process? At what points in the research does one sample theoretically? How does one know when sufficient sampling has been done?

Recollect that concepts are the basis of analysis, whether the research aim is description or theory. It is concepts that are sampled in data. Participants provide the data that tell us about those concepts. So, when researchers sample theoretically they go to places, persons, and situations that will provide information about the concepts they want to learn more about. At first glance, theoretical sampling seems contrary to everything a researcher has been taught about sampling. Researchers of conventional methods of sampling are taught to think about sampling people and controlling variables. But in theoretical sampling the researcher is not sampling persons but concepts. The researcher is purposely looking for indicators of those concepts so that he or she might examine the data to discover how concepts vary under different conditions. It is no wonder that the novice is confused when first encountering theoretical sampling.

Unlike conventional methods of sampling, the researcher does not go out and collect the entire set of data before beginning the analysis. Analysis begins after the first day of data gathering. Data collection leads to analysis. Analysis leads to concepts. Concepts generate questions. Questions lead to more data collection so that the researcher might learn more about those concepts. This circular process continues until the research reaches the point of **saturation**; that is, the point in the research when all the concepts are well defined and explained.

With theoretical sampling, there is the flexibility to go where analysis indicates would be the most fruitful place to collect more data that will answer the questions that arise during analysis. Of course, the researcher begins a study with a general target population and continues to sample from that group. In the analysis chapters that follow, Chapters 8 through 12, readers will be taken through a research project using Vietnam veterans as the target population. The reader will see that after analysis of the first interview, concepts derived from that interview and questions about the concepts became the basis for gathering more data.

Theoretical sampling is based on the premise that data collection and analysis go hand in hand. In other words, data collection never gets too far ahead of analysis because the focus of subsequent data collection; that is, the questions to be asked in the next interview or observation are based on what was discovered during the previous analysis.

Now, researchers have to be practical. There are times when a researcher has to use previously collected data or must collect data during a restricted time period. This usually occurs when a researcher has to travel to obtain participants or has to take advantage of the moment. The problem with this approach is that when analysis does occur, and questions arise as they always do during analysis, the researcher may not have the opportunity to collect additional data about a concept, leaving gaps in the research. Later in this chapter we will explain what a researcher can do if he or she cannot gather additional data. But first let us answer some important questions.

## Questions and Answers About Theoretical Sampling

### 1. What Advantage Does Theoretical Sampling Have Over Other Forms of Sampling?

Theoretical sampling is concept driven. It enables researchers to discover the concepts that are relevant to this problem and population, and allows researchers to explore the concepts in depth. Theoretical sampling is especially important when studying new or unchartered areas because it allows for discovery. Most of all, it enables researchers to take advantage of fortuitous events. Consider the following example from our study of work in hospitals. One of the main concepts derived in that study was the "workflow."

While we were collecting data, there was an earthquake of 7.4 on the Richter scale. This natural event provided us with the opportunity to return to our site to sample "workflow" in response to "contingencies" and to test our hunches about how contingencies "disrupted" workflow. Questions that we asked were: Was the flow of work altered, how, and what new arrangements were put into place to keep the work of patient care going? The results were fascinating because we discovered that the workflow was not disrupted for very long; in fact, it soon resumed, albeit sometimes in creative ways. Had we been using more conventional methods of sampling, we might not have been able to take advantage of this situation.

To make another point, theoretical sampling is cumulative. Each event sampled builds upon previous data collection and analysis, and in turn contributes to the next data collection and analysis. Moreover, sampling becomes more specific with time because the questions become more specific as the researcher seeks to saturate categories. In the initial data collection, the researcher collects data on a wide range of areas. It is kind of like fishing, for the researcher is hoping for something but does not know what will come out of that sea of data. Once the initial analysis takes place, the analyst has a greater sense of where he or she is going with the research because now the researcher has some concepts to sample for. Not every concept that comes out of the research is sampled for. The researcher has to be practical and stick to developing those categories or themes that are most important.

## 2. How Does One Proceed With Theoretical Sampling?

In doing theoretical sampling, the researcher takes one step at a time with data gathering, followed by analysis, followed by more data gathering until a category reaches the point of "saturation." In theoretical sampling the researcher has to let the analysis guide the research. The researcher has to ask questions and then look to the best source of data to find the answers to those questions. Though human subject and dissertation committees want to know in advance what persons or groups will be sampled and what questions will be asked of those participants, theoretical sampling makes it difficult to predict all of this with certainty. A researcher using theoretical sampling never knows what twists and turns the research will take. I (Corbin) didn't know when I picked up that first interview with a Vietnam veteran (see Chapter 8) where the study I was about to embark upon would lead. However, I could say with certainty, though the direction of topics of the study was largely unknown, my focus was on the Vietnam War veterans. More will be said later in this chapter about writing proposals for committees.

The procedures for theoretical sampling are simple: the researcher follows the analytic trail. Perhaps the easiest way to convey what I am trying to say is to jump ahead to the study of Vietnam War combatants that begins in the next chapter. During the first interview, the participant (a nurse) described his Vietnam War experience "as not too bad." This led the researcher to ask, "How come?" What made his experience not so bad, when all I had heard from Vietnam and other war veterans in the past was how terrible war is? Could that difference be attributed to the fact that this participant was not a "combatant"? From this insight the researcher set out to interview "combatants" to see if this did indeed make a difference in the "war experience."

The second interview was done with a combatant. It revealed a considerable difference in how the participant talked about the "war experience." What struck the researcher during analysis of the second interview was how much a combatant's war experience centered on "survival" and the degree to which unresolved issues (lack of "healing") were still present in the stories told by veterans, even thirty years later. This led the researcher to ask the question: Why so much "anger" for so long? What is there about the "war experience" that creates anger and why the "lack of healing" and "wall of silence"? Not having answers to these questions meant that I had to understand more about the war experience itself, which led me to read memoirs written by Vietnam veterans who were combatants. I also did more reading of memoirs written by veteran nurses to get a comparative base. In addition, I looked to other types of combatants, for example, pilots, questioning the nature of their experience.

After analyzing more data, the "culture" in which the Vietnam War was fought seemed to take on considerable relevance. Therefore I went back to the data, but this time instead of memoirs I went to historical data, to find out more about the "culture of war." Analyzing data about the "culture of war" identified a series of "problems" that posed a threat to "survival" through the "risks" they presented to combatants. This discovery led to a further examination of the concept of "surviving risks" and the strategies that combatants used to overcome some of the problems and reduce the risks. Based on those findings, I wanted to know more about specific "survival situations" and how persons responded to them. Then came the final question of what concept could I use to pull all of these materials together. The answer came in the form of the concept "reconciling multiple realities."

Notice that as I proceeded I was constantly following up on analytic leads derived from analysis. There was no way I could have known ahead of time the sampling path I would follow, yet all the time I stayed within my target population. It was the analysis of data and the questions about concepts that came out of that analysis that determined what kind of data I needed to collect and what I would focus on in those data. Another researcher might have gone in a different direction, yet anyone reading my analysis can follow my logic. The direction the research takes depends upon the nature of the data and the analyst's interpretation of the data, bringing the researcher and data together in the process.

## 3. How Does a Researcher Keep the Sampling Systematic and Consistent Without Rigidifying the Process?

In theoretical sampling, the researcher is not so much concerned with consistency as following up on important theoretical leads. As new analytic threads (concepts) arise during analysis, the researcher wants to be free to follow up on questions without concern of whether or not the same question was asked of previous participants. At the same time, consistency is not usually a problem because as persons tell their stories there is often much consistency between them. One of the important pieces of information that I learned from reading Vietnam War memoirs was the degree to which each memoir appeared to be written from the same basic script—the desire to survive—though many of the details were of course different. Remember, concepts that are relevant in data from one participant will almost always be found in data from other participants, though the form they take might be different. And if they are not found in the other data, then the researcher should ask, "Why not?" If at the conclusion of an unstructured interview or observation relevant topics are not covered, the researcher certainly can ask questions about these, especially if he or she feels that the topics are relevant to the study.

## 4. How Much Sampling Must Be Done?

The answer to this question is both simple and complex at the same time. It is simple in the sense that a researcher continues to gather data until reaching the level of data "saturation." It is complex in the sense that arriving at saturation is not that easily attained. Saturation is usually explained in terms of "when no new categories or relevant themes are emerging." But saturation is more than a matter of no new categories or themes emerging. It also denotes a development of categories in terms of their properties and dimensions, including variation, and possible relationships to other concepts. In other words, the aim of research is not just to come up with a list of categories. It is to tell us something about those categories. The understandings provided by the researcher must go beneath surface explanations. It would not be sufficient to say simply that the experience of combatants in war can be reduced down to survival. The researcher has to explain how and when survival takes on meaning, what survival looks like under different conditions, and what are some of the consequences of survival.

Only when a researcher has explored each category/theme in some depth, identifying its various properties and dimensions under different conditions, can the researcher say that the research has reached the level of saturation. In reality, a researcher could go on collecting data forever, adding new properties and dimensions to categories. Eventually a researcher has to say this concept is sufficiently well developed for purposes of this research and accept what has not been covered as one of the limitations of the study. Every researcher has to be aware of not concluding a study too soon. Sometimes when researchers say they have saturated their categories, what they really mean is that "they" are saturated with the data collection process. They have either run out of time, money, or energy, therefore foreclosing on the research problem much sooner than they should have, leaving gaps in the overall story. It is doubtful that five or six one-hour interviews can lead to saturation.

## 5. At What Point in the Research Does a Researcher Sample Theoretically?

Theoretical sampling begins after the first analytic session and continues throughout the research process. It ends only when the research process has been concluded. Even when writing about the findings, it is not unusual for a researcher to have new insights, discover that some categories are better developed than others, or uncover breaks in the overall logic that require the collection of more data.

## 6. How Does a Researcher Know When Sufficient Sampling Has Occurred?

A researcher knows when sufficient sampling has occurred when the major categories show depth and variation in terms of their development. Though total saturation (complete development) is probably never achieved, if a researcher determines that a category offers considerable depth and breadth of understanding about a phenomenon, and relationships to other categories have been made clear, then he or she can say sufficient sampling has occurred, at least for the purposes of this study.

Throughout the research, the investigator should take advantage of fortuitous incidents that occur. However, any data that the researcher collects through theoretical sampling must have relevance to the analysis. In other words, if the researcher is observing in a facility and an important manager dies, it is not the death of the manager per se that is relevant but what that death means in terms of "workflow" or some other concept the researcher is working with.

It is easy for novice researchers to get carried away when out in the field because so many interesting things are going on. It helps if the researcher is using concepts derived from previous observations as a guide. Let questions and concepts developed during analysis guide data collection. Just because someone says, wouldn't it be interesting to "such and such" is no reason to go off on a "wild goose chase." Stay focused. Look for situations that offer variation or different situations regarding the concept in question and sample theoretically at this point. The variation in conditions will maximize the opportunity to discover new properties and dimensions about a concept.

## 6. What If I Have Already Collected All of My Data Before Sitting Down to Do My Analysis? Can I Still Do Theoretical Sampling?

The answer is yes, though it may be more difficult. I used memoirs as data in the study on Vietnam War veterans that I present next, so using already collected data can be done. Remember, theoretical sampling is concept directed data gathering and analysis. Questions about a concept(s) serve as a guide for what incidents to look for in the next set of data. Therefore, a researcher can sample data that have already been collected or are available for incidents pertaining to a concept. It is not unusual to return to previously analyzed data and look at them with a fresh eye. Incidents or events pointing to a concept may have been overlooked earlier because the significance of these events may not have been well understood by the researcher at the time. A researcher can always return to previously collected data to see what was missed. One word of caution: gaps in the research may occur when analyzing previously collected data because there isn't the opportunity for further exploration. If a researcher is stuck with certain data, then he or she will have to make do. It doesn't mean that a study will lack significance or be superficial. A researcher can do a high-level analysis on whatever data he or she has.

## 7. Where Does a Researcher Get a Sample?

In looking for areas to sample, the analyst determines what data will best provide answers to questions and help to fill in information about categories. Let us clarify a point before continuing with this line of thought. In the Vietnam veteran study that will be presented in the next chapter, it became clear that I had to go out to find a group of combatants in order to discover what difference being a combatant made to the war experience. I did obtain a couple of interviews, but at the time was not able to obtain more. So, I went to memoirs to find my answers. However, when it was context that I was interested in, I went to historical documents because they would provide me with the type of data I was looking for to answer my questions about how the United States got into the war and why. Then, when I needed more information about the concept of surviving, I returned to the memoirs. I went back to the data and looked specifically at incidents I labeled "surviving" and analyzed these in more depth. Sometimes, in order to sample theoretically, a researcher has to collect more data and sometimes the researcher can return to data that have already been collected.

## 8. What Are Some Sampling Matters That a Researcher Must Consider Before Starting the Research?

At the beginning of a study there are many sampling matters that the researcher must consider. The initial decisions made about a project give the researcher a sense of direction and a place from which to begin data gathering. What happens once data collection is under way becomes a matter of how well the initial decisions fit what one is uncovering in the data. Initial considerations include:

(a) *Decisions made about the site or group to study.* This, of course, is directed by the main research question. For example, if a researcher is interested in decision making by executives, he or she must go to those places where executives are making decisions to observe what goes on.

(b) *Decisions made about the kinds of data to be used.* Does the investigator want to use observations, interviews, documents, memoirs, biographies, audio- or videotapes, or combinations of these? The choice should be made on the basis of which data have the greatest potential to capture the kind(s) of information desired. For example, a researcher might want to use memoirs and other written documents in addition to interviews and observations when studying executive decision making. If one is studying interaction between groups, it is only logical to observe, in addition to interviewing, because observations are more likely to reveal the subtleties of interaction.

(c) *Decisions made about how long a site should be studied.* A site is studied as long as it provides the data one is seeking. Remember, it is not sites or persons per se that we are studying, but concepts. Usually we make use of sites to collect data on concepts we are following up on. Within sites, we can always vary data gathering by looking at who, what, when, how, and where, based on concepts derived from the data. A factor that enters into the sampling decision is whether the researcher is developing formal or substantive theory. To clarify: in our study of work we confined our study on the articulation of work in hospitals to only one hospital because we were developing substantive theory. To develop a more formal theory about the articulation of work, it would be necessary to sample other types of organizations and other types of work, such as police work, construction, etc. The possibilities are unlimited. Decisions about the various sites might be made at the beginning of the investigation or chosen as the study progresses, depending on concepts. Initially, decisions regarding the number of sites and observations and/or interviews depend upon access, available resources, research goals, plus the researcher's time schedule and energy. Later, these decisions may be modified, as the study progresses, based upon questions that arise during analysis.

## 9. Can Interview and Observational Guides Be Used to Collect Data?

With theoretical sampling, interview and observational guides are not as relevant as they are to structured forms of research because they tend to evolve and change over the course of the research. However, a researcher cannot get through a human subject or research proposal committee without an indication of the questions that will be asked or the observations to be made, the purpose for which is the protection of human subjects. A student or experienced researcher usually has some basic knowledge to draw from when putting together a questionnaire or observational guide, either from experience or the literature. Putting together a list of questions or areas of observation should not be that difficult. The researcher does the best that he or she can to put together a comprehensive set of questions or areas for observation. To cover all bases, the researcher should add a sentence or two in the proposal indicating that if a participant brings up another topic that proves to be important to the investigation, the researcher will follow through on that topic.

As a practical matter, once a researcher has decided upon the target population, the place, the time, and the kinds of data to be gathered, he or she is ready to develop a list of interview questions or areas for observation. Initial interview questions or areas of observation might be based on concepts derived from literature or experience, or—better still—from preliminary fieldwork. Since these early concepts haven't evolved from "real" data, if the researcher carries them with him or her into the field, then they must be considered provisional and discarded as data begin to come in. Nevertheless, early concepts often provide a departure point from which to begin data collection, and many researchers (and their committee members) find it difficult to enter the field without some broad conceptualization of what they are going to study. That is not our style, however.

Once data collection begins, the initial interview or observational guides (used to satisfy committees) give way to concepts derived from analysis. Adhering rigidly to initial questions throughout a study hinders discovery because it limits the amount and type of data that can be gathered. It has been these authors' experience that if a researcher enters the field with a structured questionnaire, persons will answer only that which is asked, and often without elaboration. Respondents might have other information to offer, but if the researcher doesn't ask, then they are reluctant to volunteer, fearing that they might disturb the research process. Unstructured interviews, using general questions such as "Tell me what you think about" or "What happened when" or "What was your experience with," give respondents more room to explain what is important to them (Corbin & Morse, 2003).

## 10. Are There Variations on Theoretical Sampling?

As with all research there is the "ideal way" of doing things and there is the "practical way." Sometimes a researcher has to settle for the latter. Here are some variations.

(a) *A researcher may look for persons, sites, or events where he or she purposefully can gather data related to categories, their properties, and dimensions.* For example, when doing a study on the use of medical technology in hospitals, one team member noted that machinery in hospitals had several properties (Strauss, Fagerhaugh, Suczek, & Wiener, 1985). The properties included (among others) cost, size, and status. The team then proceeded to sample events and sites within a hospital, where the similarities and differences among these properties of machinery would be maximized. They went to observe the computerized axial tomography (CAT) scanner, a big and expensive machine that had been given considerable status among diagnosticians. However, CAT scanners represent just one extreme type of hospital machinery, a fact to keep in mind when collecting data. It is also important to sample machinery that might be less costly, less prestigious, and less reliable to have a comparative base. In the above case, the researchers were driven to sample by the conceptual notion that the work of patient care might be influenced by the particular properties the medical machinery brought into service as part of their care, thus integrating two categories, "patient care" and "medical technology."

(b) *A researcher may gather data very systematically (going from one person or place to another on a list) or by sampling on the basis of convenience (whoever walks through a door or whoever agrees to participate).* This is a more practical way to gather data and probably the method used most often by beginning researchers. In other words, the researcher takes who or what he or she can get in terms of data. This does not mean that comparisons aren't being made on the basis of concepts during analysis, for they are. It is just that the researcher must accept the data that he or she gets rather than being able to make choices of who and/or where to go next.

Often, differences in data emerge naturally because of the natural variations in situations. For example, when we began our study of "workflow" in hospitals we knew little about either the particular hospital or the wards or the head nurses; we simply proceeded from unit to unit, spending time with any head nurse who was willing to participate in the study. In the end, we found that each unit was different in terms of organizational conditions, the number of patients and types of work done, and how the flow of work was organized and maintained over time. Because of those differences, there was ample opportunity to sample theoretically based on concepts. But the ideal remains to let the questions about concepts that are derived during analysis guide the next data collection.

(c) *A researcher may find that differences often emerge quite fortuitously.* A researcher often happens upon theoretically significant events quite unexpectedly during field observation, interviewing, or document reading. It is important to recognize the analytic importance of such an event or incident, and to pick up on it. This comes from having an open and questioning mind, and being alert. When an analyst happens upon something new or different, he or she must stop and ask, "What is this? What can it mean?"

(d) *A researcher may return to the data themselves, reorganizing them according to theoretically relevant concepts.* An example of this form of sampling occurred during a study of high-risk pregnant women, when it became evident to the researcher that she was categorizing women according to her (the researcher's) perception of risks, which was medical, *but* that women were acting on the basis of their perceptions of risks, which was not always in alignment with the medical person's perceptions of the risks. The researcher then went back and reshuffled incidents, placing them into categories of risks according to the women's definition of the situation. Then the actions taken by the women began to make analytic sense. Note that in any one interview or observation there are often several incidents pertaining to the same concept and each is coded separately. For example, in the study of high-risk pregnant women, sometimes over the course of a week, perceptions of risks varied depending upon what was going on with the status of the chronic condition, the baby, and the pregnancy. This meant coding each incident separately because risk definitions and management tended to vary accordingly.

In data gathering and analysis, the researcher will want to sample incidents and events (either from new or previously collected data) that enable him or her to identify significant variations. By asking what difference the type of machinery makes to the type of care given to patients, the researcher is able to relate "type of care" with "type of machinery." Questions to be asked include: how the patient is prepared, how risks are managed, how the work is parceled out, who schedules and coordinates it, and so on.

Relationships between concepts, just like the concepts themselves, are compared across sites and persons in order to uncover and verify similarities and differences that will demonstrate dimensional range or variation of a concept and the relationships between concepts.

A researcher should never become upset by not being able to choose a site or obtain access to a theoretically relevant site or person(s). Rather, he or she should make the most out of what is available to him or her. When it comes to events and incidents, rarely will a researcher find two or more that are identical. Rather, there will always be something different, be it conditions, inter/action, or consequences, that will provide the basis for making comparisons and discovering variation. If the analyst is comparing incidents and events on the basis of the concepts rather than looking at data in a descriptive sense, then he or she is doing theoretical sampling regardless of how the data were actually gathered. It may take longer to uncover process and variation, and to achieve density, when a researcher can't purposefully choose persons or sites to maximize variation, but through continued and persistent sampling, eventually differences will emerge.

(e) *Toward the end of the research, when a researcher is filling in categories, he or she may return to old sites, documents, and persons, or go to new ones to gather the data necessary to saturate categories and complete a study.* Analysts are constantly comparing the products of their analyses against actual data, making modifications or additions as necessary based on these comparisons. As such, they are constantly validating or negating their interpretations.

## 11. Can I Sample Data From a Library, and If So, How?

Some investigations require the study of documents, newspapers, or books as sources of data. Just how does one go about this? The answer is to sample exactly as you do with interview or observational data, with the usual interplay of coding and sampling.

If you are using a cache of archival material, this is the equivalent of a collection of interviews or field notes (Glaser & Strauss, 1967, pp. 61–62, 111–112). However, the documentary data may not be located in one place but scattered throughout a single library, several libraries, agencies, or other organizations. Then you must reason, just as with other types of data, where the relevant events/incidents are to be found and sampled. Will they be in books about particular organizations, populations, or regions? You can answer that question by locating the materials using the usual bibliographic research techniques, including browsing purposefully in the library stacks.

A special kind of document consists of the collected interviews or field notes of another researcher. It is customary to call the analysis of such data by the term "secondary analysis." A researcher can code these materials too, employing theoretical sampling in conjunction with the usual coding procedures.

## 12. How Does One Sample Theoretically When a Team Is Gathering the Data and Still Maintain Consistency?

When working with a team of researchers, each member must attend group analytic sessions. Each must also receive copies of any memos that are written by individual data gatherers as well as those that are written during group sessions. Data must be brought back to the group and shared. The important point is that each researcher knows the categories being investigated so that he or she knows what types of questions to ask during fieldwork. Equally important is that the team meets regularly and frequently for analyzing portions of the data. Working as an analytic unit enables team members to work together and to sample theoretically. As the data pile up, it may become impossible for team members to read all of each other's interviews or field notes, so of course each has the responsibility to code his or her own materials and bring the results of coding back to the group. Everyone must read all the memos, otherwise team members miss out on the evolving nature of findings.

## 13. How Does Theoretical Sampling Differ From More Traditional Forms of Sampling?

In quantitative forms of research, sampling involves randomly selecting a portion of a population to represent the entire population to which one wants to generalize. Thus, the overriding consideration is how representative the target population from the larger population is in terms of certain characteristics. In reality, a researcher can never be certain that a sample is completely representative. In quantitative research, procedures such as randomization and statistical measures help to minimize or control for variation. In qualitative investigations, researchers are not so much interested in how representative their participants are of the larger population. The concern is more about concepts and looking for incidents that shed light on them. And in regard to concepts, researchers are looking for variation, not sameness. Variation is especially important in theory building because it increases the broadness of concepts and scope of the theory.

## 14. Is Theoretical Sampling Difficult to Learn?

Theoretical sampling is not difficult to do. However, it takes a lot of trust in self and the research process to let the evolving analysis be the guide. This comes with time and experience.

## 15. What About Research Design—What Is Its Relationship to Theoretical Sampling?

Unlike statistical sampling, theoretical sampling cannot be planned before embarking on a study. The specific sampling decisions evolve during the research process. Of course, prior to beginning the investigation, a researcher can reason that events are likely to be found at certain sites and within certain populations. Realistically, when writing proposals for funding agencies, it is important to explain both how the researcher will sample and the rationale for such.

## Summary of Important Points

Theoretical sampling is the process of letting the research guide the data collection. The basis for sampling is concepts, not persons. Relevant concepts are elaborated upon and refined through purposeful gathering of data pertaining to these concepts. It is through theoretical sampling that concepts are elaborated and as such it forms the basis for thick rich description and theory construction. Theoretical sampling continues until all categories are saturated; that is, no new or significant data emerge and each category is well developed in terms of its properties and dimensions.

## Exercises for Thinking, Writing, and Group Discussion

1. Think about theoretical sampling. What is the logic behind it? How does it enhance the research process? What are some of the drawbacks to this type of sampling as you see them?
2. In a few sentences write down how you might go about theoretical sampling in your study.
3. Discuss in a group how you might explain theoretical sampling in a proposal for a human subjects committee.

# 8

# Analyzing Data for Concepts

*But my favorite way of developing concepts is in a continuous dialogue with empirical data. Since concepts are ways of summarizing data, it's important that they be adapted to the data you are going to summarize. (Becker, 1998, p. 109)*

**Table 8.1** Definition of Terms

*Categories*: Higher-level concepts under which analysts group lower-level concepts according to shared properties. Categories are sometimes referred to as themes. They represent relevant phenomena and enable the analyst to reduce and combine data.

*Coding*: Extracting concepts from raw data and developing them in terms of their properties and dimensions.

*Concepts*: Words that stand for ideas contained in data. Concepts are interpretations, the products of analysis.

*Dimensions*: Variations within properties that give specificity and range to concepts.

*Properties*: Characteristics that define and describe concepts.

## Introduction

The first chapters provided the foundation behind the approach to qualitative research presented in this book. The next section in the text will take a different turn. The procedures and strategies will be used to demonstrate how to do analysis. I have discovered over the years that it is one thing to talk about data collection and analysis and quite another to do it. As the reader moves through the next five chapters, he or she will notice that analytic strategies that were discussed earlier in Chapters 4 and 5 are not something to agonize over but are integrated and natural parts of the analysis. The doing of analysis is fluid and generative.

I'll[1] begin the demonstration project with what is called open **coding.** Open coding requires a brainstorming approach to analysis because, in the beginning, analysts want to open up the data to all potentials and possibilities contained within them. Only after considering all possible meanings and examining the context carefully is the researcher ready to put interpretive conceptual labels on the data. Conceptualizing data not only reduces the amount of data the researcher has to work with, but at the same time provides a language for talking about the data.

Just as a bit of a refresher before beginning this section, recall that there are many levels of **concepts**. Concepts can range from lower-level concepts to higher-level concepts. Higher-level concepts are called categories/themes and **categories** tell us what a group of lower-level concepts are pointing to or are indicating. All concepts, regardless of level, arise out of data. It is just that some are more abstract than others. The process of conceptualizing data looks like this. The researcher scrutinizes the data in an attempt to understand the essence of what is being expressed in the raw data. Then, the researcher delineates a conceptual name to describe that understanding—a researcher-denoted concept. Other times, participants provide the conceptualization. A term that they use to speak about something is so vivid and descriptive that the researcher borrows it—an in-vivo code.

In this chapter, the analytic focus is upon constructing concepts out of data. First I will break the data into manageable pieces. Second, I will take those pieces of data and explore them for the ideas contained within (interpreting those data). Third, I will give those ideas conceptual names that stand for and represent the ideas contained in the data. Coding requires "thinking outside the box" (Wicker, 1985). It means putting aside preconceived notions about what the researcher expects to find in the research, and letting the data and interpretation of it guide analysis. Coding also means learning to think abstractly. The idea is not just to take a phrase from "raw" data and use it as a label. Rather, coding requires searching for the right word or two that best describe conceptually what the researcher believes is indicated by the data.

The actual procedures used for analyzing data are not as important as the task of identifying the essence or meaning of data. Procedures, you recall from earlier chapters, are just tools. The greatest tools researchers have to work with are their minds and intuition. The best approach to coding is to relax and let your mind and intuition work for you.

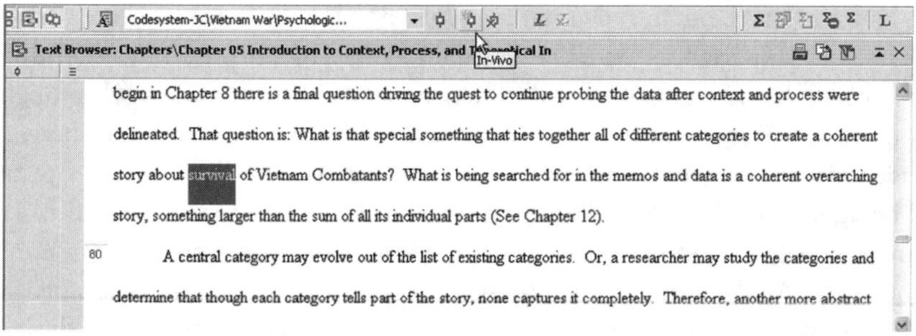

Codesystem-JC\Vietnam War\Psychologic...

Text Browser: Chapters\Chapter 05 Introduction to Context, Process, and T... cal In

In-Vivo

begin in Chapter 8 there is a final question driving the quest to continue probing the data after context and process were delineated. That question is: What is that special something that ties together all of different categories to create a coherent story about survival of Vietnam Combatants? What is being searched for in the memos and data is a coherent overarching story, something larger than the sum of all its individual parts (See Chapter 12).

80 A central category may evolve out of the list of existing categories. Or, a researcher may study the categories and determine that though each category tells part of the story, none captures it completely. Therefore, another more abstract

---

**Screenshot 8** The screenshot shows the upper menu bar of MAXQDA, situated right above the Text Brower Window, where the currently opened text is displayed and can be worked on. The example shows the in-vivo coding of the code "Survival": You highlight the word (or term) you want to turn into a code, click on the In-Vivo icon in the menu bar, and the code will automatically be created, inserted into the code system, displayed on its top, and the coding will be displayed in the code margin beside the text (which has not yet happened in this example because the In-Vivo icon has not yet been clicked). The picture also shows the icon to switch the text edit-function on and off on the left of the menu bar, which allows you to make any changes in your text without losing the position of the codings. Right beside it, you see the Quick Code bar, where you can transfer codes from the code system in order to have them right at hand to do concentrated coding for selected codes. On the right of the In-Vivo icon you find the undo function, allowing you to delete all codings within the same working session. The "L" icon besides it is the link option, which you may use to link a word within your text to any word in another place in your data set or to any other place outside your project—for example, a photo on your hard drive, a Web site, and the like.

## Demonstration Project

Beginning at this time, I would like to take readers on an analytic journey. I will begin the analysis with the first interview. Each subsequent chapter will build upon analysis derived from the previous chapter. I have chosen a topic of study that I have never researched before. Therefore, I am starting at ground level. Everything I will be doing as part of the study will be done before the eyes of the readers.

Some readers might be disturbed by the topic chosen for the demonstration study. Or, some might say that they can't learn from the demonstration project because the topic is not relevant to their discipline. I understand that some students have difficulty separating the process from the topic; that is, they can't relate to the topic so they dismiss what the author is trying to demonstrate. Remember, it is the process, not the topic, that is relevant here.

I admit that the materials I will be working with are intense. They pertain to soldiers' experiences with the Vietnam War. The purpose of this study is to explain the Vietnam War from the perspective of the soldiers who fought in that war. The relevance of this study to society is its potential to increase understanding of what soldiers go through and have to live with as a result of being in a combat zone. The study also has implications for the health and care of those men and women who served in combat after they return home from war. At the time of the Vietnam War, only men were designated as combatants. There were no women ground soldiers or fighter pilots, though in Vietnam there were women in supportive roles, such as nurses (Smith, 1992; Van Devanter, 1983). Therefore, when I talk specifically about "combatants" I will refer to them in the masculine.

I want to add a note here about the topic of investigation and the research question. Notice that I do not begin with a specific research question, but a general topic area. The question is open ended because, as I begin analysis of this first interview, I do not know where the research will lead me and what questions will evolve as I go along. When I talk about a soldier's experience, I'm *not* talking just about "inner experience" or experience from a phenomenological point of view, rather I am using experience as a general all-encompassing term to describe the entire process of being a frontline soldier or pilot in the Vietnam War, from volunteering for service, to being drafted, and to homecoming. What makes this project a grounded theory study, rather than, say, a phenomenological study, is my background and training as a grounded theorist. I analyze and interpret data differently than would a phenomenologist. My background and training also make me look at context (structure) and process (action/interaction) and lead me to go beyond description to develop a theoretical explanation. However, as readers will see, a researcher need not go all the way to theory development. He or she could stop after concept identification and development and do a very nice descriptive study, adding elements of context and process, as he or she feels competent to do.

During the analysis, I will be working with different types of data, including interviews, memoirs, and historical materials. This chapter is based mainly on interview data and focuses on open coding and concept identification. The entire transcript of the interview can be found in Appendix B. The interview took place in about 1994. It was conducted by Anselm Strauss for purposes other than this project and sat in our files for several years. It describes one man's Vietnam War experience. The interview was discovered recently when I, Corbin, was looking for materials to use for this book. I was struck by the content of the interview and requested permission from the participant to use the materials. Permission was granted.

## Analysis

The first step in any analysis is to read materials from beginning to end. (If a researcher is working with field notes, videos, or other types of documents, he or she could use the same process of going through the entire video or document to get a feel for what it is all about.) When doing that first reading, analysts should resist the urge to write in the margins, underline, or take notes. The idea behind the first reading is to enter vicariously into the life of participants, feel what they are experiencing and listen to what they are telling us. (Refer to Appendix B with Participant #1 in its entirety.)

### *Beginning Coding*

Analysts should begin the coding soon after the first interview or observation/video is completed because the first data serve as a foundation for further data collection and analysis. Once the researcher has read and digested the entire document, it is time to "go to work on the data," so to speak. I use natural breaks in the manuscript as cutting off points, and usually these breaks denote a change in topic, but not always. Then I examine each section in depth. Using this detailed approach is more tedious than just doing a general reading of a manuscript, then pulling out some themes. However, it is the belief of this author that a "close encounter" with data in the beginning stages of analysis makes the analysis easier in later stages because there exists a strong foundation and less need to go back to find the missing links. Detailed work like this in the beginning is what leads to rich and dense description and as well as to well-developed theory.

Computers can be used to do coding, but the analyst must be very careful not to fall into the trap of just fixing labels on a piece of data, then putting piles of "raw" data under that label. If a researcher does just this, he or she will end up with a series of concepts with nothing reflective said about what the data are indicating. Even with computers, the researcher must take the time to reflect on data and write memos. Thinking is the heart and soul of doing qualitative analysis. Thinking is the engine that drives the process and brings the researcher into the analytic process.

The process I will use is as follows. I'll take a piece of raw data. That piece of data will be used as a springboard for analysis. What I am thinking as I analyze data will be presented as a memo. Each memo will be labeled with a concept. Sometimes the code name changed several times as I thought and rethought about the ideas contained in each quote. Any conceptual label reflects my interpretation of what is being said, as other researchers may have their own ideas and even disagree with me. The idea here is not to quarrel with everything that I say, but to note the process that I am going through.

I will then take the first section of the interview and proceed from there with the analysis. The reader will see that each memo has been assigned a number and titled with a concept that reflects what I think the raw data are all about. Under the title are the actual data followed by the analysis. Some memos will be longer than others. Remember are this is the first analysis of the first interview. Memos will become more accurate, complex, and longer later in the study as analysis accumulates.

---

## Memo 1

### June 10, 2006

### Locating the Self: Time: Entry Into the Military

*Basically, I come from a middle-class family, very patriotic, God fearing and religious. We were a very loving family and continue to be. I have three brothers and one sister. My father is dead. My mother died in her eighties. We all get together for a family reunion at least 1 time a year. I left home at sixteen. I worked a couple of years at menial jobs, well not necessarily menial but low paying. I worked as an orderly in a hospital and that's how I became exposed to the nursing profession and decided to pursue that. I was twenty-one-years-old when I was first licensed as a nurse. Now that I'm fifty I have a long history of nursing in there. This was back in the 60s. I worked one year at a veteran's hospital in the city of X, where I was exposed for the first time to veterans, people who had been to wars. Primarily, there were elderly WW I people, some middle-aged WW II people, and a few Korean Veterans thrown in. And I was pretty much interested in listening to them talk about their experiences and all that, so in 1966 when the government finally made a commitment to Vietnam, sending lots of men and women and materials, I volunteered to go.*

In these first few lines, the interviewee is looking back "locating himself" at the time of entry into the service. He begins by explaining who he was at that time, what he was doing, and what led him to enlist in the military. I am not certain why he began the interview in this manner, for I wasn't at the interview. Perhaps he was asked to provide this information by Strauss before beginning the interview. Or perhaps the participant felt it was necessary to "locate himself" or explain who he was then versus who he is now. Under the concept of "locating" come several minor concepts—**properties** that help define who he was. The first is "family background" which includes being "middle-class," "religious," "God fearing," "close," and above all "patriotic." The second concept is "patriotism," which stands out from the rest. I'm not sure what it means to be patriotic since he does not define it for us. The third concept is "path to the war," that is, how he got to the war and his mindset and preparations for going to war. The "path to war" subconcepts include: hearing war stories, being a nurse, volunteering (which is a concept that I want to come back to), becoming a "six-week wonder," being an officer, and being quickly dispatched to the war zone. I must say that his "path" has the **dimensions** of being "straight" and "quick." There is no indication of inner conflict about going to war or negative feelings about the war itself at the time.

### *Methodological Note*

From time to time I will insert a methodological note between the memos to explain analytically what is going on. In the above case, what is important to note is that though there were several lesser concepts, like "family background," "path to the war," and "patriotism," these were not listed as topic headings. One of the mistakes beginning analysts make is to fail to differentiate between levels of concepts. They don't start early in the analytic process differentiating lower-level explanatory concepts from the larger ideas or higher-level concepts that seem to unite them. Note how the lower-level concepts fill in, explain, and tell us something about who the person was and give us some of the properties and dimensions of "locating the self." "Locating the self" is a higher-level concept because it can be applied to other interviews. The notion of "locating the self" can crosscut interviews while the specifics like being from a "middle-class family" are likely to be different for different people.

If an analyst does not begin to differentiate at this early stage of analysis, he or she is likely to end up with pages and pages of concepts and no idea how they fit together. Furthermore, if the analyst relates lower-level concepts to a broader concept like "locating the self," the concept of "locating" can be qualified more specifically such as "at the time of entry." The analyst can then look at data to determine how the same and/or other persons locate themselves later in the war experience or even later in their lives. Everyone reading this should be able to see the difference in abstraction between concepts like "family background" and "locating the self." Family background tells an analyst something about the "self" at the time of entry into the military. It is part of the explanation of who he was and why he enlisted. Locating the self is the analyst's interpretation of what the respondent is doing.

### Memo 2

**June 10, 2006**

**Being a Volunteer Versus Being Drafted Versus Being a Draft Dodger**

*Well, I kind of volunteered. I was one step ahead of the draft. So I volunteered to go. I did basic training at Ft. Sam Houston in Texas, a six-week wonder. I came out as a second lieutenant and was immediately sent to Vietnam.*

There are three concepts here. One is "being a volunteer." This concept comes directly out of the data. Drawing upon my general knowledge of the time, I also know that at the time of the Vietnam War the draft was still in place and that many of the young men who went to Vietnam did so not because they were patriotic but because they were drafted into the army and had no acceptable way out. I also know that there was a group of "draft dodgers." I have no data on draftees or draft dodgers at this time. The concept of being a "volunteer" interests me because it denotes a willingness to go to war, or at least in his case a more or less willingness to go to Vietnam. Our interviewee was young, only twenty-one. He was not in college at the time but had graduated from nursing school. He was patriotic but didn't enlist because of patriotism, and says rather it was because he believed that eventually he would be drafted. Let me spell out some of the conditions that he gives for enlisting. He came from a "patriotic family" and going to war seemed to him like "the right thing to do" at the time. Yet, he says he volunteered to stay one step ahead of the draft, a little contradiction here. As an analyst, I wonder if volunteering vs. being drafted makes a difference in the war experience. If young men volunteer for the military, are they more likely "accepting," "ready," and "committed to the war effort" at the time of entry than, say, someone who is drafted or one who doesn't want to go? Also there is another point; if a young man enlists, he is given a choice about the branch of service he enters. As I proceed with my analysis I want to see if volunteering vs. being drafted makes a difference in the overall war experience. Also, it might be important to see if there is any difference between those who are in the military for just a four-year term vs. the career people. It will be also be interesting to see if there is information out there about draft dodgers and how those who had to serve and were wounded feel about those who managed to avoid serving.

### Memo 3

**June 10, 2006**

**Being a Noncombatant Versus Being a Combatant**

*I . . . most of the time I was there I worked in transport and an evacuation hospital. We went out in helicopters and picked up people from aide stations, which were pretty much . . . it's hard to say because there were no really defined lines. The lines could change every day, two to three times a day but the aide stations were in areas of conflict. We would transport the most seriously wounded back to Saigon, which was about 75 miles away.*

Being a "combatant" and being a "noncombatant" seem to be relevant concepts. The interviewee tells us that he worked in helicopter transport and in an evacuation hospital indicating that he was a "noncombatant." Being a noncombatant is interesting because it makes him different from the guys who actually engaged in front-line fighting. However, this doesn't necessarily mean that he did not experience the horrors of war. He did fly into battle zones, that I will call "zones of conflict," and transport the wounded back into "zones of safety," which in this case was about 75 miles away.

So what does it mean to be a noncombatant? I need more data in order to understand the differences between combatants and noncombatant in war. However, I can play an analytic comparative game based on what I know from general readings about war. Being a combatant means that a person's "life is at-risk" much of the time. Combatants see their "comrades wounded" and "killed." A combatant "must kill" or "be killed" in "battle." Also, being a combatant means constant "fear and stress" during contact with the enemy. Being at war is in some ways similar to being a "hunter" but it has that added twist of being both "hunter" and potential "prey." It's like a mad video game where each side is out to get the other but in the case of war the "kill" is for "real." The "enemy" is there to kill you, and will, if you don't kill him first. I'm not certain that a person can fully comprehend the meaning of war and the constant stress and fear associated with being simultaneously hunter and prey unless one has been there.

Being a noncombatant provokes different kinds of fear and stress. Noncombatants can be "prey" if they are in the wrong location but they don't go out "on the hunt" for the enemy. A noncombatant is "not a killer;" the gun, if carried, is for "self-defense" if he somehow comes under attack. This particular noncombatant's job was to "care for the injured." His exposure to "enemy fire" was "intermittent" mostly when he flew on missions to the aide stations. The notion of "care" vs. "kill" I think is a very defining one, making the war experience different for this research participant vs. one who is a combatant. Being a noncombatant doesn't lessen the contribution to the war effort but it does frame the experience differently. There isn't the first hand experience of engaging the enemy and being shot at sometimes for hours on end as in battle, though medics have this experience. Yet, from my readings I know that many of the nurses that went to Vietnam suffered considerable stress and had many of the same problems readjusting that the combatants did (Moore, 1992; Smith, 1992; Van Devanter, 1983) because of the pressure, intensity, stress, and the sight of wounded and dead young men.

---

### Memo 4

**June 10, 2006**

**The Enemy**

*I was very much anti-Vietnamese like most of the soldiers always feel about their enemies.*

This memo ties in with Memo 3, picking up on the concept of "the enemy." He is telling us that he was anti-North Vietnamese and saw them as "the enemy." The word "always" is interesting. I wonder if soldiers always see their opponents as "enemy." The use of the word "always" kind of bothers me here. But going on, an enemy is someone who wants to harm you if given a chance, someone who wants to bring you harm, someone your country is fighting. Whether the act of defining someone as an enemy is rational or not doesn't matter if one acts upon that definition. When I think about it, the funny thing is that both sides are fighting the "enemy." In war, you are enemy to each other. Without an enemy there would be no war, no one to fight against. But, you can be enemies without necessarily fighting when tensions between two groups may be high. It is difficult for me, having never been to war, to comprehend the abstract notion of "enemy," however I think that once someone starts shooting at you with the aim to kill you the concept of enemy becomes very concrete. The "enemy" is the person who is shooting at you. I wonder under what conditions a combatant would not see the opponent as an enemy.

---

### Memo 5

**June 10, 2006**

**Zones of Safety and Zones of Conflict or Killing Zones**

This memo also picks up on an idea that came out of Memo 3, rather than addressing any specific field notes, and is important to understanding the "war experience." There are no real safe places when a person is in a war, however there are fronts, or "zones of conflict," "killing zones," or places where battles are being fought at this moment in time. And there are "safety zones," places removed from battle sites, like bases—though these are not necessarily safe and indeed were attacked during the Vietnam War. Flying into "zones of conflict" in a helicopter carries a lot of risks because helicopters were targets, especially as they got close to the ground. Even though our participant was not actually engaged in fighting the enemy, he was a potential target every time he got into a helicopter and went into the battle zone for a rescue. And as our participant says, the zones of conflict were constantly shifting from one place to another so it was difficult to know where to land the helicopter to pick up the wounded or dead soldiers. The wounded were transported 75 miles away where they could be treated. But as we know, even Saigon eventually became a battle site. So what was a safe place one day might become a battle zone on another day. That was the problem with Vietnam, or any war, for one never really knew "who" the enemies were and "where" they would strike next, keeping everyone on edge.

---

**Memo 6**

**June 10, 2006**

**The Military System**

This is still another exploratory memo coming off of Memo 3 and expanding upon it. Reading about how wounded are taken to different areas depending upon the degree to which they are injured reminds me that the military is a "giant system" of "rules" and "arrangements." The military must transport soldiers and supplies from one place to another, provide for sick and wounded soldiers, feed the troops, provide ammunition, come up with plans for attack, and so on. In the military there are policies and systems for doing everything. Also, discipline is strict and necessary for the welfare of the group. How else could you get a bunch of young men to go out and shoot at the enemy and be shot at in return? Soldiers must obey orders even if those orders are wrong or don't make sense. The logistics of carrying out a war is mind-boggling. I noticed when watching the latest war in Iraq on the television that there was a whole military system established to wage and support the war. Without soldiers there would be no war and without this whole back up and support system the soldiers couldn't fight a war. I think that how soldiers feel about a war partially can be explained by whether or not they feel supported by this back up system as well as explained by how they feel about the war they are fighting. Are the back up systems like helicopter support and additional troops there when you need them? Are supplies available? Do you get time to rest between battles? Are you adequately cared for if you are wounded? This is where participant #1 comes into the picture; he was part of this support system, providing care for the wounded.

---

## *Methodological Note*

I want to interrupt the analysis here to explain methodologically what I am doing. I am identifying concepts from data, as the reader can see. At the same time, I am making notations in memos that reflect the mental dialogue occurring between the data and me. In the memos I am asking questions, making comparisons, throwing out ideas, and brainstorming. Though this system of dialoging with the data may seem tedious, and at times rambling, it is important to the analysis because it stimulates the thinking process and directs the inquiry by suggesting further areas for data collection. Most of all, it helps the analyst to get inside the data, to start to feel them at a gut level.

Researchers often have a great deal of curiosity but often have little experience with the topics that are studying. To understand what it is like to go to war, or "the war experience," the analyst has to feel the experience through the eyes of the participants. Notice that the analysis does not seem "forced." Asking questions and making comparisons comes naturally when working with data. Though the analyst can never fully understand another person's experience, the more that he or she works with data, thinks about them, and mulls them over, the more the data take on meaning and the researcher starts to understand. One more point: If I were working as part of a team, then the team members would be having similar discussions among themselves. Working as a team is not only fun, researchers are stimulated in their thinking by the ideas of others. In teams the analysis seems to proceed at a much faster rate because of this reciprocal stimulation of ideas. It is important, though, that one person be the designated "note keeper" so that team discussions are kept on track, order is maintained, and notes are written up later as memos.

---

**Memo 7**

**June 11, 2006**

**Locating the Self: Trying to Find Meaning**

> *Let's see . . . I was pretty young, twenty-one-years-old, very patriotic and gung ho, and thought that we had every right to be there and doing what we were doing. I was very much anti-Vietnamese like most of the soldiers always feel about their enemies.*

Once more our respondent is again locating himself at the time of enlistment. He lists these personal characteristics: patriotic, gung ho, and he thought "we had a right to be there and doing what we were doing." He was "anti-Vietnamese" because this is the way "soldiers are supposed to feel" about the enemy. In making this statement, it is almost like he is looking back trying to explain why he enlisted both to himself as well as to the researcher. But this locating makes a very important point. To understand an experience from the viewpoint of those who participated in that experience, a researcher can only view it through their eyes and "looking back" at who they were then and distinguishing that from who they are in the present. Had the participant been asked how he felt about the war at the time that he left for Vietnam he would have told us, "going was the right thing to do." Between his going to Vietnam and the time of the interview he has come to see war differently. Later in the interview he makes a very important point. It is difficult to evaluate an experience when you are "living it." There is the experience as you "live it" and experience as you "reflect" back on it. Only when we look back can we put our actions and experiences into perspective. Looking back is always a construction from the present. There is another assumption made here by the participants—that soldiers are supposed to feel that way about the enemy. How are soldiers supposed to feel? I don't know.

---

## Memo 8

**June 11, 2006**

**Inconsistencies in War: Psychological Strategies for Blocking Out or Minimizing Inconsistencies**

*I guess during the time I was there I started to become aware at little nips at my conscience, inconsistencies, but don't think that I paid much attention to them. There was too much going on to have really given a lot of thought to that. And I'm not sure that it's not some sort of unconscious mechanism that keeps you from looking at what you're doing and evaluating it. I don't know if it's because you don't want to or you choose not to. I'm not sure. It's pretty hard when you're in the middle of something to be evaluative while you're doing it.*

It was during the "Vietnam War experience" that our participant first became aware of "nips of conscience" that led to his change in attitude about war. Stimulating the "nips at conscience" were events that he perceived to be inconsistent, but inconsistent with what? I presume he means with the moral standards of the society he came from? What is the meaning of this word "inconsistent"? He goes on to tell us that at the time he didn't dwell on these "nips at his conscience" because he was too busy "being in the experience" caring for the wounded. Naturally, we all have "psychological strategies" for handling uncomfortable situations, avoidance being one of those strategies. I'll code his "avoidance" as "psychological survival strategies" because, as our participant tells us later in the interview, it would be difficult to survive psychologically and physically in a war if one dwelled too much on what one was seeing or doing. He talks about "not dwelling" upon or "evaluating." Later in the interview he talks about "becoming hardened" by the experience. That is, with time, he learned not to feel things so deeply. Constant exposure to something does tend to desensitize one and maybe that is what he means by "hardened." There are now two different psychological survival strategies, "avoidance" and "hardening." I'm sure as I go on with this study I'll find many more.

---

---

### Memo 9

**June 11, 2006**

**Perceptions of the War Experience**

*Actually I can't say that my experience was all that bad. I was young and kind of enjoyed that experience. I think it's the most maturing thing I've ever done in my life to be there and realize that people would want to kill me. As far as I know I never killed anybody else even though we had to carry weapons at times. I never shot at anyone. Not on purpose anyway. It was a strange time in my development.*

Here our respondent is giving us one viewpoint, one dimension of the "war experience," the "not so bad part." I suspect that he saw his experience as not "all that bad" because he was a "noncombatant." His job was to "care for the injured" and not "kill or defeat the enemy." He also got to sleep in a bed at night in a somewhat "safe zone. " And, he was young and ready for adventure. But I know from my readings that there are many ex-soldiers who would describe the experience differently. That is why I must get data about "combatants" next to make that comparison. Two more points that he makes seem important in this section of the interview. One is that he states that he never killed anyone in the war, so that he never had to "carry that burden" on his "conscience." He also ends this section by saying it was a "strange time" in his development. What does he mean by a "strange time"? To use one of our analytic aids, I ask, what is the meaning of this term? Does he mean strange as unknown and that he couldn't quite understand what was happening to him at the time? Or, in the sense that being in war is a "surreal experience," one goes through it hoping to survive long enough to come home. I wonder how other soldiers describe this experience.

---

### Memo 10

**June 11, 2006**

**A Broader Memo About The War Experience**

I want to step outside this interview and write a broader memo about the "war experience." I want to get my mind thinking about some of the properties of the experience. It appears that the war experience can vary from "not so bad" (he doesn't go so far as to call it good) to "very bad." The experience goes on "over time" and therefore one's experience can vary over that time. The experience though "ongoing" usually takes place during youth and therefore the war experience has potential consequences for the present and future biography of the individual. It can hasten maturity by forcing one to become responsible, self-reliant, and capable. It can also have negative effects, especially if one becomes bitter, angry, and unforgiving. Most of the men who go to war are young, and somewhat innocent about what war is all about at first. "Images of war" are romanticized and derived from oral stories, and reading. Being at war, "in the experience," changes one's images to a more realistic view of what it is all about. Calling the experience a "strange" time in his development still confuses me. Perhaps he is indicating that war is an experience that one can never really be prepared for no matter how much military training one gets. You have to "undergo" the experience before you can really appreciate it. It is like stepping into a world that even in your worst nightmares you could never have imagined.

---

**Memo 11**

**June 11, 2006**

**The Culture of War and Its Inconsistencies**

*A lot of things that I hold sacrosanct such as the value of human life, I guess I saw that diminish. I was there in '66–'67 during the Tet Offensive when the North Vietnamese fought back and really won a great victory. I can remember in this one village, the village was called "Cu Chi," after they had been routed, there were dead Vietnamese, these were South Vietnamese, killed by the Viet Cong, and they were stacked along the road like racks of firewood and I can remember not having any emotion about that. It was just like "Hey this is war!" This is what kind of happens. So that kind of confused me because before that the thought of someone dying would send me into some sort of scurrying behavior. Working in a hospital, if someone is dying you really get concerned and upset about that.*

As I read these words, "a lot of things that I hold sacrosanct such as the value of human life, I guess I saw that diminish." I am struck by what might be called "the culture of war" and the personal change that occurs when one is forced to live and survive in this culture. Because killing and death are so prevalent, one begins to accept these as the norm. I see the "culture of war" as a major theme running through this interview. It is the context or backdrop underlying the war experience. The "culture of war" will probably become a category as I proceed with this analysis because the more I get into this study, the more impressed I am with this notion of a "culture of war" which definitely is different from "civilian culture." So, what is the culture of war and how does it differ from civilian culture? Civilian culture encompasses the values, beliefs, and standards of the society we grow up in. These define our attitudes and actions on an everyday basis. What is the culture of war? The culture of war is defined by the military system. It too has its own set of rules and norms. To function in a culture of war, civilian attitudes and beliefs must be put aside and new ones adopted. In a culture of war, it is okay to shoot someone designated as an enemy. The intent of war is to defeat the enemy by any means open to you within the "rules of engagement." In a civilian culture, we don't normally go around shooting someone and if someone is ill or hurt we do everything that we can to save the person. How difficult it must be, then, to set aside those values that are "bred" into you by the society. Somehow a person has to "reconcile the inconsistencies" between civilian life and war in order to function and therefore to survive. The comment "Hey, this is war," says it all because war means death and destruction. It is one of those "psychological survival strategies" used by our participant and probably others for "reconciling the inconsistencies" between civilian and war culture. Now we have another "psychological survival strategy" and that is "redefining moral values to fit with the situation."

---

## Memo 12

**June 11, 2006**

**More Psychological Survival Strategies**

*And I just really didn't feel anything about that. Like this was all well and good, that's the way thing should be in war. It was a strange feeling. And if I remember correctly, most of the people around me didn't show any emotion about that either. In fact, there was a lot of jocularity. "Well that is one less 'gook' we have to worry about." That was a common name for the Vietnamese, "gooks" . . . so let's see. . . .*

There is no doubt that physical and psychological survival within the "culture of war" depends upon a person's capacities to develop strategies that mitigate the horrors of the war experience. Our respondent tells us that he "did not feel anything." To "feel" would break through the "protective shield" you surround yourself with in order to survive. When you see bodies stacked along the road, you tell yourself "this is war" and you feel no emotion, no pity or remorse for the "enemy." Other psychological strategies, in addition to putting up a "protective shield" include "making jokes" or "making light of the situation" and "distancing oneself from the enemy"—"gooks." Psychological survival strategies help you to "block out" the sights, sounds, of war and enable you to go on. I want to hold on to this notion of "blocking out" because I think it is a very important to managing the experience of war both during the war and afterwards.

## Memo 13

### June 11, 2006

### Another Memo on the Enemy and Psychological Strategies for Dealing With the War

*For a while, then, I worked in an evacuation hospital. They kind of rotated you from job to job. The strange thing is these were Quonset huts set up like hospital units and there were . . . we would have three kinds of people in there at one time, which was strange. We would have wounded American soldiers, we would have wounded South Vietnamese soldiers, and we'd have wounded Viet Cong or North Vietnamese. So we kind of depersonalized those people. I remember when we would give report to an oncoming shift we would talk about our soldiers, use their names and stuff like that. I remember when giving report on a North Vietnamese or a South Vietnamese we would say "bed #12" or the "gook" in room such and such. It was a way of depersonalizing that person so you didn't have to feel for them. You couldn't communicate with them because you couldn't speak the language. You very seldom had a translator or interpreter around.*

In this war, the North Vietnamese and Viet Cong were the "enemy." Here is another one of those "contradictions" of war. The job of soldiers is to kill the enemy. But once an enemy is wounded he or she ends up in the same hospital ward as your own soldiers and has to be cared for. To handle the incongruity, those caring for wounded enemy developed psychological strategies for "taking care " of the enemy without coming to "care about" the enemy. No names or identities are given to the enemy. It's just "bed #" or "gook." You don't make an effort to talk to them. You don't have to because you don't speak the same language and interpreters are in short supply. Our respondent goes on to say:

*What I do remember about these men was how stoic they were. I can't remember them asking for something to ease their pain, which as I think back they must have been in. At the same time, unfortunately, I don't remember myself, or any of the other nurses or doctors ever taking the initiative to find out if they were in discomfort. The wounds of war can be terrible. I don't know. I never thought about that at the time. I don't remember ever giving a Vietnamese anything for pain.*

It is interesting that what stands out in our respondent's mind after all these years is the "stoicism" of the enemy. A little further down in the interview our respondent describes the treatment given to the North Vietnamese, in fact even the South Vietnamese who were the "friendly" Vietnamese, as "benign neglect." In the hospitals they didn't hurt the enemy or allow anyone else to, but didn't reach out to relieve pain or provide solace. "Giving solace to the enemy" is not something one generally does. It is contrary to the culture of war. One senses, however, that in looking back our respondent feels some remorse for his behavior. That is, he wishes that he had been more sensitive to the pain and psychological hurt of even the enemy.

---

### *Methodological Note*

In the above memo and several others before it, notice that I am linking concepts; that is, putting "enemy" together with "psychological strategies" for handling the enemy "under certain conditions"—in this case, having to care for the enemy. Since the analysis is still in the beginning stages, I can say that at this time I think that these two concepts have some relationship to each other, though I might not be certain yet how they are linked.

---

**Memo 14**

**June 12, 2006**

**Letting Down the Emotional Guard**

> *They were very stoic. I do remember one incident where I felt sorry for this Vietnamese person and I don't remember if he was an enemy Vietnamese or a friendly Vietnamese. It's when he woke up after surgery and looked under the covers and saw that one of his legs was missing and he was crying. Being unable . . . I don't remember anyone, myself included, being able to comfort this person in any way.*

Despite efforts to "block out" feelings about the enemy as human, at times feelings came through. The "breakthrough" moments must have been uncomfortable psychologically because survival depends on blocking out all feeling for the enemy as human. I sense that now that the war is over it's okay to allow those feelings. As he looks back, I sense he feels some degree of guilt and remorse. Of course, these feelings are being experienced in light of the present.

---

**Memo 15**

**June 12, 2006**

**More About the Moral Inconsistencies in War and Psychological Survival Strategies**

> *Hmm . . . Then again this would be abnormal behavior on the part of a medical person outside a war zone. We wouldn't let people suffer emotionally or physically the way we let these people suffer.*

War is full of "moral inconsistencies." It is difficult to judge behavior in war from the standpoint of an outsider because war of necessity requires ways of acting that are different from civilian life. Our participant compares now and then from the perspective of looking back from where he is now—older, wiser, and part of a civilian culture. He seems upset at his own past behavior. The self does not stand still; it continues to evolve or change with experience and the passage of time. Our respondent continues:

*At times there would be conflicts in the units because we would have these three groups of people. Some American soldiers or South Vietnamese would see that their enemy was in there, the North Vietnamese or Viet Cong and there would be conflict. We would always protect them from the other people. We would never allow our soldiers to physically abuse them, although I do remember a lot of verbal behaviors, threats and all, but I never saw any physical violence. There was never a question about who would get care, or who would get supplies as they were needed. Always, the Americans or the Australians came first. There was an Australian division next to ours and they would wind up in our hospital. Ah . . . they always got priority of care and supplies. Generally there was enough to go around. So ah. . . . I recall one incident where I didn't make the choice, but a choice was made to take a North Vietnamese off a ventilator and use it for an American solider because it was the only one available. That is the only time I remember that kind of decision being made. Most of the time it was more of a case of benign neglect of their needs, to see if they really did want or need something. Sometimes I can remember the South Vietnamese interrogation team came into the hospital to interrogate the Viet Cong and I can remember at times they took the people out of the hospital. I can only imagine what happened to them. They would take them out. They said they were going to take them to another hospital but I'm sure they were taken and interrogated or even killed. But again, at the time, in all reality that didn't bother me. It was war and they were just faceless people. They were just another North Vietnamese to me . . . Like I said there were times when it would slip into my consciousness, I would think about the inconsistencies.*

This long section repeats what was said before, so there is no need to take it apart on a line-by-line basis. It is important to note so that I can put this together with other similar memos. What is stated in this long quote gets at the heart of the "moral inconsistencies of war." What is normal in war is not standard behavior in civilian life. A nurse treats the enemy with "benign neglect" but he or she can't let a patient be cared for back home with benign neglect. If you did, you would lose your job. In war, a nurse couldn't let himself or herself "feel" the pain of the enemy, in other words bond or empathize with them. If a nurse did, then he or she wouldn't be able to let the South Vietnamese army take the enemy out of the hospital to be interrogated and most likely executed. So, you distance yourself from the enemy, disregard what is going on, and make them faceless people. In this way, what happens doesn't bother you. Again, the participant brings out the small "breakthroughs" of conscience, which is important. I think it is these breakthroughs of conscience that enable one to maintain one's humaneness during war. When breakthrough happens, a way to quiet the conscience is to say, "It was war and these were faceless people. They were just another North Vietnamese to me."

---

**Memo 16**

**June 13, 2006**

**Inconsistencies of Treatment Within the Military System**

*I would think about the inconsistencies. It was not only the treatment of the Vietnamese that bothered me but there was a hierarchal system within the American army system. I was an officer so I had a lot more privileges than did the basic soldier. They would have to work a 12–18 hour shift at a stretch whereas officers did not. They were the "grunts," but that's the military. That's consistent worldwide with military everywhere. I'm trying to think about my peers, to think back to see if we had any discussions about what was going on. I don't recall any. I really don't know anything about how other people were feeling while they were there, if they were having any problems with what they were seeing or not.*

All men in the military are not created equal. The military is built on a hierarchical system. That is part of the context within which all military personal must operate. The context the interviewee describes is one where officers get more privileges and better working conditions and, most of all, less physical risks than the lower-level soldiers. Other than describing how this affected him, there is little more said. He didn't see any evidence of this affecting other soldiers. Or if it did, they never said. Context is important to understanding the "war experience." But what this gets at for me, is what I sense is a growing "disenchantment" with the military and its systems. It is one of those incidents that contributes to, and acts as a condition for his evolving change in attitude about the military, being in the military, and about war and country.

---

**Memo 17**

**June 13, 2006**

**Normalizing the Situation: Another Psychological Survival Strategy**

*It amazes me how comfortable you can get in that situation. You get up and go to work and it just doesn't seem to bother you a great deal. I that guess that's part of the whole human adaptation that goes on. You just adapt to the surroundings. But life took on an almost normal feel at the time. You had parties. At times the big concern was "where are we going to get enough beer." Or "can we trade some penicillin to another group for some whiskey" or something like that. We never thought that maybe some other group needed that medicine.*

I think that what our participant is expressing here is something that I've begun to feel as I get into this study, being that "under conditions of war," there is a "moral adjustment" that takes place. Otherwise you couldn't survive or live with the moral contradictions. Occasionally, as our respondent points out, there are "breakthroughs of conscience." But these are quickly "blocked," walled off in the inner recesses of the mind, because they don't fit with every day reality of a war situation. At the same time, soldiers try to "normalize" life or maybe "escape" the situation by having parties and being concerned with having enough beer. "Normalizing" is a way of relieving stress and stepping out of the conflict for a while. Trading and bargaining are also "normal behaviors" and have been going on since the beginning of man. If one has penicillin and another has beer, why not make an exchange. But again, as our participant looks back, what seemed normal to him at the time takes on a different reality in the present. In talking about the experience, he is trying to "reconcile these different realities." "Reconciling these different realities" is what many veterans have to work through when they come home. The concept of "reconciling different realities" seems to be an important one and I'll have to keep it in mind as I go on.

---

**Memo 18**

**June 13, 2006**

**More About Moral Contradictions**

I: Were you ever attacked? Did you ever feel in any danger when you were there?

*R: Do you mean the compound or the hospital itself? The hospital itself came under fire very often and there were people killed in the encampment. When fire did come we had to move patients out of their beds onto the floor on their mattresses. The buildings, the Quonset huts, were made out of tin and when a shell would hit there would be shrapnel flying around. But we never moved the North Vietnamese. They stayed in their beds. Americans went on the floor on their mattresses out of the line of fire.*

As I work with this data, some vague notions are beginning to stir in my brain. Looking at all the inconsistencies that our participant mentions, I'm beginning to get a picture in my mind of this place called Vietnam, at least from the perspective of this participant. It is a place of "inconsistencies," moral and otherwise, a surreal environment, where wrong is right, and right is wrong. Even places of safety are not so safe because the enemy is firing into hospitals and though wounded American soldiers are taken out of their beds, the Vietnamese are left in their beds in the line of fire.

Then the respondent goes on to say:

*. . . Some of the other inconsistencies were that during the day we allowed Vietnamese to come into the encampment to work, clean up the place and that kind of thing. You don't know if at night they went out and put on their black pajamas and became Viet Cong. It's like in the daytime you are okay. We can see you. We don't know who you are at night, that kind of thing.*

In the daytime, the Vietnamese are workers at the base. At night, they are foe attacking that same base—another "inconsistency." The "face reality" is not the "known reality" but a world where nothing is as it appears. How do you give meaning to that world and maintain a sense of purpose and mental balance when there are so many inconsistencies? All one can do is focus on "survival." I keep wondering if there is something that I can compare this experience to, in order to get a better handle on it. All I can think of is Alice in Wonderland in the sense of multiple people running around in a surreal world. But that is not a helpful a comparison. Hmm, maybe a better comparison is climbing a high mountain like Mt. Everest. It calls for survival in a hostile environment. Conditions on the mountain can turn your mind to mush, confuse you, and make it difficult to get out alive, and what appears to be real might not be "real" because of the effects of altitude on the brain. On Everest, your life depends upon training, physical ability, mental strength, the proper equipment, lots of luck, and making right decisions in response to contingencies. Survival in war also depends upon having training, physical ability, mental strength and discipline, a lot of luck, the proper equipment and support, and making the decisions in response to contingencies. But in Vietnam, moral codes are also turned upside down to some extent, one is out of touch with the world one came from, and what might seem wrong at home, seems right there. Only the fittest mentally and physically can come through the experience unscathed, or perhaps a better way of saying it is only the fittest can make the readaptation to civilian life without a lot of help. I think that the most helpful readaptive strategies are what our respondent describes in the next section of the interview. These include "closing off the experience" and "putting up a wall of silence."

---

**Memo 19**

**June 13, 2006**

**Coming Home and Getting on With Life**

*I stayed there for a year. In retrospect it was not a terrible year. It went very fast. It was very maturing for me. Uhm . . . It was in '67 that I came back. That was when the peace movement was starting to be heard very vocally. I remember my first stop after Saigon was the San Francisco airport. They made us take off our uniforms and change into civilian clothes because people in the airport were throwing things at the soldiers coming back from Vietnam and calling them murderers and things like that. That made me really mad. I thought I had gone over there and taken part in something all well and good and how could they treat us like that.*

In the above words, the interviewee turns from Vietnam to "coming home." This section is fascinating because in it he describes the transition from war to home. For him, the transition is rather smooth. He goes on with his life doing those things that he had planned while he was still in Vietnam. The reason that he gives for doing this post-war planning is revealing. It provided him with something to hold on to, probably helping him to survive. He gives us another survival strategy, "planning for the future."

He also points out that when he got home he found another inconsistency. People at home did not hold the same view of the war or his participation as he did. It made him angry to think that he and others had answered the call to serve their country, and risked their lives, only to return home and be treated as somehow "unclean" for having done so.

---

### Memo 20

**June 13, 2006**

**The American Failure: A War Hostile Environment**

*This was 1967 and the peace movement was big. I was in college and I would get angry with the student marchers, groups, and stuff like that. There were still soldiers over there and I know that it hurt them to watch that, to see the news and all of that. Now looking back, as I said before, I admire the marchers. At the time I was seeing them from my viewpoint, a patriot, and they were seeing the war from their viewpoint, "this is all wrong." So looking back now I admire those people who at the time had more insight into that situation than I did at the time. It was wrong.*

I am not calling the peace marchers themselves the "American failure" but what is important is that the peace marchers served as a constant reminder that the country was engaged in a war that they seemed to be losing. In addition, our participant thought going to Vietnam was the "right thing" to do but when he returns he finds people telling him that it was the "wrong thing." Worse yet, implied in the treatment of returning soldiers is a blame game. Soldiers are blamed for fighting in an unpopular war when more often than not they were drafted and forced to go to war. Rich kids got deferments, middle-class and poorer kids went to war. He reports feeling angry with the peace marchers. With time, he changed his mind about the war, and it took on a new meaning for him. Getting on with life in this "war hostile environment" requires the strategy of "having to pass," that is, putting aside your soldier's uniform and blending into the crowd. "Having to pass" and "not responding to the masses" is not easy for someone who has just returned from war and who still believes in his country and its mission even though patriotism had a few cracks in it. In addition, many young men were still in Vietnam sacrificing their lives while others at home who had never "experienced the horrors of war" sat in judgment on them.

---

**Memo 21**

**June 13, 2006**

**Growing Disillusionment: A New Meaning of War**

*Over the years my feelings about that have changed. It was senseless for us to have been there. It's hard to lose your patriotism. It's hard to give that up.*

This is a short bit of information but a significant one. It ties together nicely his evolving meaning of war. When he went to war, he saw it as the "right thing to do," and over the years he began to think of war as futile. But a lot happened in between then and now. It wasn't only the war, but the social unrest, subsequent wars, the advent of AIDS. The war was the catalyst for change in the sense that it opened his eyes. The reality of injured men and death in war sowed the seeds of disillusionment. Events after that just helped those seeds to grow. Change begins when the first "moral inconsistencies" and "nips of conscience" happen and continue into the present. "Change of self" and "change of meaning" are important themes running through this interview. I think part of the disillusionment has to do also with peace marchers who hassled returning soldiers, and anger at the lack of recognition from society for those who fought in the war. In addition, there is anger at a government who sent the young men to war while at the same time failed to make a total commitment to fighting it. I wonder how prevalent this anger is among those who served in Vietnam? One need not have gone to Vietnam to be disillusioned with war or government. Our respondent seems to be struggling to find meaning in an experience that carries with it so much emotion and inconsistency.

---

## *Methodological Note*

In the last memo, I could have coded the interviewee's feelings as "change in attitude," or as "losing patriotism" as beginning researchers often do. But what an analyst tries to do as an analyst is get at the essence of what is being said; that is, try to understand what the underlying issues are rather than focusing on every little possible concept in data. I am also relating the "disillusionment" to the "meaning of war" and describing how that meaning evolves over time, and place, and as a result of a variety of experiences, each one feeding into the other. Also note that some of the memos put together two concepts such as "war experience" and "psychological survival strategies." These memos are examples of axial coding because they show the relationships between two or more concepts.

---

**Memo 22**

**June 13, 2006**

**War as a Maturational Stepping Stone: The Changing Self**

> *What I think that the experience did to me is give me the motivation to do something. I was maybe twenty-two or twenty-three by then. I don't remember which but by then I had formulated plans of what I wanted to do when I was discharged. I came back to X to finish my time out there. I applied to the university and received a bachelor's and master's in nursing. I was very busy. I worked part-time and went to school.*

This participant points out that going to war gave him the motivation to go on to school and do something with his life. He mentions several reasons for war being a maturational experience in the remainder of the interview, things like having to take responsibility, having good role models, and finally learning to accept himself as being gay (see full interview in Appendix A). It was only "one year out of his life," "one point" in his total development. I wouldn't say that war was as much of a "turning point" for him as it was an important "developmental milestone." So there is another process described in this data, "the changing self." In his case it was a maturational process. The war was a stepping-stone to a change that occurred in the self.

---

---

**Memo 23**

**June 13, 2006**

**The Wall of Silence**

> *I was very busy. I worked part-time and went to school. I was really too busy to think about that whole experience. I jut put it on the back burner and went*

*on with my life. I really, at this point, can say that there weren't any major negative affects of the war on my life. It's hard to know over the years how my feelings about war and killing have changed. It's hard to say what caused the change, whether it's a maturation process or whether it was just becoming aware of all the inconsistencies and feeling the futility of war. I normally have avoided situations where I would bring this stuff back into consciousness. I have never gone, never went to watch a movie about Vietnam. Those never had any appeal to me at all. I don't know why they don't appeal. I never tried to maintain any friendships with any of the people that I knew in Vietnam. I got out of the military. I knew I never wanted any more of that.*

One of the fascinating aspects about this whole Vietnam experience for me is the "wall of silence" that seems to exist about it, an internal wall built around the experience itself, and an external wall between the self and the outside world. Ex-military don't really want to talk about it especially with outsiders. When I tried to get participants for this study I was met with a "wall of silence." Only one person responded to my call for volunteer participants. Another person who responded to my call for participants but did not want to be interviewed said, "I can't talk about Vietnam to my wife, why would I talk to you?" (meaning me the researcher). All I can conclude is that for many Vietnam veterans the war was a very "disturbing experience" to put it lightly. And when the soldiers came home, the reception they received pushed them further behind their wall. They don't even like talking among themselves, as is so evident in this interview. This man never talked about the war with his partner or his brothers. This participant maintained his wall of silence intact by "keeping busy," "not talking to others," "not reading books or seeing movies about Vietnam," in other words not doing anything that would bring back the "memories." I know from my experiences when I was doing research on head nurses at the Veterans Administration Hospital that some veterans still have nightmares and "flashbacks" and some turned to drugs and alcohol to blot it all out.

---

### Memo 24

**June 13, 2006**

**More About the War as a Maturational Process: The Changed Self**

*When I think about the impact of the war on me it was a positive one. It seems strange to say that war can have a positive impact. I met some people in Vietnam, motivated people and it kind of motivated me to go on to school. (Pause.) I would say if I had to put any kind of weight on it, it was probably more positive than negative. It was a maturational process. I probably would have matured anyway but this was kind of instant maturity.*

There is no doubt that going to war makes you grow up quickly. I found that in the readings I've done about the war also. There is talk of going off to war as a youth and returning home as an old man because of the intensity of the experience. The reality of being in a war zone certainly dispels any romantic images that one might have upon going to war. Knowing that someone will kill you if given the opportunity, as our participant says early in the interview, and seeing the wounded and dead make you grow up fast. I am still intrigued by the fact that this participant says that the war had a positive impact on him, more positive than negative. I can only think that it was perceived positively because he never had to be in a battle and his time was spent caring for the sick and wounded, in other words doing something positive.

---

**Memo 25**

**June 13, 2006**

**The Evolving Meaning of War**

*I was still angry when I got out of the military. This was 1967 and the peace movement was big. I was in college and I would get angry with the student marchers, groups, and stuff like that. There were still soldiers over there and I know that it hurt them to watch that, to see the news and all of that. Now looking back as I said before I admire the marchers. At the time I was seeing them from my viewpoint, a patriot and they were seeing the war from their viewpoint, "this is all wrong." So looking back now I admire those people who at the time had more insight into that situation than I did at the time. It was wrong.*

I certainly understand his anger at the peace marchers. He had just returned home and the experience of Vietnam with its wounded and dead was fresh in his mind. He could relate to all the men who were still in Vietnam and whose lives were on the line while these guys protested from the safety of their own country. Again this quote shows the evolution of his thinking about war with time. It is quite a turn around in his thinking.

---

## *Methodological Note*

In reading the remainder of the interview, I find that much of it expands upon concepts that we've already identified. There is no point in continuing this coding demonstration on the same interview. But just to let our readers know, I will continue to code the remainder of the interview in the same manner as demonstrated here. I'll build upon concepts already delineated and look for further concepts.

But before closing off, there are two additional areas that I would like to write memos about.

### Memo 26

**June 13, 2006**

**Breaking Through the Wall of Silence**

Our respondent didn't talk to his partner or his brothers about the war experience but he did check the obituaries to see if anyone he knew was on the list. So despite his efforts to maintain that wall of silence, to keep his thoughts about the war pushed back into the recesses of his mind, it must have crept through at some level because he read the obituaries. He still had a tie to Vietnam through the men he served with.

### Memo 27

**June 13, 2006**

**Having a Future Orientation: Strategies for Surviving**

Our respondent talks little in this interview about survival, most likely because he was not a combatant. However he did have some strategies, which in a way he credits with helping him to escape the horrors of the situation and to survive. About this he says, "I think that this futuristic orientation kind of helps you separate from the reality of the situation that you're in. I thought more about the future than the present."

And then he goes on to say:

*I think anyone that did not deal with anything beyond that day . . . I just think that they would have more difficult dealing with that . . . I could see . . . And probably by formulating my plans about the future also subconsciously did tell me that I had a future, that I was not going to die, that I was going to get out.*

### Memo 28

**June 13, 2006**

**Trying to Find Meaning and Recognition: Going to the Memorial**

It seems to me that by going to the war memorial and in doing this interview, our participant was trying even after these many years to find some meaning in the war and his and others' experiences there. He went to the war memorial partly out of curiosity but more than that he wanted to find some meaning in a war that caused more than 58,000 American deaths and many more Vietnamese deaths. When he got there he wanted some recognition, he wanted to find a band, or some indication that people cared then and now. He wanted to find the recognition that was not there when he first returned from war. Reading the interview, I am not sure that he found what he was looking for by going to the memorial. In fact, I suspect that he just buried his emotions again and probably left feeling an even greater sense of loss.

---

## *Methodological Note*

By now, our readers have a good idea of how we go about our beginning analysis. It consists of brainstorming about the data in order to identify meaning, then conceptualizing that meaning by assigning concepts to stand for what is being expressed. Concepts and, most important, categories/themes that are designated are considered provisional at this point. Concepts will be scrutinized against further data and added to, modified, or discarded as the products of analysis accumulate. The following memos were derived from the memos above and summarize and synthesize our thinking to this point.

---

**Memo 29**

**June 14, 2006**

**Impressions of Interview**

After several days of working with this interview and thinking about it, I am struck by the underlying ambivalence that I feel runs through it. I still see a lot of anger and buried feelings about the war. It is difficult to explain but I'll try. I feel that so much is being covered up, perhaps not intentionally but glossed over, sugar coated. It isn't that he doesn't say things, he does. But it is the way that he responds to things like "bodies stacked like cords of wood" in Cu Chi. Everything is explained away as "this is war" and things like this are to "be expected." The interview is almost like the way the most recent war in Iraq was covered by television reporters. It was sanitized. We never saw the blood, the sweat, or the fear. Our participant worked in a field hospital and flew missions in helicopters to pick up the injured. He must have seen terrible things, soldiers with limbs torn off, their guts exposed, and body bags by the dozen. He must have come into contact with soldiers who went crazy from the fear and constant stress. He doesn't talk about this. It's the emotion that is missing, the feeling. He does say that being in Vietnam changed his feelings about things, that he was "hardened" by the experience. But the raw emotions provoked by the war are sealed off in some deep dark place. Underlying his story, one senses the anger and the guilt. There is anger at government for bringing men to Vietnam and then not fully supporting them by declaring it as a "war." There is anger at it being a futile "war that solved nothing." There is anger with the peace marcher for saying that the war was wrong because he had to reconcile that men still there were giving their lives for what some considered a "wrong war." There is anger at himself for believing in his country and for allowing himself to be what he perceives as "deluded." And there is guilt and anger for not being more compassionate or caring of the "enemy" who also was human and hurting. His trip to the Vietnam War Memorial is touching and very revealing. He wanted recognition for those who had given their lives. He wanted a band, a crowd to be there, someone to say that the war was worthwhile. But he stood there alone. It is interesting he went to the memorial looking for the name of the brother of his friend, then walked away burying any emotion, not wanting to return again. As he says later in the interview, the government never really declared Vietnam a war. This is what makes interviews done after the fact so interesting. We get that "looking back perspective." We see war through the different lenses of then and now. We also see that even after all these years it is difficult to penetrate that wall of silence.

---

## Memo 30

### June 14, 2006

### Summary Memo of Themes/Categories

In this interview I see several themes/categories evolving. By this, I mean threads or ideas that run throughout this interview. These have already been identified as concepts, but at this time they will be elevated to the status of category/theme not only because they seem to run throughout the entire interview but also because they seem to be able to pull together some of the lesser concepts.

1. The first theme/category is the "culture of war." By that I mean that war has a culture all its own, a culture where things happen that often come into "conflict" with "civilian" norms and standards of behavior. These conflicts are experienced as "nips at conscience" or better still as "inconsistencies." In addition, the culture of war is a surreal one, taking place in a country so foreign to one's own. There is an "enemy," who if given a chance will kill or capture you. It is a culture of rules set up and enforced by the military machine that fights wars, a machine that each soldier must obey—going where they tell him or her, when they tell him or her, and doing what they tell him or her to do. If they tell you to go into the jungle to battle with the enemy you have to

do it even if you're frightened. It is a culture of battles, death and destruction, and sometimes overpowering fear calling for psychological and physical survival strategies. It is a culture that just by the very act of taking responsibility and surviving accelerates the maturational process. Under this heading I would put concepts such "the enemy," "zones of conflict" and "safer zones," "military systems," "combatants" and "noncombatants," and "psychological and physical survival strategies."

There is also the theme/category that at this point I will call "the changing self." I don't quite know what to do with this category at this time, but it seems to have to do with a gradual change in the self by undergoing the war experience. There is a patriotic, gung ho individual starting out, many of the men very young and through the experiences that take place during that year or more in Vietnam, the person is transformed sometimes for better as in the case of our participant and sometimes for worse. For our respondent, the Vietnam experience was a maturational process. It helped him to recognize who he was and to set up plans for the future. Other soldiers might be affected very differently and that remains to be seen in future data. Our respondent grew up quickly knowing that someone "out there" wanted to kill him. I think under this theme would come concepts such as "path to the war," "self-locating," "war as a stepping stone," "experiences during the war" and "experiences immediately after the war" such as "getting on with life."

The third theme has to do with the "evolving meaning of war." Many of the young men, just as our respondent did, entered the war full of enthusiasm and with romantic notions of war. In the end they were disillusioned by the futility of this war. "Nothing changed, nothing was accomplished." Under this heading I would put concepts such as "volunteering," the "peace movement," and the "wall of silence." I find silence throughout this interview. Ex-soldiers don't want to talk about this experience. They don't go to see movies or read books about the war. Among the reasons presented in the literature for this wall of silence are the nature of the experience itself, that is, the viciousness of some of the battles and tenacity of the enemy, and the despondency of seeing so many dead and wounded bodies, especially when those bodies were of comrades. Added to this experience, as Isaacs (1997) makes it clear that there was the lack of recognition of their sacrifice and effort when soldiers returned home. There were no bands, no parades for them as in other wars. In fact, soldiers were often blamed for the war and the destruction that occurred there. Also, some soldiers like our respondent, felt that the government never really committed itself to the war. They sent men off to fight without a clear purpose other than some vague ideological reasons of a fight against communism then didn't adequately support them once they got there. They didn't understand why the enemy were fighting and underestimated them as a fighting force.

**Memo 31**

**June 14, 2006**

**Questions and Directions for Theoretical Sampling**

Coding this interview has left me with a series of questions that will be used to guide theoretical sampling or further data collection. This participant was in the army and a "noncombatant." After all the analysis that went on in relationship to this interview, I am left with one very important, and I think most relevant, question. That question is, what was being in a war like from the perspective of combatants?

### *Methodological Note*

From a methodological standpoint, the question raised above is the most important thing to come out of the analysis to this point. It is the question that will direct the next data gathering. Analysis began with an interview that provided the concepts of "combatant" and "noncombatant." The interview was with a noncombatant. What came out of the analysis is that this man's experience with the war was certainly very different in many ways from what I would expect in the sense that he said it was "not such a bad experience." It is the description or dimensionalizing of the experience as "not so bad" that led me to the question, well then, what would that experience look like for a combatant. Would he describe it in the same or a different way? This is how theoretical sampling comes about. The concept of "the experience" is dimensionalized, in this case as "not so bad." My intuition tells me that the reason he thinks it was not so bad is because he was not in the front lines actually fighting the war (a kind of hypothesis to be verified or disapproved through further data collection). This hypothesis led me to gather my next set of data from combatant(s) (theoretically sampling) in order to determine if "combatants" describe their experience the same or differently from "noncombatants." In doing so, I am not only extending my understanding of the "war experience," I am also looking at how the concepts "combatant," "noncombatant," and "war experience" relate to each other.

## List of Concepts/Codes

At this point I have a list of concepts/codes and perhaps some suggestion as to categories/themes. I'll list these now so that I can keep them in mind as I proceed with analysis of the next interview.

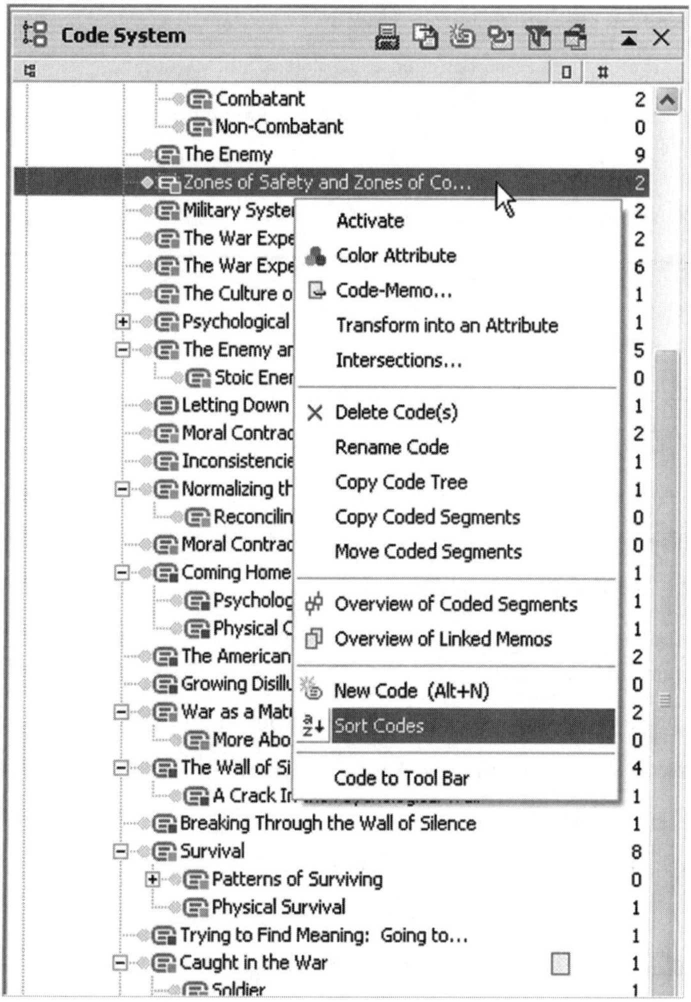

**Screenshot 9** Here you see how the codes in MAXQDA can be managed. You may arrange them in a hierarchical order. They are organized similarly to Windows Explorer—you see if a code is a subcode by the little + in front of a code. All codes can be moved around by dragging and dropping. The little rectangle in the code icon indicates the code color you have chosen. All code stripes will be displayed in this color and the colors are also used for the visual functions like the TextPortrait, the CodeStream, and others. The color choice is completely up to the researcher and can be chosen out of a range of several hundred colors. The numbers on the right side of each code indicate how many text segments have been currently assigned to the code. Right mouse clicking on any of the codes will bring up the context menu. So, you can easily see which options you have to manage your Code System.

1. Locating the Self: At Time of Entry
2. Volunteering Versus Being Drafted Versus Draft Dodging
3. Being a Noncombatant Versus Being a Combatant
4. The Enemy
5. Zones of Safety and Zones of Conflict or Killing Zones
6. Military Systems
7. The War Experience and Strategies for Blocking Out or Minimizing Inconsistencies
8. The War Experience
9. The Culture of War and Its Inconsistencies
10. Psychological Survival Strategies
11. The Enemy and Psychological Survival Strategies (An axial coding memo)
12. Letting Down the Emotional Guard
13. Moral Contradictions of War and Psychological Survival Strategies (An axial coding memo)
14. Inconsistencies Within the Military System
15. Normalizing the Situation: Another Survival Strategy
16. Moral Contradictions
17. Coming Home and Getting on With Life
18. The American Failure: War Hostile Environment
19. Growing Disillusionment: A New Meaning of War
20. War as a Maturational Stepping Stone: The Changing Self (An axial coding memo)
21. The Wall of Silence
22. Breaking Through the Wall of Silence
23. Survival
24. Trying to Find Meaning: Going to the War Memorial

### *Methodological Note*

The purpose for listing these concepts here is to provide readers with a memory refresher as we move into the next chapter. If the researcher is using a computer program, the concepts or list of codes are readily available. But remember, just listing concepts is not what we are about. It is the thought that goes into those concepts and their development in terms of properties and dimensions that is important. Those properties and dimensions are not always spelled out in our memos but they are there in our words.

## Summary of Important Points

This chapter demonstrates early coding. The researcher began by breaking the data down into manageable pieces, reflecting upon that data in memos, and conceptualizing what she thought the data were indicating. To arrive at an understanding of what the data were stating, there was a lot of brainstorming going on with questions asked about the data, comparisons made, and a lot of reflective thought. Some of the memos expanded upon the concepts by including some of the details or subconcepts contained in the piece of data. There was some attempt to do some beginning axial coding, or relating minor concepts to broader level concepts such as "locating the self" at time of entry and listing subconcepts under it. Also, a couple of possible themes or categories were delineated, though at this point the categories remain unverified and undeveloped. Almost of all the analysis in this chapter provided direction for the next set of data collection. Directing the next data collection is the question of how does being either a "combatant" or a "noncombatant" influence the "war experience"? In the next chapter, I will pick up with the analysis where I left off, building upon previous analysis using the next data set.

## Activities for Thinking, Writing, and Group Discussion

1. Sit down and think about what the researcher did analytically in this chapter.
2. Write a detailed memo about the analytic process and what you learned from it. If you do this using MAXQDA, think about a meaningful symbol for your memo. Make use of the possibility to link codes to your memos, if your memos relate to any existing codes.
3. Discuss your memo with the group, pointing out what you learned from the above demonstration.
4. As a group, take a piece of data, an interview, or an observation provided by one of the group members or the instructor and go to work analyzing it. Come up with some concepts and write memos that explain and expand

upon those concepts. If you are working with MAXQDA, decide on using an "individual" color for each person of the group, then put aside all other code colors. When finished coding, switch all colors on and compare the coding done by the different members of your group. Discuss the differences of the coding.

## Note

1. The pronoun I, rather than we, will be used in the following five chapters because the analysis was done by Corbin, who takes full responsibility for it.

# 9

# Elaborating the Analysis

*On the part of the researcher, creative and solid data analysis requires astute questioning, a relentless search for answers, active observation, and accurate recall. It is a process of fitting data together, of making the invisible obvious, of linking and attributing consequences to antecedents. It is a process of conjecture and verification, of correction and modification, of suggestion and defense. (Morse & Field, 1995, pp. 125–126)*

**Table 9.1** Definition of Terms

*Axial Coding*: Crosscutting or relating concepts to each other. Though this is not specifically addressed in this chapter, note that when two concepts are discussed in the same memo I am using what was called in previous editions of this book axial coding.

*Comparative Analysis*: Comparing incident against incident for similarities and differences. Incidents that are found to be conceptually similar to previously coded incidents are given the same conceptual label and put under the same code. Each new incident that is coded under a code adds to the general properties and dimensions of that code, elaborating it and bringing in variation.

*Conceptual Saturation*: The process of acquiring sufficient data to develop each category/theme fully in terms of its properties and dimensions and to account for variation.

*Open Coding*: Breaking data apart and delineating concepts to stand for blocks of raw data. At the same time, one is qualifying those concepts in terms of their properties and dimensions.

*Theoretical Sampling*: Data collection based on concepts that appear to be relevant to the evolving story line.

# Introduction

As an introduction to Chapter 9, I (Corbin) thought I would pass on the following story. After finishing the draft of Chapter 8, I went into the kitchen to prepare dinner. Somewhere between preparing the main dish and the salad I had a sudden insight (which often happens when one has spent the day immersed in data analysis). I went back to my computer and wrote the following memo.

---

**Memo 1**

**June 1, 2005**

**Methodological Note on the War Experience**

I missed an important category/theme in what I thought were the major ideas contained in the interview with Participant #1. It was "the war experience." Yes, I had defined the "war experience" as a concept but subsumed it under the category/theme of the "evolving meaning of war," rather than defining it as a category/theme on its own. After stepping away from the data it became clear to me that, though related, the two concepts are analytically different. Furthermore, both concepts are different from still a third concept and that is the "culture of war." The "evolving meaning of war" category pertains to the feelings, emotions, and attitudes that one has about war, based on one's experiences there. It is very subjective. The "culture of war" pertains to the context in which war takes place. It is less subjective in the sense that it has to do with the purpose of war, the actual events that happen in the war zone and back in the home country in relation to the war. It also includes the norms of war and how the realities of war differ from civilian norms. It also has to do with the whole military system that is established to fight and maintain a war. The "war experience," on the other hand, has to do with "perceptions" of those actual events, such as how one experiences battle, the death of friends, and seeing bodies along the roadside. It too becomes part of the context but is the more personal level of the conditional matrix. Even before sitting down to look at the next set of data I revised and refined my thinking a little.

---

The moral of the above little story is that the more the analyst works with data, the more he or she is likely to have "aha" experiences or sudden insights into possible meanings of the data. Insights can happen at any time and in any place, so the researcher must always be prepared to jot down those ideas before they are lost. Revising and re-revising the emerging analytic scheme, especially during the early stages of analysis, should not be cause for alarm. Interpretations are not set in stone but are subject to revision as data accumulates. In fact, new insights and subsequent changes in the analytic scheme often occur right up to the end of the study. Seeing data differently later in a study does not indicate that earlier analysis was wrong. It only points out that understandings evolve and that subtleties, previously overlooked, take on meaning the more one works with data.

In this chapter, I will continue to build upon the data analysis started in Chapter 8. The question that is directing data collection in this phase of analysis is one that was based on analysis of the previous interview and determined by me, the analyst, to be of major relevance to the evolving storyline. The question directing this phase of the analysis is this: In what ways was the Vietnam War experience different for combatants versus noncombatants? In other words, it is the concept of "combatant" that is directing the analysis. The researcher is interested in knowing in what ways, and why, the war experience was similar or different between combatants and noncombatants.

Data collection and analysis will proceed as follows. I will obtain data from combatants and compare that data to previous data from Participant #1, a noncombatant. Comparisons will be made at the concept level. To be more specific, data will again be broken down into manageable chunks. Each chunk of data will be examined closely. If a chunk of the new data is conceptually the same as data from the previous interview, then it will be coded using the same conceptual name, but this time I'll be asking about what else is being learned about this concept that will further extend understanding of what it is like to go to war. For example, if an incident in the second interview is coded as "locating the self" at the time of entry, I will then want to know where this participant locates himself. That is, what does he tell me about himself that might be the same or different from our first respondent? Anything new he tells me will be added to the list of properties and dimensions. Also, the researcher will be looking for new concepts that might not have been in the previous data, adding them to the list of codes.

In addition to making comparisons along conceptual lines, I will continue to ask theoretically based questions that will lead to further data collection (**theoretical sampling**). Research is a continuous process of data collection, followed by analysis and memo writing, leading to questions, that lead to more data collection, and so on. In this approach, the original question(s) is modified over and over again in light of what is being discovered during the analysis. This entire data collection and analysis process will go on until I am satisfied that I have acquired sufficient data to describe each category/theme fully in terms of its properties and dimensions, and until I have accounted for variation (**conceptual saturation**), and most of all until I can put together a coherent explanatory story.

I want to make an important point here. Though this phase of the analysis is focused on the concept of "combatant" versus noncombatant" because I deemed this concept to be important, another researcher might have gone in a different direction. Focusing on this concept does not mean that other concepts such as the self and images of war are being ignored. They are still important categories. It is because I have this analytic hunch that differences in the self, images of war, the culture of war, the war experience, and homecoming might be related to whether or not an individual was a "front-line combatant" that is engaged in some direct manner with the enemy. Before I began analysis of the first interview, I had only a general and very open-ended question. That question was, "What was the Vietnam War experience like for military persons who served in Vietnam during the war?" I had no idea where I was going with the study. I let my interpretation of the data from that first interview guide me on where to go next.

Allowing the data to guide you is one way of working with data, and perhaps too open for some researchers. Some researchers prefer to stay much closer to their original question, though I would venture to say that more experienced researchers are more willing to "go with the flow" of data and let the data guide them. But then more experienced researchers and less likely to have to answer to committee members and more willing to trust their intuition about what is important. Notice, however, that I am staying with my target population, Vietnam veterans, and I still want to know what going to war was like for them.

## Axial Coding

In previous editions of this book there was mention of something called **axial coding.** In the 2nd edition, axial coding (the act of relating concepts/categories to each other) was presented as a separate chapter as though it occurred separately from **open coding** (breaking data apart and delineating concepts to stand for blocks of raw data). But as you probably noticed from the memos in Chapter 8, open coding and axial coding go hand in hand. The distinctions made between the two types of coding are "artificial" and for explanatory purposes only, to indicate to readers that though we break data apart, and identify concepts to stand for the data, we also have to put it back together again by relating those concepts. As analysts work with data, their minds automatically make connections because, after all, the connections come from the data. For example, when I say in a memo that "blocking" is a "psychological survival strategy," I am relating these two concepts, "blocking" being the lesser concept and "psychological survival strategies" being the broader, more encompassing one. Though I came up with the terms "blocking" and "psychological survival strategies," these labels and the connection between them were based on data provided by the interviewee. Then, if I say that using "psychological survival strategies" is essential to "surviving" the "war experience," I am making a more abstract hypothesis linking two categories, a hunch to be checked out against data, where it is either verified, invalidated, or amended depending upon what I find in the data.

As I link categories, I am also elaborating them. When I note in my analysis that "blocking" is a "psychological survival strategy," it becomes one of the explanatory descriptors under psychological strategies, explaining how persons survive the war experience, thereby elaborating that concept. Linking occurs at various levels, lower-level concepts such as blocking to a higher-level concept such as psychological survival strategies, and psychological strategies to the "war experience." It is like putting together a series of interlinking blocks to build a pyramid. The pyramid represents the entire structure, but blocks, and how they are arranged are the components that make it what it is. What the reader will notice in this chapter is that very often the memo titles (which are essentially codes) contain two or more concepts with the memo spelling out the links between the concepts.

### Analytic Strategies

I want to remind readers that the analytic strategies of asking questions and making comparisons continue to be major analytic strategies for elaborating the analysis. For example, if I wanted to think more about "psychological survival strategies," before going on to analyze the next interview, I could ask "who" uses psychological survival strategies, "when" are they used, "why," "how," and "with what consequences"? The first participant tells us that he used those strategies essentially to help him survive the war experience and handle the moral conflicts that it the war aroused in him. Then, I could ask, do other soldiers use these strategies for the same reasons or different reasons? Thinking through comparative situations makes the analyst more sensitive in the sense that it alerts him or her to what to look for in data.

## Something About the Interviews Used in This Chapter

The interviews used in this chapter are somewhat different from the one that was used in Chapter 8. In that chapter, I worked with an interview that can be described as an "unstructured interview." The participant was allowed to tell his story as he saw it and only when he had finished his narrative did the interviewer ask questions about points brought up in the interview that the interviewer felt needed further elaboration. The next two interviews are different, and what might be called "structured interviews," because the participants responded to a set of questions derived by the analyst from previous data. Doing the interview in this manner was not a matter of choice. I prefer doing unstructured interviews. But doing unstructured interviews proved to be impossible in this case. How I acquired the additional interview material is interesting. After analyzing the first interview I realized that as part of theoretical sampling, I had to interview "combatants." Unfortunately, I didn't know anyone who had served in Vietnam as a combatant well enough to ask for an interview. I decided to go to the Internet to see if I could make contact with Vietnam veterans and perhaps find some potential research participants. I found a chat line for Vietnam War participants and made a request for research participants. There was only one response.

At first I was disappointed that persons didn't jump at the opportunity to participate in this book. In fact, I received a very cold shoulder to my request. One veteran wrote, "If I can't talk to my wife about this experience, why could I talk to you?" Good point! The one Internet person who agreed to be interviewed for this study, Participant #2, said he would be willing to answer questions about his experience as part of his desire to educate others about the Vietnam War. In fact, he often speaks to groups about the war. Participant #2's responses to questions are brief but very honest and powerful.

Additionally, I want to assure my readers that I did take measures to maintain ethical standards when conducting interviews via the Internet. The site is a closed one and the participants had to contact me. I could leave a message at the site but I could not chat with anyone. For the two responders, I did fully disclose the reason for the interviews and the use to which the materials would be put. I even sent the participants a copy of the chapter in which the materials were used so that they could respond and raise any objections. I also obtained a signed consent and took measures to ensure confidentiality and anonymity (Flicker, Haans, & Skinner, 2004; Hamilton & Bowers, 2006).

There are two parts to the interview with Participant #2. Both parts of the interview can be found in Appendix C.

---

**Memo 1**

**June 20, 2006**

**Locating the Self: Entry Into the War**

> *I was twenty-one when I went to Vietnam. I came from an average Southern family in X, my father being a schoolteacher, coach, and athletic director. My mother was a homemaker and I had one sister nineteen months younger than me. I wasn't married or engaged. My father was a World War II combat veteran*

*flying fifty combat missions on a B24 out of Toretta, Italy. My family was supportive of my choices, not necessarily of the war in Vietnam.*

The concept of "locating the self" applies to this interview also. In this interview, we know that the biographical information was given in response to a direct question posed by the researcher. This respondent's "family background" is quite similar to that of Participant #1. Participant #2 came to the war from a middle-class, close, and supportive family. His dad also served in the military during World War II. The fact that both our participants came from middle-class, intact families is quite interesting because one often hears that the men who served in Vietnam were mainly minorities or from low-income families. Though there may have been a disproportionate number of men and women in Vietnam from minority or low-income backgrounds, obviously not everyone fits that profile. What is different between the two men is that Participant #1 had training as a nurse, therefore was not likely to become a combatant. He didn't volunteer to become a marine. Becoming a marine puts you out in the front lines as a "combatant." Turning back to the concept, of "self at the time of entry" in our analysis, we have identified properties that pertain to "age," "education," and "family background." From this case we can extrapolate "young" as a dimension of age (twenty-five or less). Dimensions of family background include "middle class," "intact," "patriotic" or "having a sense of duty," and "supportive family." Dimensions of education include "having some college."

### *Methodological Note*

During analysis a researcher is carrying a lot of different ideas around in his or her head. Computers can be used to keep a running list of concepts and a log of memos. They help the analyst shift concepts around, retrieve memos, and provide easy access to what already has been done. In that sense, computers are an excellent analytic tool and can be added to the other analytic tools we have already identified. But computers don't do the thinking needed to move a study along. Only a person can do that. That is why the human element is such an important part of doing qualitative research. Computer programs are exciting, and they add another dimension, but analysts should not be fooled into thinking that if they use a computer program they can omit the thinking and memo writing. With these few words of wisdom behind me, I want to explain where I am going with the analysis.

Remember from the last chapter that the analysis focused on the concept of the "war experience." I hypothesized that the "experience" might be quite different if the person were a "combatant" versus a "noncombatant" and therefore I purposely gathered data from a combatant. The idea was to check out my hunch that the description of the experience would be quite different.

What is interesting from a comparative standpoint is that our two participants share much in common in terms of who they were at the time of entry and family background. Therefore, any differences are probably not related to "self at time of entry," though this needs to be checked out further. The question that remains when examining these data is determining what made a difference in the experience, and could this be related to the fact that one was a combatant and the other a noncombatant? As we proceed with this chapter the reader will notice that when I do the analysis, I often use the terms "we" or "us" or "our." These words refer to the reader and I, because I am taking the reader along on my analytic journey.

---

**Memo 2**

**June 20, 2006**

**Being a Volunteer"**

*I was a volunteer as all Marines were when I entered service in 1964. I did not serve with any draftees in Vietnam.*

Again, our respondent is a "volunteer." He served in an all-volunteer Marine Corps group, thought of by those who join it as an "elite" and "well-trained" corps. This stands in stark contrast to Participant #1 who described himself as a "six week wonder." This participant gives the date that he joined the Marines as being 1964. From a war standpoint, this is important because at the time he volunteered, the only U.S. military persons being sent to Vietnam were there in an advising capacity. Their job was to train and support the South Vietnamese army. This tells us that the participant didn't join the Marines expressly for the purpose of going to Vietnam. It was by chance that he ended up in Vietnam. He said it was not the "cause" per se that attracted him to the marines, it was what the Corps stood for, the defense of our country and protection of our rights guaranteed under the Constitution. We have some contrast here from our first respondent. Participant #1 volunteered for the Army Nurse Corps to stay one step ahead of the draft. The U.S. had increased its involvement in Vietnam and he knew he would probably be sent to Vietnam. He didn't mind because "at the time" he thought that the country "had a right to be there" and "it was the right thing to do." If #1 had not thought that he would be drafted he probably would not have volunteered for military service. Participant #2, on the other hand, went to Vietnam because he was already a Marine and it just so happened that the war was escalating. The "paths to war" were different but they both ended up in the same place, Vietnam.

---

---

**Memo 3**

**June 20, 2006**

**Being a Combatant**

*I was a combat Marine rifleman also certified in 3.5-inch rocket launchers.*

Now we get to the heart of the matter. This participant was a "combatant," in fact he defines himself as such. He was a front-line soldier who fought the "enemy," a contrast from our first participant who only came into contact with the enemy when the enemy happened to be wounded. It will be interesting to follow the data and note how his "being a combatant" shapes our understanding of the experience.

---

**Memo 4**

**June 20, 2006**

**The Enemy, the War Experience, and the Culture of This War**

*The Viet Cong were a very well-trained and disciplined military force who gained footholes in local villages by terror, killing, and torture.*

An interesting comment, "the war experience," is partly defined by the "enemy" one is fighting. This participant defines the enemy as "well-trained" and "disciplined," (these are properties of the concept "enemy" as this participant describes them). He even qualifies, or dimensionalizes, the discipline and training for us, stating that the enemy was "well-trained." He tells us more about the "enemy." He says that it "controlled villages and villagers" often through a "campaign of terror." Part of the "culture of this war," then, is that ordinary people were "caught up in the war" providing sanctuary and support to the "enemy" often against their will. This made it difficult for the soldiers to define who the "the enemy" was because every citizen (man, woman, or child) was potentially an "enemy." So combatants not only encountered a "well-trained" enemy, they also had to contend with villagers, who also posed a threat to the soldier's well-being.

---

## *Methodological Note*

Notice the crosscutting, or relating, of concepts/categories above through statements of possible relationships. These relationships will be checked out against incoming data and accepted, modified, or discarded with further analysis. In the next memo, we are also crosscutting concepts. It is equally important to notice that we are still doing open coding. That is, as we talk about the concept of "enemy" we are qualifying it by saying it was a "well-trained" and "disciplined" enemy, and when we ask who the enemy was, the answer we receive is that it could be "anyone."

---

**Memo 5**

**June 20, 2006**

**The War Experience, Military Systems, and the Culture of War**

> *Marines like myself were extensively trained to follow orders, no question why or the politics of the situation. I could kill without hesitation as that was my job and I was trained to do just that. It doesn't take long for one to get into the groove seeing friends wounded and killed. The killing becomes a habit and self-defense as time goes on and you survive. Marines fight for other marines and the corps, not necessarily the cause.*

I should note here that he tells us that "combatants" who were marines were "extensively trained" to follow orders describing for us who at least this group of combatants "were" and "what" their job was. The "culture of war" is a culture of "killing," and that was this man's "job." I guess this really defines what constitutes a "combatant" from, say, a noncombatant. They not only come into contact with the "enemy," they are trained to "kill" the "enemy." The notion of kill or be killed was a large part of the daily "experience" of "combatants." The fact that killing becomes taken-for-granted should not come as a shock to anyone. As a soldier, you may not like killing but, as this participant states, you do what you must do to "survive." It is this participant's reaction to his dead and wounded comrades that is interesting to me. Does one really get "into the groove" of seeing death happen? Or, is "getting into the groove" one of those "blockers," a "psychological survival strategy" necessary for survival? Another point this participant raises is that he was not fighting for "the cause" as such. "The cause" is too abstract and too far out there when you come into battle. "In a battle soldiers are fighting for their lives and for the lives of their comrades." It's that basic.

---

---

**Memo 6**

**June 20, 2006**

**Caught in the War**

> *I was always supported when I served. There were a few of us that did not want to be there but no one wants to be in a life or death situation of combat if they have a choice.*

As he says, few people want to go to war, to be shot at or to shoot at others. The problem is that young men join the military for a variety of reasons, some ideological and some for the sake of having an adventure, to get a skill or training, or to get away from home. They have no idea when they set off what being in war entails. When the reality of war hits, it must be a real shocker to a young man who lived a middle-class life where people are basically nice to each other.

---

## Memo 7

### June 20, 2006

### The American Failure and the Impact on the Self

> *As far as the anti-war movement was concerned, that's one of the reasons GI's fight. The right of free speech, right to protest and right to live free. However, when that movement attacks GI's due to their choice to serve, call them "baby killers" just to mention one name, and to have never served this country in anyway with the exception of running their mouths about things they know not or will never know anything about I detest to this day and to my grave. These groups will be the downfall of the United States, as we know it. The anti-war movement did nothing but gain a dishonorable peace and disrespect 58,000 Americans who paid the ultimate price for the rights of its citizens. The GI's of the Vietnam War were treated like traitors to the student and activist anti-war movement of that era. That should never again happen to an American GI.*

As a collective society, we also failed our GI's during and after the Vietnam War. I can see myself as a person intruding into this analysis but I can't help it. It's my reaction to this data. Protestors saw a war, a war they felt was morally wrong. They failed to distinguish the war from the soldiers who were forced into fighting it by their government and the society that sanctioned the government. These two concepts, war and soldiers, though they go together, are very different. I really don't believe that it was the "individual soldier" that protestors disliked even though they "spit" at them and called them "baby killers." It was the symbolic meaning represented in "soldier" that they responded to.

What this participant is trying to tell us, I believe, is as follows. We have certain freedoms because there are men and women who are willing to serve in the military in order to protect and defend those freedoms. But protestors use that freedom, bought at the price of soldiers' lives, to attack the very soldiers who are out there in the battle zones defending the protesters's right to free speech. It's all so ironic. Of course, there are the true pacifists, those who believe that all wars are wrong regardless of the circumstances. But these persons I suspect are relatively few in number and even they would probably be willing to serve in some capacity if their country were attacked. Most Americans believe that having some sort of military is important for defense against aggression. But having a military and having a war are two different things. It wasn't the military that started the war. It was the elected officials. Where we failed as a society was in sending young men off to fight a war, then blaming them for doing what they had to do when they got there. Isaacs (1997) makes a good point. He says that after World War I and World War II, returning soldiers were treated as heroes. There were parades and recognition for their sacrifices. The effect of this recognition was a sort of collective sharing of guilt for any atrocities that might have occurred as a result of war. The soldiers who served in the Vietnam War didn't receive recognition for their valor when they came home, rather they were held responsible for the war by persons who had never been to war and thus had little understanding of the conditions of war. Fifty-eight thousand men were killed. It took years before their sacrifice was recognized in a Vietnam War Memorial. One more note, the "meaning of war" for this participant is a "dishonorable peace." To him, settling for a dishonorable peace does not justify the 58,000 lives that were lost.

---

### Memo 8

**June 20, 2006**

**Carrying the Burden**

*Every combat veteran, and some who were not, are affected for a lifetime by the killing, carnage, loss of friends and family. Some carry the burdens easier than others. Outwardly anyway.*

Here our respondent clearly describes it for us. The killing and carnage place a burden on combat veterans that they carry for life. "Carrying the burden" is an in-vivo code. The men who fought in the Vietnam War ended up "carrying the burden" of war. They fought to survive when many of their buddies did not, and many to this day carry the "burden" for having lost the war. It is interesting that some ex-soldiers have been able to come to terms with the war experience and the many losses they suffered. Perhaps it is because they had, and still have, a greater repertoire of "psychological survival strategies." Or perhaps it is because they've been able to talk about it, and in talking, let go of some of that burden. I want to mark this concept "carrying the burden" because for me it helps explain some of the residual anger and the "wall of silence" some Vets are still carrying.

---

### Memo 9

**June 20, 2006**

**Patriotism as a Motivation for Being a Volunteer: The Post-War Experience and the Meaning of War**

*In closing, I joined the Marine Corps by choice out of State University. At that time we only had advisors in Vietnam. Myself as well as my entire unit did not join the*

*Corps especially for the Vietnam cause. I joined, as John Kennedy said, "Ask not what your country can do for you. Ask what you can do for your country." I wanted to give something back to the country and people I so love. Myself, and the tens of thousands of others were in the same boat when the leaders of this country who were elected by the people took us into the Vietnam cause. I'm a true American patriot and believe that those who choose to serve or are required to serve should do just that in an honorable way. Those who choose to attack us for our service, those who ran away to other countries, are not the foundation this country was built on. These attitudes carry to this day with many and never should have been tolerated or excused by the American people. The difference with Vietnam compared to World War II or World War I [is that] we weren't attacked by a foreign force. The GIs of all those eras are no different in their service to the United States. Just the cause.*

This participant, and many like him, joined the military for ideological reasons. They accepted the Marine code of "service to country." This man was in the military for five years and achieved the status of sergeant. I don't know how many of those five years were spent in Vietnam. The usual tour of duty in Vietnam, I believe, was thirteen months. Yes, mistakes were made in the war and innocent civilians were killed. But most soldiers were honorable men, as he states, doing their duty and many were maimed or were killed. What makes the difference when looking back at war is the "meaning" given to a war by individuals and societies. This war came to be viewed by society as a "mistake" but by many Vets as a "dishonorable peace." The sad thing is that the GIs suffered for society's "mistake." Some Vets continue to suffer today from flashbacks and worse. For Participant #2, the young men who fled the country rather than go to war were the dishonorable ones, because they fled their country in the time of war. President Carter later pardoned them. The pardon was another blow to the Veterans. They did what they thought was the "right thing" and paid with their lives. I think another reason why Vietnam is so difficult to talk about is the sort of collective guilt that society feels for engaging in the Vietnam War.

---

**Memo 10**

**June 20, 2006**

**Summary Memo**

I am struck by the "burdens of war" that this man carries within him to this day. Along with that burden is an unresolved anger. I am placing "carrying the burden" in my evolving list of categories. I also see that I am developing some concepts further such as the concept "meaning of war." For this man, the Vietnam War came to be viewed as a "dishonorable peace." It is his perception that honor comes from winning a war and not pulling out. As far as he is concerned, in accepting the negotiated settlement, America turned its back on the 58,000 that lost their lives and approximately 300,000 that were wounded.

As a researcher, I find working with these materials very difficult. I feel the pain of these men. I know that only they can "come to terms" with the experience. I feel very inadequate at conveying the depth of feeling contained in the words above. I know that as a researcher I have a very deep responsibility to those who trust me with their stories to present those stories accurately and fairly. It is their side of the story that I'm trying to capture in this study. To really bring out the complexity of their experience, it must be placed in the context of everything else that was going on at the time, including the peace marches.

## *Methodological Note*

Based upon analysis of Participant #2's statements, I had more questions for him. I present the questions next.

### Memo 11 Part II of Interview

**June 10, 2006**

**The Psychological Aftermath of War**

R: In the first interview [with Participant #1], which by the way was done with a good friend of mine, several themes came out and I wonder if you could respond to them. I think in some way you have, but wonder if you might say more. One is about the "culture of war" and how what goes on in war conflicts with normal standards behavior. Because of that moral conflict, there were times in Vietnam that my friend had pangs of conscience about what he was seeing and doing. But the only way to survive that was to push those thoughts aside, see the enemy as the "enemy"—one who would kill you if given the chance, call them "gooks" to distance oneself from them being human, and just not talk about it. In fact, he had never talked about the war with anyone during or after the war, up until the time of the interview. He just blended into the college campus when he returned home, avoiding all antiwar activities and discussions on campus. Did any of that haunt you then or afterwards and how did you deal with it?

*P: It has haunted me everyday of my life. Not a day passes that I don't remember something about that era. I never mentioned or talked about Vietnam to anyone including my wife of thirty-seven years until the late 90s.*

These are pretty powerful words. This participant makes it very clear that one does not kill and watch one's fellow soldiers be killed and then walk away from the experience unscathed. There is pain, remorse, regret, and terrible memories that one lives with. One way to function in the everyday world is to bury those memories deep within the "self" and avoid anything that will trigger the pain and suffering. But I also wonder if part of the reason for not talking about one's experiences in Vietnam is the fear that others won't understand and therefore make judgment on your actions. How do you explain what you went through, the visions that haunt you, the things you had to do, and the fear that returns in the night? To carry that experience for so many years, and not be able to let go, really points to the depth of the experience and its lingering effects. The pain is almost paralyzing in the sense that it is difficult to talk about, even with those whom one holds most dear.

---

### Memo 12

**June 20, 2006**

**Survival: A Matter of Chance**

R: I guess what I'm getting at is that you say that you thought of it [the war] as a survival experience, but what were the strategies that enabled you to survive?

*P: Surviving the war was a matter of pure luck. You happened not to be in the wrong place at the right time. That was merely luck. You could not survive the war by being careful, a coward, or trying to stay in the rear with the gear. I know guys who served an entire combat tour without even a briar scratch and then I knew others who were there less than thirty days and [were] nearly blown in half.*

Our participant here is talking about "physical survival," which he describes as a "matter of chance." This war had a high death and casualty rate precisely because the enemy was "well-disciplined," "well-trained" and described by some as "cunning." In fact, Moore and Galloway (1992) state, "From that visit I took away one lesson: Death is the price you pay for underestimating this tenacious enemy" (p. 49). According to my readings, the enemy often hid in villages and placed mines and booby traps along the paths that they knew marines would pass. Though villagers often passed along these same paths, they were not injured indicating that they probably knew where the mines were placed (Anderson, 1981). When a soldier was wounded, the enemy used him as bait because they knew that a marine would never leave another marine, dead or wounded, behind (Waugh, 2004). The enemy hid snipers in trees and it was difficult to see them because of all the foliage, hence the defoliation with Agent Orange. If you can't see an enemy or a trap, then your "survival is threatened" and you have to get rid of the foliage either by burning it out with Napalm or dropping Agent Orange. It makes sense from a military standpoint. Unfortunately, there were consequences of the military strategies to the innocent villagers who were caught in the war.

### Memo 13

**June 20, 2006**

**Psychological Survival Strategies**

R: How did you deal with the death that was happening all around you?

*P: Death and mutilation is all around you in war and it becomes a matter of acceptance and habit. You mentally try to remove yourself from all the carnage and put our mind in another place and another time. Your mind spends hours upon hours at home in a warm, dry, clean, safe bed with family and loved ones. It's my opinion that marines were better trained than some of the other services to deal with the carnage. Not better GIs just better trained and much closer to each other.*

Death, destruction, and mutilation "make up the culture of war." If one is going to "survive a war," then one has to accept the "realities" of it, and come to terms with killing and death. I think it is interesting that this participant tells us something very similar to what Participant #1 told us. Our first participant stated that "planning ahead for the future" and "psychologically removing himself from the scene" helped him survive. He daydreamed about what he would do when he got out of the military. This participant also projected himself at home, thinking about returning to his safe warm bed and to family. I guess "thinking about a future" and about home not only helps you to escape mentally but it also gives you the "mental steel" to keep going despite the bloodshed and difficulties. I code this psychological survival strategy as "mental escaping" but this label hardly does justice to the complexity or profoundness of it. This quote does tell me that both combatants and noncombatants had to use psychological survival strategies to keep themselves going.

### Memo 14

**June 20, 2006**

**More on Psychological Strategies: Mentally Removing the Self**

R: How do you turn that off?

*P: I was able to mentally remove myself from the carnage. I always felt if I dwelled on it and allowed it to consume me I would be the next one hit.*

This little quote says it all, verifying the above hypothesis about the role of psychological strategies. One must mentally remove oneself from what is going on. To focus too much on the present would severely impact one's ability to physically and psychologically survive the "war experience."

### Memo 15

**June 20, 2006**

**Healing: A Summary Memo**

There is one thing that I have not heard much about in either of these two interviews, or from my readings of various memoirs, and that is "healing." I suppose this is because, for some, there has been no "healing" and for those who have "healed" there is no need to talk about it. After going through such a profound experience, how, when, or can healing even take place? To what degree does it occur? And, if healing doesn't happen, then what happens? Are the pain, loss, and memories brought out as anger, rage, and depression? This is an interesting series of questions. I need to follow-up on this question in my next round of data collection.

### Memo 16

**June 20, 2006**

**Aftermath: Unresolved Rage, Anger, and Depression**

R: Then and now?

*P: Since Nam and now I put it completely out of my mind with friends, family and loved ones. I avoided drinking completely as booze would bring on the most vivid mental attacks of rage, anger and depression. I would not be talking about it today unless a great friend of mine through boot camp and Nam found me after forty years and all the memories flooded back into my mind. Talking with a brother you served with is easy but not the general public. This guy was a machine gunner in my weapons platoon and now we see each other regularly, which allows us to dump all the memories on each other, which is like taking a drug. I've been so lucky to have a woman in my life who never pushed the issue, never asked questions, held me quietly when the nightmares came and gave me her unyielding support.*

There is not only a "wall of silence" between himself and others regarding Vietnam, there is a "wall put up within the self" to blot out memories and feelings about Vietnam. Alcohol lets down those defenses and the anger, rage, and depression come flooding through. Other Vets have turned to alcohol to blot out memories. Again this gets back to the question of "healing." Participant #2 tells us about this underlying anger and rage. Though not expressed directly by Participant #1, I couldn't help but feel that he too had many unresolved issues, though certainly not to the same degree as Participant #2. I hypothesize the difference had to do with being in "combat." Conceptually, though, there are similarities between #1 and #2 in the sense that each has buried memories of Vietnam and had nips at conscience, which they carry to this day as "burden."

Participant #1 has some good memories mixed with the not so good. Participant #2 doesn't seem to have any good memories. I wonder what is it about the war experience that creates such intense feelings? In the memoirs written by Vietnam veterans that I've been reading, there is a lot of rage felt towards the enemy because of the ferocity of the fighting and their perceived viciousness. There is rage at seeing death and mutilation in your buddies and a strong desire to get revenge (Waugh, 2004). In fact, Bird (1981) states, "The casualties taken in the fighting really got to us and uprooted us. That also incited fighting in a way that when somebody was hit or killed, it made the others that much angrier wanting revenge" (p. 43). The rage expressed here also seems to be at the peace marchers, who are felt by many Veterans to have brought the U.S. government to a point of accepting a "dishonorable peace" (Sar Desai, 2005). It seems amazing that after all these years memories are still so vivid and the nightmares so real. Civilians go on with their lives after a war but for the GIs it seems that their lives are affected forever. It is good that this individual has gained some relief through talking with his marine "brothers" and that he has a loving and supportive wife, both conditions for handling the memories upon "coming home."

---

**Memo 17**

**June 20, 2006**

**Maintaining Contact: A Healing Strategy**

R: Just the name of your Web site intrigues me, "n. g. a."

*P: N.g.a. as you have guessed has to do with the ghost of war and Vietnam. The name popped into my head in 1996, thirty-one years after Nam. Several dozen Nam vets use to gather at a Web site put up by a lady and Vietnam vet supporter who was never associated with a Veteran or Vietnam in any way. It became too much for her to deal with over the years so I put [up] a chat room and Web site to honor my unit and maintain contact with many Vietnam veterans I've met over the years. Mostly marine combat vets but we have a few others from other services including the Air Force, Army, and Navy who join us weekly. We're a very tight knit group and stay to ourselves for the most part. During our gatherings online we try to avoid the ghost of Vietnam. Therefore the name . . .*

The bonds that tie these guys together have to do with their shared experience in "combat." There is something comforting in the camaraderie of a group of men and women who have had the same experiences as you and that understand what you've been through and are still going through. The group must be therapeutic in the sense of having a knowing brother/sister to talk to without having to bring up all the details. One can see why they want the ghosts to stay buried. The dead are always with them.

---

### Memo 18

**June 20, 2006**

**The Ghosts That Haunt**

R: Did you ever have contact with the enemy and what was that like?

*P: The contact that I had with the enemy was with the dead or dying. I watched several last breaths and can see each one today as I did then. We had intimate contact with ARVN (Army of the Republic of Vietnam) [South Vietnamese Army], which in some cases I'm convinced were VC, the enemy. There were no differences in the Vietnamese friend or foe as far as the people were concerned. They were of a different culture and religion but human. I never view friend or foe as nonhuman or villains.*

These words are so powerful. To think that all these years later he can still visualize the sight of dead and dying enemy, as well as the death of friends. These are the "ghosts" that haunt you forever. Friends, enemies, they are the same when injured or dead. They bleed, have pain, suffer and they too are afraid of death. Many Vietnamese friends and foes were lost in this war, a fact that is important to remember.

---

### Memo 19

**June 20, 2006**

**The Enemy Is Everyone**

*Like your medic friend, I did not trust any of the Vietnamese, friend or foe. You never knew what they were from one day to the next. Under the right pressure of being killed or tortured, your friend on Monday was your foe on Tuesday. They were still human, just the enemy. You depended on your GIs who came from the same land as you.*

To this participant, the enemy is someone you can't trust and have to fight in war. One of the sad things about Vietnam is that South Vietnamese civilians were forced to become "the enemy" under fear of torture and having their families killed. I suppose that some villagers became "enemy" because they believed in the "cause" or unification of their country. Since you never knew who was friend or foe, to "survive" you must treat all Vietnamese as foes. The only persons you could trust were your fellow soldiers.

---

**Memo 20**

**June 20, 2006**

**Meaning of War**

R: Would you say that the war hardened you, made you more sensitive and feeling, disillusioned you about war?

*P: Unfortunately war has become a necessary evil of the world, as there are cultures that want to murder us, each and everyone. I'm not against war under the right circumstances and Vietnam for sure did not make me a pacifist. I viewed myself as hard nosed before Vietnam, owned my first gun when I was seven. Hunted alone before I was nine. Things that our parents would go to jail for today. Not then. Vietnam showed me how many Americans really are in their attitudes about God and country. I learned they are all about themselves and will kill Americans to have their own way or force their views on society. Whatever one wants to call [it] these people need to give this old GI a wide berth in life. If you want to brand that hardened, yes, I'm hardened. It's my feeling [that] the elected leaders of this country should put GIs in harms way only as a last resort. World War II was a last resort. I'll have to say I'm not sure about Vietnam, Korea or Iraq. The average American does not have the information at hand as our elected leaders have to make the determination of war. History will prove whether these other wars made a difference in the world or the well being of the USA. I wish I would be here for those answers. I detest seeing humans abused, tortured and killed now and before Vietnam. I think we are blessed as a people, which puts us in a mindset to help others. Is this a justification of war? I'm not sure and don't have all the answers.*

There is a lot in this long quote and it really gets at what this man struggles with to this day. Is war ever justified? This, of course, is an issue that has been debated probably since man began. He makes a distinction between World War II in which there was an aggressor that threatened a continent and the wars in Vietnam and Iraq. In terms of the later, he questions the meaning of those wars and leaves it to history to tell us if they made a difference. This man has not become disillusioned about his country or war in the way that Participant #1 has. The betrayal that this man feels is not betrayal by country, but a disillusionment with the civilians who criticized those men and women who were willing to fight for the ideals that Americans hold sacred.

---

## Memo 21

**June 20, 2006**

**A Crack in the Psychological Wall**

R: Have you been to the war memorial and how did that affect you?

*P: Yes, I've been my one and only time. No way can I explain how seeing those 58,000 names, many being GIs I served with as well as friends from high school and college, affect me. I will say I never want that feeling again.*

There is a profound sense of grief upon seeing the names on the wall because each name represents a real person. Going to the memorial was difficult for Participant #1 also. I think that it is not only the loss that affects them so, but that going to the wall brings back the whole "war experience." It breaks through the "wall of silence" and lets out "the ghosts" out to "haunt" them.

## Memo 22

**June 20, 2006**

**Keeping Up the Wall of Silence**

*Ms. Corbin, in closing I just want to warn you if you don't already know, asking these questions of some Vietnam vets will bring on aggressive responses and sometimes verbal attacks including guys who patronize my Web site, I would say most of them as a matter of fact. I choose and never have edited the message board and the guys know it. We offered our lives for freedom of speech as well as all other GIs who have served. Who am I to censor free speech? I've tried to accommodate teachers and students like yourself over the year with basic input to enable those who were not involved to the views of many, especially the views of veterans in a feeble attempt to create an understanding of their views. Just don't take it personally, if some tell you to take a hike.*

This participant is telling me that it is better not to disturb that "wall of silence." The hurt that these veterans feel must be deep and profound. Talking about the experience revives old memories and brings out the unresolved anger. But I still don't understand, why so much rage so long after the war? It seems to be a nonspecific and diffused type of anger, not focused on a specific person or thing. Why the terrible memories and why can't the GIs let go of the memories and the anger? What is there about war? This question needs to be examined further and this is where theoretical sampling comes in. I say that veterans have the right to not speak about the war. After all, they earned it.

### Memo 23

**June 20, 2006**

**A Summary Memo**

What this part of the interview makes me realize is the degree to which the aftermath of war lingers and how little healing has taken place in some of these vets. It appears that the "ghosts of war" haunt men who served in combat and perhaps those who served in war support roles as well. The ghosts must be more than just ghosts of the dead. The ghosts must also include memories of battle, the noise, the fear, and the chaos. Though many vets live ordinary lives, beneath the surface in some there is a volcano that if disturbed can erupt as rage. O'Shea with Ling (2003) calls this "The Beast Within," a beast that he says has robbed him of a good deal of his life.

## *Methodological Note*

Note that with each interview, knowledge and understanding about the war experience from the perspective of those who lived through this experience expands. In coding this interview I came up with some new concepts and a new probable category, "carrying the burden." Notice, also, that analysis leads to questions that lead to further data collection that lead to further analysis. That is why I am such a strong believer in alternating data collection with data analysis. I know that this is not always possible, but when it is possible, the process enriches the findings considerably. A researcher cannot possibly know all the questions to ask when beginning a study. It is only through interaction with data that relevant questions emerge. Questions based on evolving concepts such as "anger," "war," and "healing" come out of the data. Sometimes the only way to answer the questions is to go back to the field and gather more data.

The other important methodological point to bring is that though the memos don't specifically state that something is another property or dimension of a concept, that information is in the memos. For example, though not specifically stated, "combatant" and "noncombatant" are dimensions of "type" of soldiers. It was also learned from Participant #1 that another dimension is status for rank with dimensions ranging from "officer" to "enlisted man," commonly referred to as a "grunt." In other words, researchers do not always specifically spell out in memos that "this" or "that" is a property or dimension, but the information is there in the memos.

Analysis leads to theoretical sampling, or sampling on the basis of concepts derived from data. Though the idea of doing theoretical sampling sounds rather complicated, it is not. It is sampling that follows a line of logical thinking. For example, I was struck by the concept "residual anger." I wanted to learn more about it. Fortunately, I was able to ask about it with the next participant who I will call Participant #3.

## Participant #3

Some months after my interview with Participant #2, I received an e-mail from another marine at the same web site who said he was willing to speak with me. He did not serve in Vietnam but did serve in Bosnia and Grenada. I asked him, why the anger?

---

### Memo 1

**June 21, 2006**

**Residual Anger and Coming Home**

*The anger comes from several avenues. It starts in Boot camp. They are training you to protect, defend, and to kill if it comes to that.*

*They frustrate you, intimidate you and irritate you because any sane person would not make you do the things that they do. Then if you do go to war and experience it, our anger splits, like an atom does and creates heat. Anger is volatile. You are sent someplace to defend your way of life, to protect your country, her women and children and her divine right to exist. You get mad because you don't understand why the other guy hates you [enemy] because you're an American. It builds and builds because everything you were told as a child you have to protect. You are afraid it will be taken away. This adds to the anger. You do your job. You win and you get to go home and everyone has been protected. No one loses sleep while I'm protecting you.*

*You come home and no one cares that you fought for them. They didn't feel the pinch, the lead flying around, the bullets, smell the death, smell diesel fuel, the napalm, the gun powder, the smells that are burned into the soldier's brain. Because they didn't experience it nor did they actually feel that their liberties were in jeopardy, they don't think that you did anything for them. So their freedom was never really challenged in their eyes so quit overreacting. You didn't do anything for me. This reaction from an ungrateful person adds to the anger; it continually compounds. Now remember you are still young and you do not have the coping skills because so much happened to you so fast that the coping skills are short circuited in the process.*

*Now you begin to think it was a waste and your buddies died for nothing and you got shot for what? More anger, you're like an atomic bomb with its atoms splitting. It is a continual reaction. Add alcohol to this already explosive mixture. You are in hell, you don't understand. You did it right. You were a Marine and defended America. You did what you were supposed to do, why does life hurt so bad and why do I not want to be here anymore? You can't think it through; there is no logical thought pattern that will help you put this together. Now add the hormones, the dopamine, the epinephrine, because you were in a constant state of excitement and fear, your hormones that flow in the brain to maintain emotional stability are all screwed up and stuck high. You can't process it now if you wanted to.*

*The anger is actually a chain of events, then it goes to a chemical reaction in the brain, then add the Jack Daniels to this, the anger does not go away till one of these chains are broken. That's why it takes years to "come home."*

*I hope this answers the question for you. Take this info and help more guys to be able to "come home." You will help me by bringing all of us home.*

This is pretty heavy stuff. How representative are the reasons that he gives for all young men who serve in combat—I have no idea. I am certain that what this participant is describing pertains to some others as well, because the men do share a lot of information between them. To his list I would add the "sense of loss" that comes from seeing your buddies killed or wounded. I wonder how many Vietnam vets were diagnosed with post traumatic stress disorder? This is another thread I have to follow up on. I know from my experiences with veteran's hospitals here, that many of the vets being treated in the hospitals were on the drug and alcohol units. I like the way he describes the anger as a process that begins in boot camp and grows with the war experience. Many young soldiers go through boot camp but not all come out angry. My educated guess is that a major condition for the anger is the intense pressure and stress of the war experience. Added to this, when a combatant comes home, there is apparent indifference to his or her experience on the part of society. I wouldn't say that the general population doesn't care, but that they are caught up in their own lives. I also really like his "in-vivo code" of "coming home." There is the "physical coming home" and just as important is the "psychological coming home" in the sense of "readjustment to civilian life" and this is a process that takes time. I don't think the vets are asking for bands to play, but merely a little understanding and respect for what they have been through.

"Coming home," though, seems to be a time when problems arise. The men are too busy when in the midst of combat to think about what is happening around them. Thoughts are more dreams about the future and what they will do when they get out. But when they get home, the war experience comes back in the form of flashbacks and other memories (the ghosts) to haunt them.

**Memo 2**

**June 20, 2006**

**Coming Home: A New Category**

I like the concept of "coming home" and think that it should be a category. It marks transition back to civilian life. Upon coming home, many vets "carry a heavy burden" that manifests itself as flashbacks, bad memories, and anger. Some vets are able to let go of that "burden" as they take up satisfying lives once more. Others seem to be carrying the burden of Vietnam even to this day. I see "coming home" as a process that happens at different rates, and to different degrees, depending upon the individual, probably depending upon youth, whether or not one serves in combat, and experiences in the war. It is more than a physical returning. It has to do with psychologically "letting go" of the "burden of war," which requires "burying the ghosts" that haunt them once and for all, if that is possible. Though the anger begins in boot camp, boot camp alone is not sufficient to produce that degree of anger. Lots of guys go through boot camp and not all come out of the military angry. Boot camp has to be mixed with and tied to one's "war experiences." Degree of "healing" and "letting go" of the bad memories and anger are likely to be related, among other things, to the support system that one has at home (a hypotheses on the part of the researcher to be checked out against incoming data).

### *Methodological Note*

More theoretical sampling is indicated at this time in regard to the concepts of "coming home" and "letting go." I then asked Participant #3, why is it so difficult for some vets to "let go" of that anger and to "come home" psychologically? Here was his reply.

**Memo 3**

**June 20, 2006**

**Residual Anger**

*What I have found to be true is that a veteran goes through a grieving process—denial, bargaining, anger, and acceptance. After the "imprint of horror," a video is imbedded into the memory of a soldier. The video often replays continually until the coping skills are exercised and the imprint is reduced. Anger stays as the primary emotion because this is where everything is stuck—anger at loss of life, loss of innocence, loss of the "fun years," loss of power, loss of any number of things. The average age of a service man is eighteen to twenty-five. What do you remember of those years and why do you remember it? College, spring break, friends, all nighters, etc . . . these are fond memories, in contrast of war for the veteran. The secret is to get the veteran to use coping skills they don't know they have because they were never been taught to use them as you were with "critical thinking" in college. Emotionally, 'til the veteran uses coping skills, they can't advance in emotional or cognitive age. They are stuck with thoughts, hormone imbalance, etc. Some need not only counseling but medication also to help maintain psychological homeostasis. I learned how to use coping skills with meds, counseling, support network of other veterans, and my wife. That's why I am finally back in college going after what I wanted to be fifteen years after the normal age of doing that. Regret is another hang up. Have you ever done anything that you regret because you didn't think it was you really doing it? Regret turns into confusion emotionally and it in turn creates anger. It's a cycle that continues until you break it.*

Though Participant #3 represents a sample of one, his words do give me a lot of insight into why anger seems to hold on so long. There are not only the experiences of war, but also the fact that while one is at war, the lives of civilians are going on quite normally. While a young man (or woman) is at war, he is too busy to think about this. But when he or she returns home there is "lost time" and "lost experiences" that he or she can never make up. Add that to the loss, the terrible memories, the fear that returns at night, the hormonal imbalance (strong doses of adrenalin and other stress hormones?) and the perceived uncaring on the part of civilians. Then there is "youth." Being young usually means being physically strong but not necessarily psychologically strong. Until the "cycle of anger" is broken through intervention, the anger is likely to remain.

---

**Memo 4**

**June 20, 2006**

**Summary Memo**

What I take from analysis of these two interviews is as follows. First there was greater development of the concept of "combatant" in terms of understanding what it is and identifying some of its properties and dimensions. These two men tell me a great deal of what the experience was like for them as "combatants." (Though I did not include all of the analysis of the interview with

Participant #3, readers can go to Interview #3 in Appendix D to read the complete interview.) Combatants, especially marines, are well-trained and disciplined soldiers. Their job is to engage and kill. During battle, they fight not for a cause per se but to survive, their own survival and that of their fellow soldiers—the only persons they can really trust. It can be said that the whole focus of combatants is on surviving and they develop a number of psychological and physical strategies to help them do so. Second, the experience of "fighting" really changes combatants and places a burden on them that they carry for the remainder of their lives.

Though a combatant is trained to kill, killing really goes against all the social mores that one grows up with, leaving one not only with a bad conscience but a great deal of anger at having been forced into the position of having to do this. Third, there is a major adjustment that must be made upon coming home, just like the major adjustment that must be made when going into a war zone. But because of the ghosts that haunt these combatants the remainder of their lives, these adjustments are not easy. Making it more difficult for the Vietnam vets was the reception that they received at home. No one understood what combatants had been through. Upon coming home, they were forced by men and women who had never been to war to carry the burden of an "unjust war" and a "dishonorable peace." The rage that began in boot camp was fostered in Vietnam (I mean being angry at the enemy makes it easier to kill them), and was reinforced upon "coming home."

My concept of "combatant" was developed considerably through this analysis. I also know quite a bit about the "anger" upon coming home, the whole lack of recognition, accusations, and so on. But there still remains for me the question of what happened there in Vietnam? Yes, there is the killing, but what occurs in battle? How does a combatant feel in battle, how does he act, and think? How or why does battle foster anger? I now have direction for further theoretical sampling. I am directed to look for data on the concept of "battle" or said a little differently, what it is like to be in "combat".

---

### *Methodological Note*

Having determined that the war experience was in many ways very different for those who were combatants versus the noncombatants, I am now curious about "what is combat?" I want to learn more about this concept, and flush out its properties and dimensions. I realize from my readings that "combat" can be many things from face-to-face engagement with the enemy to being a pilot out on a bombing mission and being shot at from below. I guess the main descriptor of "combat" is engagement with the enemy. I need more data to continue with my analysis. I could try to reach more vets via the Internet or put out the word among friends, but at this point

I don't have the time, as getting this book about method out is more important that searching for vets who may or may not want to be interviewed. So where do I turn to find my answers to my questions?

### *Examining the Concept of Battle*

Needing more data and not certain where to find it, I turned to memoirs written by Vietnam veterans. Once I discovered the memoirs, I found a treasure of information. I realize that the vets who wrote the memoirs were probably the articulate ones and not representative of all vets, but writers tend to be insightful and perhaps for my purposes they can be most helpful. I think that many wrote the memoirs not only to explain what Vietnam was like, but also to help the writers bury some of the ghosts that haunted them. In reading and analyzing the memoirs, I am theoretically sampling, following up on questions that were derived from previous analysis. I want to know: What is combat? What emotions does it generate? How do combatants react to it, and describe it?

Methodologically, it is important to point out that when coding memoirs, the analytic process is the same as it is with interview data. From the point of the reader, what is important is to note how I use the materials to sample theoretically and follow up on analytic questions that were raised in previous analyses. The first memoir that I coded was titled *Hill 488* by Ray Hildreth, a former Marine. The coauthor of this book is Charles W. Sasser, a writer. The book was published in 2003 by Pocket Books and describes Hildreth's account of the battle that gave the book its name. Of the men who participated in the battle with Hildreth, one man received the Medal of Honor. Four men received Navy Crosses. Thirteen men received Silver Stars and eighteen men were given Purple Hearts. Some of the decorations were given posthumously. Though I'm certain that some of the events described in the book were dramatized to make a point, for the most part the overall story seems genuine. It is similar in tone to the other memoirs that I've read about Vietnam. Rather than using direct quotes, I've paraphrased the description by Hildreth.

Describing a battle is not easy even for someone who has been through it. For the researcher trying to capture this experience second hand, it is even more difficult. However, to understand why the war experience has such a profound and long-lasting impact on a young soldier it is important to know what goes on in battle. I'll try to recapture aspects of this experience.

## Memo 1

### June 21, 2006

### The Recon Mission

While out on recon (reconnaissance) patrol the recon team came upon a village with considerable enemy activity. Usually villages consisted of men, women, and children. But this village was different in its make up of residents. It appeared to be an enemy base of operations. The recon team relayed this information back to headquarters and soon there were American air strikes on the village. The recon patrol remained in their position gathering further intelligence on enemy activities that continued to go on despite the bombings.

Commentary. What this section of the memoir tells me is that the enemy did use villages and villagers to fight from. It also explains why villages were bombed, despite the civilians who were "caught in the middle" between the Viet Cong and the Americans. Interestingly enough, the enemy did not withdraw or run away when the bombing began but continued their operations. Also interesting about this piece of information is that some combatants had the job of reconnaissance, or spotting the enemy for military strikes. However, being out there, a small group alone in the jungle, made them very vulnerable to attack themselves and they had very little equipment with them as they had to keep things light as they were usually on the move.

## Memo 2

### June 21, 2006

### Discovery and the Waiting

The recon team would have been surprised by an enemy attack if one of the sergeants with them had not overheard a conversation on his satellite phone between some Green Berets on patrol and a South Vietnamese military group who also happened to also be in the area. The warning was to be wary because a group of enemy were in the area and seemed to be looking for something or someone. Hildreth's (2003) recon team upon hearing the message realized that they had been spotted by the Viet Cong and were the probable targets of the advancing army. Unfortunately for the men it was getting dark and too late to call for a helicopter airlift out of the area. The recon team would have to hold their position and wait until morning for rescue. Traveling light, they had no heavy artillery with them. For defense, all they carried were their rifles, a little extra ammunition, and a few grenades. Preparing for a possible assault, the men were placed in two man teams and placed in locations around the perimeter of the hill. For cover there was only the tall grass.

Commentary. This tells me that combatants are not always prepared for "battle." Sometimes it comes upon them and they are unable to escape it. Fortunately the recon team was warned through the radio conversation that they overheard and was able to prepare to some degree. Missing was the heavy equipment necessary to fight a difficult battle. At this point though, the men did not know how many "enemy combatants" there were or how intense the attack would be.

---

### Memo 3

**June 21, 2006**

**The Assault Begins**

The men waited for the enemy to come. Finally one of the soldiers saw movement and fired his gun in that direction. Upon hearing the signal that the battle was beginning Hildreth says, "I froze in place, unexpectedly and totally scared to death" (Hildreth, 2003, p. 197). One of the American soldiers was wounded. His screams pierced the night air. Hildreth found the screaming unnerving. Finally, a medic reached the wounded soldier and gave him a shot of morphine. The screaming quieted down. But the battle was just beginning and this was the first of the wounded. This was the recon team's first experience with battle. About the experience Hildreth (2003) states, "Battle was difficult to grasp the first time you experienced it" (p. 199). Using the satellite phone, the platoon called for air support. However, because it was so dark and the enemy was so near, sending planes to bomb was thought to be too risky to the Americans. The planes however dropped flares so that the American soldiers could more easily spot the enemy hiding in the tall grass. Hildreth saw the man next to him killed. The death of a close comrade, he says, had a strong psychological affect on him. It made him realize that it could just as easily have been him.

Commentary. Hildreth is giving us some insight into what happens in battle. Despite combat training, a soldier is never sure how he will respond. Hildreth's first reaction to battle was fear and to freeze in place. As Participant #2 told us, to do nothing is not a good strategy if a combatant wants to survive. It only puts him at greater risk of injury or death. Yet, I can imagine the fear and confusion that occurs during those first few moments of battle. There is bound to be confusion, a few moments when a combatant is not sure what is happening or what to do. The data above also tells us that death is never far away for combatants during a "battle." Survival is just as much a matter of chance as it is skill.

---

## Memo 4

### June 22, 2006

### The Second Assault

Despite being outnumbered by enemy soldiers, the recon team managed to temporarily repel the enemy. The American soldiers knew, however, that the enemy was just using the opportunity to regroup and plan their next attack. Hildreth (2003) says, "Nothing could be as bad as the waiting" (p. 213). Finally, the soldiers on the hill heard the clicking of bamboo sticks (the enemy's way of communicating with each other) and knew that the second assault was about to begin. The second assault was fierce. There were heavy casualties. A helicopter tried to reach the battle zone to pick up the dead and injured but was shot down and the pilot killed. Other rescue helicopters in the area aborted the mission. The situation looked hopeless. The enemy brought in a 50-caliber machine gun. Hildreth states that when the gun went off, it felt like the whole hill was being ripped open. The platoon sergeant was wounded. Yet, every man, wounded or not, who was alive and able to fight continued to do so. At this point Hildreth states that for him the battle took on a surreal quality. He was certain that they would all be killed. Of that time, Hildreth (2003) says, " . . . I hadn't moved since the guy fell dead next to me with the top of his head blown off. I lay among the dead in a graveyard of the still unburied . . . " (p. 307). Again, despite their advantage, the enemy was repelled a second time.

Commentary. It is interesting to note how combat training comes into play during actual battle. Though frightened and somewhat disorganized at first, the men automatically go into the mode of doing what they were trained to do, that is, fight back. Driving them is that survival instinct. There is, though as Hildreth tells us, that feeling that you too will die, just like those around you. What is most interesting are the strategies men use, the mental escape—the surreal quality battle takes on, that is almost standing outside the action, watching it go on, yet all the while participating in it. I guess that this is a very important strategy for surviving the horrors of battle. The fear, the horrors, and smell of death are what vets want to bury deep within, not only at the time of battle but later also. The experience is a "nightmare" but an experience one can never fully escape, just like a battle.

---

## Memo 5

### June 22, 2006

### A Turn in the Battle

Preparing to make the third assault, the enemy started yelling, "Marines, you die tonight . . ." (Hildreth, 2003, p. 261). Rather than demoralizing the marines, the cries of the enemy enraged and energized them. The marines started yelling back at the Viet Cong. This went on for some minutes, until the marines who remained alive and able to fight realized the absurdity of the situation. Here they were in the middle of a battle having a yelling match with the enemy. But somehow the act of yelling lifted them out of their sense of hopelessness and renewed their will to survive. Hildreth (2003) states, "It was a turning point in the fight" (p. m263). The marines' resolve was bolstered by the aid that they finally received from Tactical Air Support. Seeing the intensity of the battle, and even though it was still dark, American bombers moved into the area and started to drop bombs on the enemy. The bombs did not deter the enemy. They continued to advance. The marines still able to fight continued to do so. The battle ended around daylight. The marines were still holding the hill. But the price the recon patrol paid was high. Six out of the eighteen men in the platoon were killed and twelve were wounded. Of the wounded, only three were able to walk without help. Of this experience Hildreth (2003) states, "None of us would ever be 19 years old again . . ." (p. 324). Waiting for the dead and severely wounded to be evacuated, Hildreth says, "I spotted movement among a group of VC corpses and I whirled and starting shooting in a rage . . ." (p. 325). He goes on to say, " I went down and looked at the gooks I killed. They were the ones that killed Adams [his team member]. I looked at them and felt nothing. I didn't know if I would ever be able to feel again . . ." (p. 325). At this point he says, "I was one sick, confused Marine. Fucked up from the shock and horror" (p. 265).

Commentary. "Being a combatant" seems to define the war experience in very definite ways. In the space of a few hours, a night in this case, a young man changed from a youth into someone much older. So what is "battle"? I can only state that a battle is a "fierce struggle for survival" though the military might define it differently—i.e., in terms of a cause. But a "struggle for survival" is how the participants of this study define it. Being in battle seems to engender a range of emotions that can vary over the course of that battle: fear, horror, sorrow, hopelessness, shock, rage, and a determination to live. Being in battle can "fuck" one up. I suppose by "fucked up" Hildreth means mentally confused, dazed, left trying to understand how easily one can kill and be killed, and most of all enraged. Perhaps rage is an emotion necessary to overcome the built in sanctions one has against killing someone else. It takes rage to "kill." As noted, once rage is set in motion and the adrenalin is soaring, it is difficult to turn off. Rage can lead to actions that a combatant might be sorry for later when he or she has time to cool off.

### *Methodological Point*

I now have analyzed several interviews and memoirs. As I continue to read memoirs, I see the words expressed by Hildreth (2003) echoed in the memoirs of other ground soldiers, marines, and pilots. Those who fought or flew in Vietnam seem to be telling the same general survival story, though the specifics may be a little different. Memoirs that I analyzed include Alvarez and Pitch (1989), Bell (1993), Caputo (1977), Downs (1993), Foster (1992) Herr (1991), Marrett (2003), Moore & Galloway (1992), Nhu Tang (1986), Rasimus (2003), Santoli (1981 & 1985), Terry (1984), Trotti (1984), and Yarborough (1990).

The memoirs validate my interviews and also demonstrate that the men (and women) who served in Vietnam carried the war home with them so much so that they felt a need to write about it, some years later. At this point in the analysis, I have a lot more analysis-derived questions that need to be answered. Questions such as: Who exactly was this enemy and why were they such persistent adversaries? How did America become involved in war way over in Vietnam? And what were the conditions—political, social, and environmental—under which men like Hildreth (2003) had to fight? I realize that these questions are contextual related questions, which means that I have to extend my investigation and examine some of the broader, more macro, issues if I am to understand the war better from the perspective of the combatants that fought in it. I already have some of the more micro contextual factors in my existing categories, for example "changes in self" from "time of entry" to "homecoming" and "beyond" and "psychological survival strategies." What I am missing at this juncture in the analysis is the larger sociopolitical picture and its impact on the combat experience. I turn to the macro issues next in my data collection and analysis.

## Summary of Important Points

This chapter explored and developed the concept of "combatant," extending it to include the notion of "battle." Though exploration and development of the concept of combatant was the main focus of the analysis, new concepts were derived from the additional data, concepts such as "ghosts that haunt," "homecoming" and "carrying the burden." The most important point for readers to take away from this chapter is the importance of memos to analysis. A researcher can readily see that it would have been impossible without the use of memos to maintain in my head all of the ideas explored in this chapter—very few of us have that good a memory—or to be able to follow up on important analytic questions through further theoretical sampling of memoirs. The analysis became too complex and loaded with conceptual development to be retained without writing those thoughts down. I also want to point out that though computers are very helpful to analysis, even when using a computer the analyst must not approach analysis in a mechanical way. It is the freedom to think, the ability of the researcher to change his or her mind, to check out ideas, and to follow the data trail wherever it leads that makes the findings derived through qualitative research so compelling and relevant and the process of getting there such an exciting voyage of discovery.

## Exercises for Thinking, Writing, Group Discussion

1. Examine some of the memos contained in this chapter and think about what was taking place analytically; that is, how concepts are being extended and developed and how relationships between concepts are evolving. If you use MAXQDA, make use of MAXMaps in order to visualize your ideas.
2. Think about the role of the researcher in this analysis and how the analysis might look different if the researcher had been different. How would you analyze the same data? What would you get from this?
3. Jot your ideas down and bring them to class for discussion.
4. As the reader will notice, I did not place a list of concepts at the end of this chapter. As part of this exercise, go through the chapter and make a list of the new codes (concepts) and add them to the list from Chapter 8. Now, having read the field notes and memos from both chapters, think of how the concepts might somehow fit together to form categories. What major new themes emerged? If you work with MAXQDA, define your (eventually) defined new concepts in the code system. Use the Teamwork Export and Teamwork Import function to share your changes in the team.

# 10

# Analyzing Data for Context

*The U.S. debacle in Vietnam can be attributed primarily to the incorrect diagnosis of the reasons for the insurrection. The conflict was not as much pro-Communist as it was anti-Diem and later anti-Ky and anti-Thieu, all of whom failed to initiate and implement the much-needed political and socio-economic reforms. . . . The conflict called for a political rather than a military solution, probably to satisfy a widespread urge to reunify the country. (Sar Desai, 2005, p. 120)*

**Table 10.1** Definition of Terms

*Conditional/Consequential Matrix*: An analytic device to stimulate thought regarding the wide range of possible conditions and consequences that can enter into context.

*Context:* The sets of conditions that give rise to problems or circumstances to which individuals respond by means of action/interaction/emotions. Context arises out of sets of conditions ranging from the most macro to the micro.

*Paradigm:* A model for integrating structure with process.

*Process:* Ongoing responses to problems or circumstances arising out of the context. Responses can take the form of action, interaction, or emotion. Responses can change as the situation changes.

# Introduction

In the last chapter, it was established that the main issue for combatants in Vietnam was "surviving." An analyst might say that the war threatened the very existence of combatants as well as their psychological well-being and their moral integrity. But where did the threats to survival come from? In order to more fully understand the concept of "survival," as researchers we are faced with two analytic tasks. The first task is to explore the **context** in which the war was fought because it will reveal the circumstances or problems that presented threats to survival. There are two parts to context, the more micro conditions, which in this research are the immediate set of conditions faced by combatants on a day-to-day basis, and, second, the macro or larger socio, political, and historical conditions that led to the more "immediate" set of conditions. The second analytic task is to explore the **process** through which persons responded to the life threatening conditions through action/interaction/emotions. Process will be taken up in Chapter 11.

### *Methodological Note*

Before exploring the "context of survival," I want to divert the readers' attention for a moment with a methodological note. Normally, at the end of a research investigation the products of analysis are presented as a set of findings. There is little or only brief mention of what the researcher went through in the process of arriving at those findings. In writing this book I wanted to give readers something more than just a book about procedures. I wanted to let readers inside the entire analytic experience. Next is a memo that I wrote between the last chapter and this one. I place it here so that readers can obtain some insight into my analytic journey.

---

**Memo 1**

**June 21, 2006**

**Theoretical Sensitivity**

It is interesting to note how much I've changed since beginning this research project. It's not that the interviews and memoirs that I'm reading are any different than they were at the beginning of the study, but that I am far more sensitive to what they are saying. It takes being immersed in the materials for some time before the significance of what is being said comes through. Sensitivity grows with exposure to data. I might say that analyzing data is like peeling an onion. Every layer that is removed takes you that much closer to the core. This is what is meant by "theoretical sensitivity," being more in-tune to the meanings embedded in data.

In addition to increasing sensitivity to data, what I find interesting are the changes in myself since beginning this project. I can't help but be touched by the war stories I've read. I've seen movies and read books about war. But there is something about the interaction that occurs between analyst and data during analysis that has altered how I think and feel about combatants. It has to do with taking the role of the other, feeling for a short time what it is like to be a soldier. I know that I'll never look at a Veteran of any war in quite the same way. I notice that as I work with data, I feel anger at the circumstances that brought young men into combat in Vietnam and the rules of engagement that made it so difficult for them to fight that war. I'm saddened by what I perceive to be the suffering of some men both during and after the war, and I also feel for the enemy soldiers and the civilians who were "caught up in the war." It was a difficult period for all. At the same time I know that I can't let my emotions get in the way of doing the research. It's okay to feel, but at the same time I must retain the ability to do justice to the stories of participants and not get so carried away by anger and other emotions that I become ineffective as an analyst.

There is still much I don't understand about how things were for combatants in Vietnam. To gain a better understanding of the problems, I have to delve deeper into the contextual factors that shaped experience, that is, the political, social, and historical conditions leading up to, during, and after the Vietnam War. This is the direction the research is leading me. The questions driving the analysis at this stage of the research are: What were the conditions that combatants had to fight under? And what are the historical/political/social factors that led to those conditions? I want to know for my own sake as well as for the research.

## Linking "Culture of War" With "Survival"

Returning to the analysis, one of the concepts derived earlier in analysis was that of the "culture of war." The "culture of war" was described as the context in which combatants were operating, or said another way, in which their war experience took place. In examining data for context, I am now "opening up" the concept of "culture of war," exploring what that context consisted of and at the same time I am linking "culture of war" to the concept of "survival" by showing the relationship between context and survival actions (to be taken up in the next chapter). I return to the data on Vietnam to uncover the conditions that threatened the survival of combatants. It is obvious they were being shot at. But in addition to being shot at, what other conditions of combat put the men at risk and threatened their physical, psychological, and moral beings? After analyzing Interview #3, I turned to the memoirs for clues as to just exactly were the problems that threatened survival and made it the major issue. One memoir that I found particularly insightful was titled *A Rumor of War,* by Philip Caputo (1977), because it spells out those problems so clearly. Caputo says that his book is a "story about war," "what men do in war," and "the things that war does to them" (Caputo, 1977, p. xiii). In other words, it was perfect for our purposes of discovering the link between our two concepts "culture of war" and the "survival experience."

Caputo joined the Marines in 1963. He was sent to Vietnam March 8, 1965, as a first lieutenant, where he served for about thirteen months. He returned to Vietnam ten years later in 1975 to cover the fall of Saigon as a journalist. The fact that he wrote his book from the perspective of both soldier and journalist makes him a valuable informant. In reading the data, I was specifically looking for conditions that increased vulnerability and threatened survival. The conditions were presented as a series of problems because that is how combatants experienced or talked about them.

---

**Memo 1**

**June 23, 2006**

**Problem #1: Idealism and Youth**

> *We went overseas full of illusions, for which the intoxicating atmosphere of those years was as much to blame as our youth.* (Caputo, 1977, pp. xiii & xiv)

Patriotic and idealistic young men joined the military with the notion of signing up to defend their country. Some enlisted, some were drafted. Since young soldiers' understandings of war were derived mainly from movies, most were unprepared for the sight of bloodshed, the suffering, and hardship that they found when they arrived in Vietnam. The "realities" of war hit hard and survival called for a dramatic shift in attitude and action, a shift that took the romance out of the situation and replaced it with, if not an "acceptance," at least a recognition that "war is about killing or being killed" and a soldier has to harden to the notion of bloodshed and death and work hard to survive.

---

**Memo 2**

**June 23, 2006**

**Problem #2: A Sense of Powerlessness**

> *Men were killed and wounded, and our patrols kept going out to fight in the same places they had fought the week before and the week before that. The situation remained the same.* (Caputo, 1977, p. 182)

In reading the various war memoirs, and especially that of Caputo, I sensed a general feeling of powerlessness in combatants—a powerlessness to change the conditions they found themselves in. At the time of the Vietnam War, the draft was still in place. Though early combatants, especially marines, were volunteers, as time wore on there was a greater need for soldiers but fewer volunteers, and more and more soldiers were draftees. They were sent to the front lines with only basic military training. Yet it was the "combatants" who saw the predominance of enemy action. And it was they who suffered the greatest number of war casualties. The name given to the front-line soldiers was "grunts," a word that says a lot. The grunts did the "dirty work of killing," were at the bottom of the military hierarchy, and had little power to control the military situations that they found themselves in. They had little control over where they went or what they did and no sense of accomplishment because they didn't seem to go anywhere.

The notion of powerlessness is an interesting one analytically. Perhaps, it explains some of the anger the veterans carried with them after the war. At least I can hypothesize that and later check it out. There was little a front-line soldier could do except go where he was told, when he was told, to do what he was told even if it didn't make much sense to him and even if he knew that his life was on the line.

---

## Memo 3

**June 23, 2006**

**Problem #3: Confusing Rules of Engagement**

The "rules of engagement" that defined this war can be called confusing at best. Listen to what Caputo has to say:

*According to those rules of engagement, "It was morally right to shoot an unarmed Vietnamese who was running, but wrong to shoot one standing or walking; it was wrong to shoot an enemy prisoner at close range. . . . "* (Caputo, 1977, p. 218)

How does a young and inexperienced combatant find his way through the confusing mass of overt and covert rules called the "rules of engagement"? How can he maintain his sense of "right and wrong" under conditions where killing is sanctioned in some situations but not in others? It is no wonder that young men were confused, disillusioned, and disturbed by what they saw and did, and in turn did things that they should not have done. Furthermore, if one perceives one's life to imminently be in danger, does one stop and think about things called "rules of engagement" or question the morality of things? Or does one just act?

### Memo 4

**June 24, 2006**

**Problem #4: Fighting Someone Else's War**

When Caputo first arrived in Vietnam, he was told he was being sent there to defend the air base in Da Nang and not undertake an assault on the North Vietnamese Army. Early in '65 the war was still defined as a Vietnamese war. Caputo was told when he landed:

> . . . *Okay, listen up. When you brief your people, make it clear that our mission is defensive only.* (Caputo, 1977 p. 35)

Yet, with time and it became clear that the South Vietnamese Army alone couldn't hold back infiltration of the North Vietnamese military the purpose of the U.S. military in Vietnam changed. It went from its early focus of providing reconnaissance, training, and support to becoming the major combatant force in the struggle to prevent a North Vietnamese takeover of the South. Caputo (1977) says, "The war would no longer be only 'their war' but ours as well; a jointly owned enterprise" (p. 68). But it was difficult for young men thousands of miles away from home to understand the purpose of why they were fighting, especially for something so vague as communism in someone else's country.

### Memo 5

**June 24, 2006**

**Problem #5: Entering a Frontier Between Life and Death**

When the American military became *the* primary combatant force in Vietnam, they entered a world that might be called the "frontier between life and death." In a war zone, the danger of death or injury is always present. "Training" for combat, no matter how good, is very different from "being in combat." There is a certain amount of learning that can only be accomplished by "living" the experience so to speak. If a soldier is to have a chance at survival, he has to very quickly make the transition from "novice" to "seasoned soldier." Caputo provides some of the properties of "being seasoned" and also describes some of the consequences.

Caputo explains:

> *We began to change, to lose the boyish awkwardness we had brought to Vietnam. We became more professional, leaner and tougher, and a callus began to grow around our hearts, a kind of emotional flak jacket that blunted the blows and stings of pity.* (Caputo, 1977, p. 90)

## Memo 6

**June 24, 2006**

**Problem #6: Fighting a Determined and Experienced Enemy**

When American troops took over combat in Vietnam they believed that it would be a short conflict because of perceived U.S. "superior" military strength. This turned out to be a false assumption. The North Vietnamese army was well trained and highly motivated. They knew how to use the terrain to their advantage. They dug tunnels where they hid supplies and even had living and medical facilities underground. They put booby traps and mines in and around villages and along trails. They were heavily armed and equipped by the Chinese and Russians. And they put their supply trails in Cambodia and Laos out of reach of American soldiers. The Americans may have had military technology but the North Vietnamese had skill and motivation.

## Memo 7

**June 25, 2006**

**Problem #7: An Inhospitable Terrain**

The Viet Cong were not the only enemy the men had to fight in this war. Though the American soldiers received some jungle training, no training could quite prepare them for the terrain they encountered when they got to Vietnam. Soldiers had to make long marches in the heat and rain, and also walk and camp out in dense brush or jungle. They were subjected to mosquitoes, leeches, and jungle rot. It was an environment that can only be called "inhospitable."

## Memo 8

**June 25, 2006**

**Problem #8: A War With No Apparent Sense of Direction**

The "policies" and "rules of engagement" established in Washington did nothing but constrain the military's ability to act in a manner that would enable them to defeat the enemy. North Vietnamese soldiers could attack U.S. soldiers then retreat to the safety of Cambodia, Laos, or North Vietnam, knowing the soldiers were not supposed to pursue them there. (Planes flew secret missions over these territories but even they were constrained in their ability to carry out a war.)

Another apparent issue was that there was no interest in permanently securing territory. American soldiers fought a battle then retreated to their bases once the battle was over, leaving the enemy free to reclaim the land they had just fought so hard for. Sometimes the U.S. soldiers had to go back and fight for the same territory again. This baffled the young soldiers who could see no purpose for what they were doing. Even pilots wondered why they were called upon repeatedly to bomb the same areas at considerable risks to themselves. After the first bombing attack, the enemy was prepared and waiting for the planes. It seemed to combatants that there was no strategic plan. The constant loss of lives and high number of injuries coupled with no apparent military gain was demoralizing. As Caputo (1977) states, "Men were dying but it wasn't accomplishing anything" (p. 213).

---

## Memo 9

**June 26, 2006**

**Problem #9: Measuring War Success Through Body Counts**

The U.S. war policy adopted for the Vietnam War was to break the morale of the enemy and drive the communist Viet Cong and North Vietnam Regular Forces back into North Vietnam. Military effort was directed at "search" out and "destroy." Success in battle was measured by "body counts." After each battle soldiers had to go around and count the number of enemy and U.S. soldiers dead and wounded.

---

## Memo 10

**June 26, 2006**

**Problem #10: The Cultural Divide**

The young men fighting this war had little knowledge of Vietnamese culture or history. Many of the young soldiers didn't even know where Vietnam was before going there. The soldiers expected the Vietnamese villagers to react to situations as they would. When the Vietnamese villagers didn't behave in a manner that fit American expectations, it made the villagers seem less human. Dehumanizing the Vietnamese people as well as the enemy made it easier to excuse atrocities. On one of the patrols, Caputo's platoon came upon a village that had been burned down. At first, Caputo says, he felt shame seeing what another group of American soldiers had done. As he passed through the village, Caputo expected the villagers to show some emotion, like hate or anger, at the passing soldiers.

However, the villagers exhibited no outward emotion. Caputo states that the lack of emotion caused his pity to turn to contempt. Only years later could he understand why the villagers reacted the way that they did. Caputo (1977) says, "They had suffered so much—endless war, bad harvest, disease that they had learned to accept and endure" (pp. 124–125).

---

### Memo 11

#### June 26, 2006

#### Problem #11: A Constant Turnover of Troops

The tour of duty for combatants during the Vietnam War was more or less one year depending upon the branch of service. For some pilots, the duration of service in Vietnam was set at a year. For other pilots, duty was fulfilled after completing 100 missions. For marines, a tour of duty in Vietnam was thirteen months but for army soldiers it was twelve months. Regardless, whether the normal tour of duty was a year or a little over a year, the frequent turnover meant that just about the time a soldier became "seasoned" and able to protect himself and others, his tour of duty was over. He was then replaced by a "novice," who was more of a hindrance than a help to the group. In addition to the normal rotation plan there was a high casualty rate in Vietnam necessitating a need for replacements, again with "novice" or inexperienced soldiers. Being "a novice" had its drawbacks. It increased the risks of death from enemy fire because the soldier lacked survival skills. It also increased the chances of accidents or "friendly fire incidents."

---

### Memo 12

#### June 26, 2006

#### Problem #12: A Desire for Revenge

War not only generates fear and anger, it can bring about a desire for revenge. After several months in the battlefield as a platoon leader, Caputo was rotated out of the field and given a job back at headquarters. His new position consisted day after day of recording casualty rates of both Americans and Vietnamese. After some months of performing this task, Caputo states that he began to suffer nightmares. Fearing that he was going insane, he requested a return to combat even though it would mean an increased risk of death or injury. He wanted to inflict on the enemy some of the suffering that he had endured.

**Memo 13**

**June 26, 2006**

**Problem #13: A War That Did Not Stand Still**

Over time the intensity of the fighting increased. Battles were more ferocious and more frequent than they were when Caputo first arrived in Vietnam. When he returned to the front lines he discovered that while he had been away the nature of the war had changed. There was an increase in the number of North Vietnamese Regular Army troops, as opposed to the less well-trained Viet Cong, operating in South Vietnam. The North Vietnamese were entering South Vietnam by means of the Ho Chi Minh Trail. The North Vietnamese soldiers carried AK 47 rifles, hand grenades, and had Claymore mines supplied by China and Russia. When Caputo returned to combat duty he found, "It was not really a guerilla war any longer"(1977, p. 207).

**Memo 14**

**June 26, 2006**

**Problem #14: A Physical Wearing Down**

With the intensification of the war, American soldiers made more frequent patrols to "search for" and "destroy" the enemy. There was less time to rest between forays into the jungle. The long marches in the heat, the lack of quality sleep, the meals taken on the run, the stress, strain and fear, took a physical toll on the men and increased the danger they faced. Caputo (1977) says, "The company had run nearly two hundred patrols in the month I had been with it, and then there had been all those nights on the line. The men were in a permanent state of exhaustion" (p. 237).

**Memo 15**

**June 26, 2006**

**Problem #15: Psychological and Moral Breakdown**

Despite complete exhaustion, Caputo and his men were forced to spend more and more time searching out the enemy. One day Caputo and his platoon came upon a village that was supposedly a Viet Cong stronghold. As the platoon made its way towards the village, Viet Cong soldiers attacked the approaching

U.S. soldiers. The Viet Cong soldiers then fled, leaving the villagers to face the angry soldiers. Caputo says that when he and his men arrived at the village, they just "lost it." They whooped through the village setting fire to every building they came across, while the villagers looked on in horror. Caputo says he felt powerless to stop the men and powerless to stop himself. Burning of the village seemed "almost an emotional necessity" and a means of letting go of months of fear, frustration, and tension. "We had to relieve our pain by inflicting it on others" (Caputo, 1977, p. 288).

**Memo 16**

**June 26, 2006**

**Survival in a Context of High Risks**

So what does all of the above mean for our analysis? It means that the culture of war was one defined by many threats to individual and collective survival. The threats presented themselves as a series of problems that combatants had to overcome if they were to come out of the experience physically alive, psychologically undamaged, and with their moral integrity intact. Engaging in combat day after day is difficult enough but if one sees no gain in territory, no seeming purpose to the war, no end in sight to the fighting, and a constant supply of enemy combatants ready to kill you, the only thing left for a combatant is to endure and survive until he can get out, hopefully alive.

## Context as Structure

So, the question that comes up now is why these conditions or problems and not others? What made this war and the conditions combatants experienced particular to Vietnam? This is not to say that combatants in other wars have not experienced similar conditions, but Vietnam was different in that the threat to the country as a whole was rather vague (communism) and certainly not immediate, while the treat to the individual combatant was very high because of the rules of engagement and the duration of the war. When we talk about structure or context, we mean more than writing a chapter citing historical events as background material. What we are interested in is how the problems faced by combatants have their foundation in the historical, political, and social conditions that set the tone and policies for the war. Though often miles apart figuratively and actually, decisions made in Washington and Hanoi, based on their individual and collective ambitions, understanding of past events, and national interests established the culture for this war and created the conditions that combatants on both side were operating within. In other words, all of the conditions spelled out above as problems can be traced back to the historical/social/political events. This tracing back will increase our understanding of why combatants found themselves caught up in a directionless and seemingly endless war.

Our analysis of the socio/political/historical will take the form of memos. They answer the questions why, who, what, and where and round out our exploration of the concept "culture of war."

### *Methodological Point*

I must admit that I've been bogged down by the analysis for this chapter. There is so much material on the Vietnam War that sorting through it has left my head spinning. I feel like I have reached a point of oversaturation. After putting this project aside for a while, I came back to it realizing that I'd been too caught up in the descriptive aspects of data and not thinking as conceptually as I should have been. By refocusing on the concepts rather than the details I know I can get going again. Data for this section on context came from the following sources: Ellsberg, 2003; Isaacs, 1997; Langguth, 2002; McMaster, 1997; McNamara, Blight, Brigham, Bierstaker, and Schandler, 1999; Nhu Tang, 1986; Santoli, 1985; Sar Desai, 2005; Sheeham, 1988; Sallah and Weiss, 2006; Summers, 1999; Tucker, 1998.

---

**Memo 1**

**June 27, 2006**

**The Culture of War**

The "culture of war" is one fraught with physical, psychological, and moral risks. The risks arise from the problems that framed the combat experience and that threatened survival. To find out why the problems and therefore why the risks I have to go back and look at *who* brought the U.S. into the war, *why* at this time and place, *who* set the policies for the war, *who* were the enemy? *What* were the limitations set by the rules of engagement, *why* did the war last so long and *what* finally brought it to an end? It's time to delve into some of those historical books.

This is a long but important story. It explains the combat conditions: why Vietnam became America's war, why troops were fighting in Vietnam (a culture so foreign to the U.S. at the time), why the U.S. used body counts instead of gaining territory as a measure of success, and why and by whom rules of engagement were established and why the enemy were so determined. There is a lot more material that could be covered but it is not in the interest of what we are trying to teach in this chapter to put it all in here.

Why were U.S. combatants fighting in Vietnam? For many years Vietnam was a French Colony. A revolt against the French by the Vietnamese led to defeat of the French in 1954 and the development of a strong sense of nationalism. French departure did not result in a unified country. As part of the Geneva Accords, Vietnam was divided into two parts at the 17th parallel, a Communist north and a non-Communist south. According to the Geneva Accords, the issue of reunification was to be resolved through free elections within two years. However, The South Vietnam government, with U.S. backing, refused to sign the Geneva Accords. Behind this refusal was a fear that if free elections were held, control of the country would pass into the hands of the Communists. The U.S fear was that if South Vietnam fell into communist hands, Cambodia, Laos, and the rest of Southeast Asia countries would follow in a sort of "domino effect" (Sar Desai, p. 68). It was this fear of communism and the "domino effect" that set the stage for U.S. involvement in Vietnam.

Meanwhile, South Vietnam had its political problems. In 1954, Ngo Dinh Diem was designated president of South Vietnam. He was a Catholic, originally from the North, coming into a traditionally Buddhist South Vietnam. Right from the start Diem struggled militarily against religious and political opposition. Rather than using the financial aid he received form the U.S. to rebuild his country, Diem used the money to build and equip his army. He filled government posts in Saigon and administrative posts in local villages with relatives and friends, disrupting Buddhist traditions that went back for centuries. The Viet Cong promised relief from the corruption of their South Vietnamese government. It is for these reasons that many villagers were sympathetic to the Viet Cong fighting against or at least not being helpful to U.S. troops whom they saw as propping up the corrupt government. Though many of the Communists went north when the country was divided, several thousand Communists (the Viet Cong) remained in the south in preparation for eventual reunification. When the Diem regime fell, political instability followed. As various military leaders fought for control, the gap in leadership opened the doors to infiltration by the North Vietnamese Army, who came to aid the Viet Cong. North Vietnamese troops and supplies moved south via the Ho Chi Minh trail. A good portion of the trail passed through neutral Cambodia and Laos, thus was supposedly out of reach of U.S. and South Vietnamese Armies.

President Kennedy sent military advisors to the South Vietnamese Army. The basis for this decision was that the Viet Cong could easily be defeated if the U.S. provided military assistance in the form of training, support, and intelligence to the South Vietnamese Army. When Kennedy was assassinated in 1963, and Johnson assumed the presidency of the United States, Johnson had his own agenda for the presidency and wasn't quite sure what to do with Vietnam. He relied on Kennedy policy makers to advise him.

McNamara, one of the young advisors left over from the Kennedy administration, became the chief "architect" of the Vietnam War. He had no military background. Rather, he worked for the Ford Corporation before becoming a presidential advisor and applied his business sense to war. McNamara's plan for the Vietnamese conflict consisted of the application of what he called "graduated pressure" to North Vietnam. Under this plan, there would be no major assault. Rather, slow pressure would be applied through a series of controlled and limited military actions (bombing missions) conducted against North Vietnam. The purpose of the bombings would be to cripple the infrastructure of North Vietnam so that they would be willing to negotiate a peace agreement more favorable to South Vietnam. McNamara's plan was flawed from the start. First, it was a plan for "containment" and not a plan for winning a war. The plan was based on a civilian understanding of war and how war should be conducted, rather than based on military strategy and conduct of operations. Every proposed military action had to be relayed and approved by Washington before it could be implemented, often delaying action until the strategic advantage of a military action was lost to the enemy. In other words, the Joint Chiefs of Staff, those responsible for carrying out the war, were given little opportunity to provide input into how the war should be conducted, nor were they given the freedom to conduct the war in a manner that would allow them to win, leading to a sense of powerlessness. Second, McNamara's plan of containment and strategic bombing ignored the fact that there was little infrastructure in North Vietnam to bomb. North Vietnam was not an industrialized society. Most of their arms and ammunition either came from China or Russia and from what they were able to capture from South Vietnamese troops (American supplied). Third, the plan failed to consider the determination of the North Vietnamese to reunify their country as was intended when the Geneva Accords were written. Finally, the plan failed to take into consideration the impact that bombing would have on the Vietnamese people. Rather than breaking their will, it intensified it, generating a determination to fight regardless of the costs, making them a very determined enemy (McMaster, 1997, pp. 323–334).

In the early years, there were few, if any, actual combat troops in Vietnam. By 1965 it was becoming more and more obvious that the South Vietnamese Army, despite U.S. support, was no match for the Viet Cong, thus it became America's War putting the troops into the frontier between life and death. In order not to alarm the American public at the number of troops being sent to Vietnam, a piecemeal approach was taken to deployments, with the number gradually increasing over time. The first deployment consisted of 18,000 soldiers sent to South Vietnam to defend U.S. bases that had come under attack from the Viet Cong. In 1968, the Tet offensive occurred. Though Tet was a military defeat for the Viet Cong—they were almost totally destroyed—it was a political victory for the North Vietnamese because the ferociousness and boldness of the attack turned American public opinion against the war. The North Vietnamese used the death and mutilation of U.S. soldiers as propaganda to fuel antiwar sympathy in the U.S. and for demoralizing American troops in Vietnam (Nhu Tang with Chanoff and Van Toai, 1986). Yet, the U.S. continued to send troops into this inhospitable terrain, preparing the troops to fight but not to understand the culture and providing them with very confusing rules of engagement.

## Memo 2

### June 27, 2006

### The Enemy

Who was the enemy and why did this enemy put up such a determined fight? In brief, for over a thousand years, beginning with the Chinese, the Vietnamese fought to maintain their sovereignty and their identity. The last occupying force was the French government. A desire to gain their independence from France gave rise to a strong Nationalist Movement. Inspired by the revolution in China, many Vietnamese youth, both Communist and non-Communist, went to China to train as revolutionaries in order to overthrow the French. The inspiration behind the revolutionary fever was not so much communism as it was nationalism. The Vietnamese, including Ho Chi Minh, were "nationalist first and communist second" (Sar Desai, 2005 p. 53). This sense of nationalism is what made them such a formidable enemy.

Who were the combatants? The early 1960s were defined as the Kennedy years and were marked by a sense of idealism with service to country and mankind. At the same time, there was the Cuban Missile Crisis and considerable fear of Communist aggression that would do away with the American way of life. Furthermore, a generation had passed since World War II and images of war were not based on any reality, but derived mostly from movies that focused on heroes rather than death and destruction. Thus, there were many young men ready to serve their country, to fight the Communists. They were patriotic and not yet cynical about the war. That came later as the war dragged on and on and the lack of progress wore the troops and the country's morale down.

Where was the war fought? It was fought in Vietnam, a country unknown to many Americans at the time. It was a country with a climate and terrain very different from the U.S. Combatants fought in the jungle, heat and rain making combat conditions very difficult. They found parasites, mosquitoes and all sorts of other disease carrying organisms that took their toll on health. Furthermore, they found themselves fighting an enemy often in villages filled with residents who often gave support to the enemy, sometimes more out of fear than sympathy to the cause. Soldiers were often confused as to who was enemy and who was not.

How did the U.S. involvement in Vietnam come to an end? Finally, the American public had had enough and through peace marches and other means brought support of the war to an end. A peace agreement was reached in Paris in 1974, resulting in the withdrawal of Americans troops and ending the war without a victory. Following the withdrawal, the North Vietnamese Army launched a massive military campaign against the South Vietnamese Army. With the fall of Saigon in 1975, the communist takeover was complete. It is interesting to note that many non-Communist members of the South Vietnamese National Liberation Front who fought so hard against the Americans felt sold out after the war. Thousands of South Vietnamese were persecuted or sent to prison camps for repatriation. Those who were able to flee the country did so, and are now living as ex-patriots in France and other countries far away from the country they loved and fought so hard for. It seems that no one really won that war. The following words written in the dedication of the book *A Viet Cong Memoir* by Nhu Tang with Chanoff & Van Toai (1986) make the point:

> To my mother and father.
>
> And to my betrayed comrades,
>
> who believed they were
>
> sacrificing themselves for a
>
> humane liberation of their people.

---

The above socio/political/historical conditions represent the various levels of the **Conditional/Consequential Matrix** and answer to the questions, who, what, where, when, and why. Not every condition has been presented but there is sufficient background to understand why combatants were in Vietnam and why they faced the particular set of conditions or problems that they did. The problems included: idealism and youth, a sense of powerlessness, confusing rules of engagement, fighting someone else's war, entering the frontier between life and death, a determined and experienced enemy, inhospitable terrain, a war with no sense of direction with success measured by body counts, a culture divided, constant turnover of troops, a growing desire for revenge, a war that didn't stand still, a physical wearing down, and the potential for psychological and moral breakdown. All of these problems taken together characterize the "culture of war." If different decisions had been made in Washington and Hanoi, if different policies and rules of engagement had been established, if the enemy had not been so committed, if personal agendas had not colored judgment, then the war might not have occurred, or if it did it might have looked a lot different from the perspective of combatants. The next question from an analytic standpoint and the question directing the next round of subsequent theoretical sampling and analysis is, how did combatants overcome the obstacles or manage those problems to increase their chances of surviving?

### *Methodological Note*

I did not arrive at this point in the analysis easily. I had to do a lot of reading, thinking, and writing and rewriting of memos. One of the most salient points that I wish to convey is the importance of setting aside time to think. Young researchers are often constrained by time and work. They don't have the large grants that established researchers often have allowing them to put aside their regular teaching or other duties. I can't emphasize strongly enough that qualitative research can't be rushed. A qualitative researcher has to allow time for sensitivity to grow and for the evolution of thought to take place. I am distressed when I hear that students say that they don't like doing the detailed type of analysis presented in this book because they think it is too much work. They want shortcuts or easy ways of doing qualitative research. In the end, it's the "quality" that a researcher puts into qualitative work that sets it apart and that gives findings significance and freshness. When doing research, the researcher should never short change the data, the participants, the profession, or the self.

## Summary of Important Points

The purpose of this chapter was to identify the context in which the combat experience took place and in doing so open up and expand upon the "culture of war" concept and, at the same time, link that concept to that of "survival." The combat experience was characterized by a series of problems that could be traced back to the larger social/political/historical conditions that were particular to the Vietnam War. The problems that combatants faced posed physical, psychological, and moral risks. Opening up and elaborating upon the concept of "culture of war" brought the analysis forward, bringing it to the next logical step of theoretical sampling for the strategies used by combatants to handle the conditions or problems they faced in combat.

## Activities for Thinking, Writing, and Group Discussion

1. Using this chapter as a model, explain how knowing about context enriched understanding of the combat experience. If you are working with MAXQDA, make use of MAXMaps in order to reflect about the relationship between "context" and "experience."

2. Follow along with the analysis and try to discern how the author went about extending and elaborating our concept of the "culture of war."

3. Jot your ideas down and bring them to the group for discussion.

4. Add the new concepts from this chapter to your ongoing list of concepts. Did any new categories/themes emerge? If you work with MAXQDA, insert the new concept as codes into the code system. Discuss your codes with your group.

# 11

# Bringing Process Into the Analysis

*The plane banked and headed out over the China Sea, toward Okinawa, toward freedom from death's embrace. None of us was a hero. We would not return to cheering crowds, parades, and the pealing off of great cathedral bells. We had done nothing more than endure. We had survived, and that was our only victory. (Caputo, 1977, p. 320)*

**Table 11.1** Definition of Terms

*Process*: An ongoing flow of action/interaction/emotions occurring in response to events, problems, or as part of reaching a goal. The events, problems, and/or goals arise out of structural conditions and the actions/interactions/emotions that are taken in response lead to outcomes or consequences, in this case, survival. A change in structural conditions may call for adjustments in activities, interactions, and emotional responses to promote survival. Action/interaction/emotions may be strategic, routine, random, novel, automatic, and/or thoughtful.

## Introduction

Once again, I take up the case of Vietnam and move the analysis forward by exploring and elaborating in more depth the concept of "survival." This time I am specifically looking for the action/interaction/emotional responses made by combatants to survive their experience as combatants. The concept of "survival" appears in some form in every interview and memoir. Despite all the talk about patriotism, desire for adventure, wanting to be a pilot or marine, in the end it all comes down to wanting to "survive" deployment to the Vietnam War. Every memoir I read was essentially a "survival story" because the individual(s) lived to tell about it. Pilots talk about surviving the risks they took each time they went out on a mission. Ground troops talked about surviving battles. When I sat down to think about what I knew about "survival," I found that I had a great deal of data floating around in my head and thus was ready to think about bringing process into the analysis. But pulling all the data together as process isn't necessarily easy even for an experienced analyst.

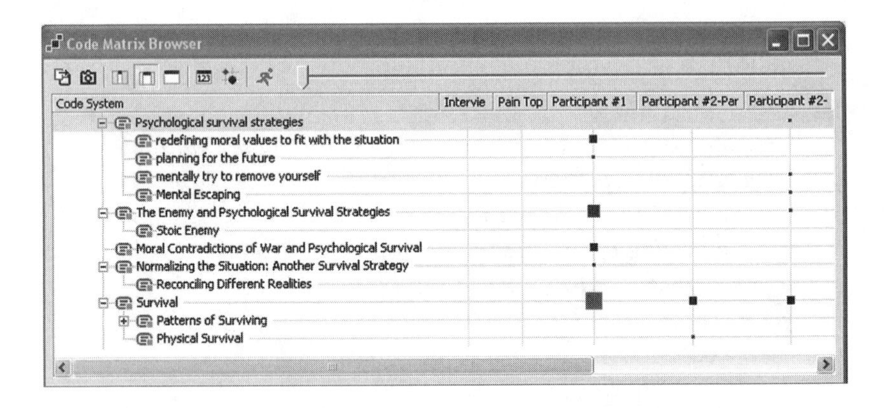


**Screenshot 10** This screenshot shows one of the visual tools of MAXQDA: The Code-Matrix-Browser allows a clear and differenciated overview of the topic "Survival" and its allocation in the different interviews: Participant #1 talks much more about this topic than Participant #2. Color and size of the squares indicate the number of coded segments; moving the mouse over a box shows the numbers of contained coded segments. The symbols may be changed from squares to numbers; the table can easily be exported to quantitative softwares like MS Excel or others.

## Surviving as a Process

The first step that I took in looking for process was to reread the memos from previous analyses. Based on my rereading I wrote a summary memo to keep the main issues in front of me while I worked with data. Usually, by the time a researcher has reached this point in the analysis, he or she has some intuitive sense of what the process might be. But it does take stepping back and thinking about the larger picture in order to hypothesize how to fit **process** (the action/interaction/emotions) with structure. Since this is not a research report but a book about method, I present my thoughts in a series of memos.

---

## Memo 1

### July 8, 2006

### Summary Memo

Combatants wanted to survive, but survival was complicated by the presence of physical, social, psychological, and moral risks. These are risks that presented themselves as a series of problems, or said another way, as a series of obstacles to survival. These problems or obstacles included youth and idealism that left combatants unprepared or "unseasoned" as combatants; a sense of powerlessness over the conditions they were forced to fight under, including rules of engagement that hampered the ability to defeat the enemy. Once in Vietnam, combatants found themselves in an inhospitable terrain fighting "someone else's war" against a determined enemy that put them in a frontier between life and death. After a time, combatants came to the conclusion that they were fighting a war with no sense of direction, and for which the only measure of success seemed to be "body counts." It was a guerilla war fought not only in the countryside but in villages with a culture so different that it made it difficult to discern friend from foe and to understand the behaviors and fears of civilians who themselves had been caught in the war. Increasing the risk factor was the constant turnover of troops with its influx of "novice" soldiers. Then there was the war that didn't stand still in the sense that it seemed to those who were doing the fighting that as more North Vietnamese regular army entered the South through the Ho Chi Minh trail, the battles became more brutal and intense, resulting in high casualty and death rate. Seeing their friends blown up or maimed created a lot of anger along with a desire for revenge. The constant stress of living and fighting under such conditions eventually led to a physical wearing down, demoralization and high potential for psychological and moral breakdown. Survival had its consequences however. A combatant might survive the physical risks of war but often carried home the psychological and moral scars of war. Some returning combatants developed post-traumatic stress disorder while other veterans, though not diagnosed, had difficulties adjusting to civilian life or turned to alcohol and/or drugs to drown out the nightmares and keep the ghost of war from surfacing.

What becomes noticeable in data is that just having a repertoire of survival strategies doesn't automatically guarantee survival. Just as important as having strategies is the ability to use them and use them effectively. For example, a combatant might be highly trained and have a gun ready to shoot at the enemy, but if he freezes under fire, his chances of surviving a battle is diminished. Thus there were a series of what might be called "intervening conditions" that entered into the survival picture, some enhanced the possibility of survival while others diminished the chances.

Some of the intervening conditions that enhanced a combatant's ability to make use of survival strategies included: being a seasoned soldier, having strong bonds with fellow soldiers, working as a member of a team, having good leadership, remaining focused during battle, having adequate and well-maintained resources and equipment, and having backup support during difficult times. Though each of these "intervening variables" are important, perhaps the two most important to survival were "becoming a seasoned soldier" and "having good leadership." Being a seasoned soldier enabled combatants to read the situation accurately and act quickly and decisively under duress. Having a good leader was essential because good leaders kept up morale, provided guidance and discipline (very important to maintain especially under conditions of duress), and coordinated action thus increasing the chances of individuals and the platoons surviving.

Obstacles to the ability to make use of protective strategies included being a "novice" at fighting, the inability to control fear and stress in the heat of battle, inept leadership, the lack of backup resources and a "wearing down" over time, especially if fatigue is accompanied by being demoralized and a temporary psychological and moral breakdown. Perhaps the most important of these obstacles include "inept leadership" and "wearing down." With wearing down physically, the ability to be alert to danger and respond quickly diminishes, while inept leadership can actually put a combatant in greater harm's way.

A person can't talk about survival in war without talking about the interactive or collective aspect of it. Survival not only necessitates individual action, it requires that persons work together as a team or act in behalf of others to enable others to survive. We see this often with heroes, medics, rescue pilots, and backup and support staff like doctors and nurses, engineers, technicians, and supply persons. Even in a battle, soldiers have to work together to fight off the enemy. Some soldiers, regardless of the danger, put themselves in the line of fire to save a wounded comrade. Medics, army and navy (navy medics were assigned to marine units as the marines didn't have medics of their own) went into battle zones with combatants and took the same, if not greater risks, than did the soldiers with whom they were marching. The enemy often used the opportunity provided by an injured soldier to shoot at others coming to his rescue. There were pilots whose main job in Vietnam was to fly behind enemy lines to rescue downed pilots or a platoon trapped behind enemy lines, or come to the scene of an active ground battle to provide backup assistance to ground troops that might have been caught in an ambush.

---

## *Methodological Note*

Since "being a seasoned soldier" and "good leadership" were so relevant to survival and "wearing down" a major hindrance, additional memos were written on these concepts.

---

**Memo 2**

**July 9, 2006**

**Being a Seasoned Combatant**

I want to look a little more closely at the notion of "being seasoned." Being a seasoned combatant didn't guarantee survival but it sure helped. What does it mean to "be seasoned"? Being seasoned is not necessarily equivalent to being an expert in the sense that an expert excels. A soldier may be seasoned at survival but not excel at waging war. Here are some of the properties of being a "seasoned combatant" that I've discovered so far.

It indicates that a combatant has acquired a set of strategies and skills that can be called upon to solve or manage problems that threaten survival. The strategies and skills come with exposure to situations and from handling difficult situations successfully. Becoming seasoned necessitates that a combatant has let go of fantasy images of war and accepts a more realistic viewpoint, i.e., "there is an enemy out there that wants to kill me and will if given the chance." It indicates that a combatant has acquired a respect for the enemy and his capabilities and knows how to take protective action as necessary. It also indicates that a combatant has sharpened his intuition about situations and learned to read the cues in the environment, and that he has developed routines for guarding his health and for maintaining the equipment that he is so dependent upon. It also means that a combatant has become physically and psychologically hardened to the extent that he can endure, and not become too emotional about death and destruction. Also, being seasoned means that a combatant has learned to harness his fears and continue to perform under duress. Finally, being seasoned indicates that a soldier has proven himself under fire and has come to view the self as a man. Becoming seasoned is a transformation that can happen very quickly, say, during the first battle. Or it can take time depending upon the individual and the types of experiences he encounters.

---

**Memo 3**

**July 8, 2006**

**Wearing Down**

There is another concept that seems to be directly related to survival. It is the concept of "wearing down" that can lead to physical and emotional exhaustion and moral bankruptcy. "Wearing down" happens over time. It appears to be a response to constant exposure to conditions such as fear, stress, environmental elements, enemy fire, and continued exposure to battle conditions. It is characterized by uncontrollable anger at the enemy, the desire for revenge, and a sense of being demoralized by the conditions. It also includes overwhelming fatigue that is unrelieved by rest, the inability to differentiate right from wrong, fear to a degree that it interferes with functioning, not caring anymore what happens, being physically plagued by a variety of ailments like diarrhea, boils, and having that vacant stare or far away look in the eyes.

**Memo 4**

**July 8, 2006**

**Leadership**

Though leadership has not been mentioned much in memos, from reading the memoirs it appears to be a very significant factor in survival. Good leaders maintain discipline and order. They keep the morale of their troops up. They foster teamwork. They know how their personnel react under pressure of combat. They know whom they can trust to lead and support others, who they can't trust, and how to problem solve and get their men out of difficult situations with minimum losses. They can make things happen and make their men respect and follow them. Inept leaders make mistakes that put their men at risk. They don't have the ability to maintain order or discipline or to keep up the morale of their troops. With inept leadership there is likely to be chaos rather than teamwork under pressure, increasing the risks to individuals and the group.

## *Methodological Note*

For persons who are accustomed to thinking of process in developmental terms, or in terms of "psychosocial process," the following section might seem strange because, as stated above, the process of "surviving" here does not follow a developmental form. I wasn't studying "becoming a survivor" or even "becoming a seasoned soldier." The main issue or theme that kept coming out in the data was "survival" and surviving was a constant day-to-day affair. Each set of problems or obstacles that a combatant encountered in Vietnam had to be solved or overcome through individualized or collective action in order to increase the chances of coming out of Vietnam alive.

---

**Memo 5**

**July 8, 2006**

**Patterns of Surviving**

After much thought about how I might conceptualize the process as it pertains to surviving, I decided upon the use of patterns of "surviving." By patterns here I mean both a combination of routine and novel ways of acting/interacting/emoting in response to the various problems that combatants encountered in Vietnam and that threatened their ability to survive. Some of these patterns denote routine ways of acting. Obviously, even in war, many routines are established for handling the known risks.

---

As Strauss (1993) states:

> Repetitive goal-directed action requires a patterning of action that does not need to be invented on the spot each time that a person or collectivity acts. During the course of encountering situations, unless an actor could quickly classify most with a standard definition, he or she would unquestionably become exhausted. (p. 195)

Yet, since planned action can be interrupted by contingency (an unplanned-for situation or change in conditions) or because patterned ways of acting might not solve a particular problem, novel ways of acting are called for. Strauss goes on to say:

> When the situation can be defined as slightly different, novel, or unusual, then although appropriate patterns of routine action are called upon, these will be supplemented with new actions or a slight adaptation of the routine. Even in the most revolutionary of actions, the repertoire of routines does not vanish; at least a lot it becomes utilized in combination with the new. (p. 195)

Combatants in Vietnam made use of both routine and novel strategies to increase the chances of survival. Two of the patterns described next denote routine ways of acting, while the two other patterns describe more novel or adaptive responses to contingency. The first pattern is the "routine and individualized pattern." This pattern includes strategies taken in response to the "everyday problems" posed by virtue of being in Vietnam and in a war zone. The strategies were carried out by the individual and were part of the daily routines combatants used to enhance the chances of physical survival and to help maintain their psychological and moral equilibrium.

The next pattern of "survival strategies" was designated as the "institutionalized pattern." Many of the risks of war are known and can be planned for and handled by institutionalized procedures and arrangements that involve persons working together for the good of the whole.

The third pattern of survival strategies is a contingency response pattern called the "rescue pattern." Rescue operations were dangerous operations that often did not go according to plan. Rescue operations were put together to remove individuals or groups from situations of very high risk and required the coordinated effort of many individuals working together in novel ways to carry out the rescue.

The fourth pattern is called the "escape and evasion" pattern and pertains to individuals or small groups purposely working behind enemy lines or trapped unwittingly in an ambush or behind enemy lines. There is no group available to rescue them. To survive, combatants must rely on their cunning and novel ways of acting. However, notice in the example provided below how past experience and training contribute to the ability to get out of a situation alive.

Not every strategy will be mentioned under each of these patterns. What is important to note is that there are certain problems associated with being in a war zone and that combatants take an active role in overcoming those problems and thus promoting their survival through the use of various physical, psychological, and moral protective strategies.

---

## Memo 6

**July 8, 2006**

### Individualized and Routine Patterns of Surviving

A war zone is dangerous. A combatant never knows when or where the next enemy attack will take place, if he will survive the next battle or mission, or what event might trigger a moral, psychological or emotional response that could lead to his demise. Though a soldier can become "seasoned" with time and experience in war, there is always the potential for "wearing down" over time because the risks are constant and the emotional and moral demands high. Even when on base, a zone of relative safety, combatants could never really "feel safe." There were nightly raids, especially on small bases by the enemy. There were snipers outside the base perimeters just waiting to take a shot. There were risks from the rigors of the environment in the form of fungus, mosquitoes, contaminated water, heat, rain, humidity and so on. There were patrols to go out on, nights spent out in the bush, mined paths, and jungle growths to hack through. And when a buddy was killed in front of you or failed to return from a mission it wasn't always easy to cope with the loss.

"Overcoming the problems of inexperience," of fighting against a determined enemy, an inhospitable environment, confusing rules of engagement, a sense of powerlessness, and so on (all of the problems that defined the "culture of war") called for "individualized and routine" patterns of strategic action/interaction/emotional response aimed at handling the problems and reducing the risks. These strategies were characterized by their repetitive nature taken over the course of deployment and by the fact that though they were often institutionalized, the strategies were also individualized as to when, where, and how they were carried out. Another characteristic of these strategies was that they were aimed at maintaining psychological health and moral integrity as well as the physical wellness.

Among the strategies and tactics aimed at reducing the physical threats of living in a hot humid climate were the taking of malaria pills, the use iodine pills to purify contaminated water, regularly checking the feet for fungus, carrying extra socks when going out in the bush, reporting physical problems to the medics before they became debilitating, keeping the canteen filled with clean water, maintaining adequate hydration, wearing a helmet, wearing clothing to protect against the sun, maintaining military equipment in proper order so that it works when you need it, always being on guard, and responding quickly to calls to take cover. When out on patrol or when flying a combat mission, combatants had to be especially on guard, reading cues that would indicate danger, not only watching out for the self but also for fellow soldiers and other pilots (if a combat pilot).

To handle the psychological stress and strain associated with war there were other strategies. Positive strategies included taking pride in winning a battle or accomplishing a mission and forming bonds with peers. Not all strategies were constructive, however. Others, though perhaps psychologically helpful, were detrimental to physical health. For example, some combatants turned to excessive drug use or alcohol to numb the realities of war. These activities may have blotted out "reality" temporarily but tended to be destructive and addictive. Though not always used excessively, alcohol was often at the center of socializing because of its relaxing effects and social associations. Memoirs mention the use of alcohol when relaxing at beaches, in the "hooches" (living quarters for officers) and at parties. A tradition of pilots who were rescued after being shot down was to buy a round of drinks for the men that rescued them. Pilots that completed a tour of duty celebrated their survival with parties that usually included alcohol. On leave, combatants often went into the nearest town to party and to whore. They looked to any activity that might "normalize" their lives a little and relieve the stress.

Not every soldier or pilot drank, used drugs, or whored. Many used the off-duty time to engage in sport activities, write letters home, engage in social acts like playing cards, read or study, do charity work in villages, or just rest up before the next mission. Other psychological strategies used included distancing of the self from Vietnam by daydreaming about a girl at home or planning what to do with life after Vietnam. With time, many combatants were able to harden themselves to the sights and sounds of war or put the worst aspects of war into the back of their mind at least "temporarily." A frequently used strategy was to demean the enemy, by calling them "gooks" and using other derogatory terms that somehow psychologically lessened the threat that the enemy posed.

To handle the ongoing moral dilemmas, the "nips at conscience," and the atrocities that combatants witnessed, combatants often reframed things that happened, referring to negative events as "the nature of war." There were religious services that combatants could attend to find renewal and solace. To maintain their sense of humanity, some of the medical personnel and some combatants worked in orphanages, shared their C-rations with the Vietnamese villagers and soldiers from the South, and reached out to people in the villages engaging in gestures of friendship.

---

## Memo 7

**July 8, 2006**

**Institutional Patterns of Surviving**

Over the years the military established "institutionalized" strategies for managing the increased risks associated with combat. These strategies often take the form of policies and procedures and are carried out through arrangements involving a division of labor between different levels and types of personnel. Considerable training is often involved in preparing persons to carry out these strategies and they are time and place specific.

Though there are many examples of patterns of "institutional" strategies carried out on an ongoing basis throughout the war, the complicated nature of these strategies can be seen in the activities that go on each time a pilot is sent out on a mission. Notice the "team nature" of the work and the necessity of an "alignment of action" if a mission is to be carried out successfully.

To prepare to go into an inhospitable terrain and engage in combat with a determined enemy under rules of engagement established by persons a long way from Vietnam, a pilot had to undergo several years of training in high performance airplanes long before going to Vietnam. Pilots were also required to attend jungle survival school, just in case they are shot down. Then when the time came for the mission, considerable planning went into every mission, from the top military officials involved down to the men who work on the airplanes. For the pilot, the day began before dawn with a briefing on the details of who would be flying which airplane, where, in which position, to carry out what activities. After breakfast, came "suiting up." Suiting up was necessary because of the risks presented by flying at high altitudes and at high speeds (Trotti, 1984, pp. 22–25). The first piece of clothing a pilot had to put on is the flight suit. This was made of a type of material designed to withstand life in the jungle should a pilot be shot down. On top of the flight suit, went the g-suit, an inflatable girdle that covers the vulnerable areas of abdomen, thigh, and calves and designed to prevent black outs. After the g-suit, the pilot put on a harness designed to keep him from being thrown around the cockpit during maneuvers and to spread the shock throughout the body should a pilot have to eject. Finally, the pilot added the survival vest. Its purpose was to provide the pilot with those objects needed to survive if he is shot down. Among the objects contained in the vest was a gun, knife, code book, maps, shark repellant, repelling ropes, radio transmitters, canteen of water, a pound of rice, fishing line, and even morphine for pain.

Only after putting on all of those articles of clothing was the pilot ready to get into the plane. Once in the plane, the use of institutionalized procedures continued. First, the pilot had to be strapped into his seat with leg restraints to keep the legs tight against the seat in case of ejection. Next the pilot donned an oxygen mask and helmet, and after that he plugged in the g-suit. Of course, between missions, the plane had its institutionalized series of checks. Planes were checked over completely after each mission for gun shot holes and other damage. Each plane had an assigned crew whose job it was to maintain the plane's flying ability and to make certain that the plane is fueled and ready for take off at any time. Then there were the step-by-step procedures that each pilot had to carry out before and during take-off. The purpose of the procedures was to ensure that the flight controls and communication systems were working.

Pilots in Vietnam often flew with a lead pilot and a wingman. When flying over enemy territory a pilot had to focus on the target therefore couldn't be watching out for enemy fire. The wingman's role was to cover the pilot while he dropped his bombs. Then the lead pilot covered the wingman while the wingman dropped his load of bombs. During a mission, air coordination is essential and there are procedures for linking up with planes from different bases (each having a different role to play in a mission) and for refueling. If a pilot's schedule is off the time, delay could place all of the pilots and the mission in jeopardy. Fighter planes use a lot of fuel. Pilots usually have to refuel once or twice as part of a mission and this too had its set of safety procedures. Upon reaching the target area, pilots use maneuvering tactics like rolling, banking, and what pilots called "jinking" or changing altitude and direction every few minutes to avoid being hit by enemy gunfire. All of the above action/interaction is necessary to carry out one mission. As Trotti said after completing a mission to Hanoi where one airplane carrying two persons was hit and crashed (the pilots were rescued):

> That was it. A hundred-man hours of labor for fifteen seconds over the target. We had expended 30,000 pounds of fuel and $2.5 million of airplane to put six tons of explosives on ten acres of real estate, and that's probably as much as we'll even know as to the value received for the price paid. (Trotti, 1984, pp. 96–97)

---

## Memo 8

**July 9, 2006**

**Patterns of Rescue Operations**

Rescue operations were characterized by their intensity and the need for alignment of action and were aimed at providing air cover to a group of combatants caught in an ambush, or to retrieve wounded soldiers from a battle area or bring back pilots shot down behind enemy lives. Though there were institutionalized procedures established for rescue, often rescues were complicated by contingency calling for "on the spot problem solving" if survival was to occur. Perhaps one of the most dramatic patterns of rescue operation that occurred in Vietnam were those that involved the retrieval of American pilots whose planes went down behind enemy lines. In these situations the rescue operations, though planned, often didn't go according to plan. More often than not, rescues were complicated by contingencies. Defining features of rescues were the need for "coordinated" types of novel action. Rescue operations required highly trained individuals problem solving together. Usually, there were considerable risks to all parties involved, the rescuers as well as the person(s) being rescued. The sequence of activities and coordinated interactive nature of rescue operations can be seen in the following example. Notice the variety of problems the rescuers encountered and how they used "on the spot" problem solving and novel ways of acting.

During the Vietnam War there were special teams established to carry out rescues. One team was established to rescue pilots shot down within enemy lines—usually over North Vietnam, Cambodia, or Laos. One aircraft defined especially for rescue was the Skyraider, a single-engine, propeller-driven WW II vintage airplane. The usual procedure was for two Skyraiders to respond to a call of a downed pilot, the senior Skyraider pilot (senior in terms of service in Vietnam) and his wingman. The Skyraider pilots did not do the actual "rescuing," that was left to helicopters. Rather, a Skyraider's role was to cover the downed pilot and rescue helicopter so that the rescue could be carried out. Air force pilots who flew the Skyraiders were known as Sandys, as their call sign was "Sandy." Another aircraft involved in rescue was the helicopter known as the Jolly Greens. Its crewmembers were known as "Jollys."

Marrett (2003, pp. 156–161), a Skyraider pilot, tells the following rescue story. A phantom F-4 jet was shot down over the Ho Chi Minh Trail in Laos during a bombing mission. There were two pilots in the plane. One pilot was able to eject, while the other pilot, the senior pilot, went down with the plane. The surviving pilot suffered multiple fractures during the ejection but remained sufficiently conscious to activate his transmitter and notify the home base of his position.

Soon after the F-4 was shot down, two Skyraider pilots and a helicopter crew were dispatched to the rescue. As the planes approached the target area one of the rescue Skyraiders was shot down. The pilot managed to eject safely from his plane. Now there were two live pilots down in enemy territory, the original pilot and the rescue pilot. Since it was getting dark, rescue operations were suspended until morning.

At daybreak, two more Skyraiders and Jollys were dispatched to the area to complete the rescue. The badly hurt pilot was the first to be rescued. The other pilot shot down remained in a tree, surrounded, but unseen, by a group of North Vietnamese soldiers. After locating the second pilot shot down, but before the rescue could be completed, another Skyraider was shot down. The Skyraider pilot died in the crash. The pilot shot down earlier had to remain undetected in enemy territory while still waiting for rescue. A third rescue team was sent to the area, located the pilot, and prepared to descend for the rescue. While the pilot was being hoisted up into the helicopter, small gunfire erupted. One of the helicopter crewmembers moved to the open back of the helicopter where he could get a better shot at the enemy. As the helicopter was lifting, the helicopter gunner assigned to defend the helicopter from incoming fire, and who was shooting at the enemy from an open area in the rear of the helicopter, was badly wounded in the leg and the helicopter itself was hit. Attention was turned from the pilot being rescued, who by now was in the plane, to caring for the wounded gunner who was bleeding profusely. Meanwhile, the helicopter crash-landed. Miraculously, all six men aboard the helicopter including the injured man and the rescued pilot survived the crash. (This was the second crash in two days for the pilot.) Another helicopter was dispatched to rescue the six downed men and eventually all were returned safely to the base.

Not only did this survival require many resources, it took a lot of readjusting of action or contingency management to finally carry it off. When one plan failed, another backup plan had to be put into place immediately. The hallmark of rescue operations is the coordination of equipment, personnel, and communications and most of all the use of novel strategic action/interaction to handle problems as they arise.

## Memo 9

**July 9, 2006**

**Patterns of Escape and Evasion**

There were times in Vietnam when combatants were in "extreme danger" such as when they were sent on a mission behind enemy lines, or were caught in an ambush, or an individual found himself behind enemy lines and no one knew quite where, or even if, he was still alive. The defining characteristic of the escape and evade pattern is the drawing upon past experience and knowledge to come up with novel strategies to handle conditions or risks as they arise. Unlike rescue operations, where others do the rescuing, in escape and evade it is the individual or group that must save himself or themselves. Here is just one example of decisive and extraordinary strategic action taken to escape and evade the enemy.

This is the story of Specialist 4 James Young of Alpha Company, 1st Battalion, 5th Calvary as told by Moore and Galloway (1992). While out on patrol, a group of American soldiers were ambushed by the enemy. A fierce battle ensued. During the battle, it was noted that an American machine gun had been taken by the enemy and was being used to shoot at American forces. Specialist Young volunteered to leave the safety of the group and locate the position of the gun so that an air strike could be called. While making his way through the tall grass, Young was hit by a bullet in the head. Wounded, he started to return to his unit only to discover during the few moments that he had been away from the unit, that the enemy had surrounded the American troops. Young found himself in a situation where he was wounded and cut off from his unit. All he had on his person were a few rounds of ammunition, a few hand grenades, a rifle, two canteens, and a small mirror. Unsure of where he was or where he should go, Young tried to recall where his unit had come from and decided moved back towards that point. To protect himself from enemy fire while running away, Young shot into the trees towards snipers and threw his few hand grenades in the grass dogging and zigzagging to avoid enemy fire. He could hear the enemy pursuing him. To outwit his pursuers Young drew upon the skills he had learned while hunting as a young boy. He came to a stream and waded upstream. He filled his canteens in the stream for later use and drank all that he water that he could hold. He left the stream in a rocky area so he wouldn't leave a trail that others could follow. He went into a valley where he would have a clear view of the trail behind him. It was getting dark. Young took refuge behind some rocks. There he took out his diary and wrote a note to his family with the hope that they would find the message if he did not return. By now, his head was hurting badly and he was vomiting every time he tried to drink.

All through the night, Young lay awake sleeping for only a few moments at a time. Toward morning he heard the sound of helicopters overhead and knew they must be American. He tried to signal the helicopters using his hand mirror but was unsuccessful. To aid his escape, he followed their flight path, a strategy that took him close to American lines. There, the battle was still raging. Young hid behind a large log and waited for the battle to be over. American planes started dropping bombs into the area to dispel the enemy. Fearful that he would be injured by friendly fire, Young looked for safety in the tall grass. By now it was dark again. Young knew that if he crossed the grass separating him from his troops while it was dark, he would be shot by either his own men or by the enemy. So he spent another night in the bush. He covered himself with brush to stay warm. The fighting continued through the night. At daybreak, Young cautiously approached the perimeter of American forces. He made it back to the safety of American lines just before the last American troops were airlifted from the area (Galloway & Moore, 1992, pp. 318–321).

In order to survive, Young called upon the repertoire of survival skills he had learned as a boy, adapting them to the terrain and problems in Vietnam. His skill as well as strong motivation to survive enabled him to escape and evade the enemy until he could make it make to American lines.

### *Methodological Note*

I could stop here with the analysis. I have brought process and context into the analysis and certainly this would enough for some research projects. But I, myself, as an analyst, am not satisfied. There are categories such as the "change in self" and "images of war" and "homecoming" not yet accounted for. That means more analytic work is necessary to pull this whole story together. The question concerning me now is, how do I put this all together? Chapter 12 will focus on integration.

## Summary of Important Points

This chapter explored the concept of survival, and in doing so, brought process and structure into the analysis. The researcher looked at patterns of ongoing strategic action/interaction/emotion in order to discover how combatants managed to overcome the problems or threats to survival they encountered. Patterns are one way of conceptualizing process and putting it together with structure. The patterns emerged from data but had to be recognized by the researcher. There are different ways of conceptualizing process. A researcher might think of process in terms of phases, stages, levels, degrees, progress toward a goal, or sequences of action. In this case, process was much more of an ongoing day-to-day, in fact at times minute-to-minute, activity aimed at increasing the chances of surviving. There is no magic trick in identifying process in data. The researcher has to study the memos and raw data and look for how the main issues or problems of the research are handled or managed over time. Once analysts have uncovered process in data, they are able to paint conceptual pictures that add to the understanding of an experience.

## Activities for Thinking, Writing, and Group Discussion

1. From the discussion above, think about what you've learned about process. Do you have any additional thoughts on the topic?
2. Can you see how process enriches the analysis and increases understanding? If you work with MAXQDA, make use of MAXMaps in order to visualize the impacts of process.
3. Can you think of any other ways of putting process and context together?
4. Add any new concepts from this chapter to your list of evolving codes. Think of how you might begin to group concepts into higher-level categories. Perhaps you can come up with a scheme different from mine. If you work with MAXQDA, use the features of copying and moving codes to rearrange your code system. Switch the view on your code system from the hierarchy to the linear listing. Discuss the use of the option to leave the hierarchical structure of the code system in the group.
5. From your reading of the data presented in the various chapters, do you see any other patterns that this researcher might have missed? If you work with MAXQDA, make use of the lexical search to support your search for pattern.

# 12

# Integrating Categories

*What remained was sorrow, the immense sorrow, the sorrow of having survived. The sorrow of war. (Ninh, 1993, p. 192)*

**Table 12.1** Definition of Terms

*Integration:* The process of linking categories around a core category and refining and trimming the resulting theoretical construction.

*Negative Case*: Though a researcher can continue to collect data searching for the negative case, finding that negative case does not necessarily negate the analyst's conceptualization. Often the negative case represents a dimensional extreme or variation on the conceptualization of data.

*Theoretical Saturation*: The point in analysis when all categories are well developed in terms of properties, dimensions, and variations. Further data gathering and analysis add little new to the conceptualization, though variations can always be discovered.

## Introduction

Not every researcher is interested in theory development, as we have been stating throughout this book. However, for some researchers, including myself, theory building remains an important goal. This chapter is for persons interested in theory building, though other readers might find it interesting for the following reason. Though the analysis of combatants in Vietnam has progressed and an interesting story is emerging complete with context and process, to me the analysis remains incomplete, and in fact, unfinished. Perhaps it is my training and bias as a grounded theory researcher. Perhaps it is my need to have all the various analytic treads come together, but I feel compelled to go on with the analysis. I want to reach the point of final **integration**. The previous chapter ended with the comment that though survival seems to be the main theme or phenomenon to be derived from analysis of the study, there remains that nagging feeling that it does not tell the whole story. Something seems to be missing from that explanation. The purpose of this chapter is to search for that missing piece, and in doing so, pull all of the research threads together to construct a plausible explanatory framework about the experience of combatants in Vietnam. Again, the format used to bring the study its logical conclusion is memos. Only note that here the memos are more like summary than exploratory memos. They pull together many different ideas.

---

### Integrative Memo 1

**July 23, 2005**

**The Descriptive Story**

I've been thinking a lot lately about the central notion of the study. The main theme that keeps coming through to me in the data is "survival." I guess if I asked the question of what the Vietnam experience was like for combatants, I could reply that after being in Vietnam for a while and especially after engaging the enemy in combat, it came down to the basic story being "one of survival." Everyone who returned from Vietnam alive was a survivor and the major goal of combatants there was to survive. But surviving, though central to this story, doesn't tell the whole story. There are the changes in "the self" and changes in "images of war" and "homecoming" that somehow contributed to that survival or were a consequence of it. I mean, people definitely changed as a result of going to war. They were not the same people when they returned home that they were before. And images of war based on civilian understandings underwent profound change, changes that have endured even years later. Going back and rereading all the memos, it seems to me that there is something deeper than just "surviving" going on here. By deeper, I mean that if someone climbs Mt. Everest, yes the climber wants to survive, but along with surviving (even if it is not overtly articulated) is the notion of getting to the top in order to "fulfill a dream" or "prove something." In the data about the Vietnam War, survival is paramount. It comes down to something that basic. But survival seems to depend upon something deeper than just drawing upon survival strategies. Young men must be able to make the necessary adjustment in self, and in images of war, so that they can face the risks of war—that is, shoot the enemy when necessary, but at the same time draw upon inner courage and strength necessary to maintain psychological equilibrium and moral integrity.

Young men go to war out of a civilian culture with one set of standards, values, and activities. That was the "reality" that made up their lives. Once in Vietnam, the young soldiers are forced into the role of combatants where they are confronted with a whole different "reality." This "reality" is comprised of a series of problems or situations fraught with physical, psychological, and moral risks; and depending upon how they perceive or define the situations, they find themselves in calls for a whole new set of standards, values, and activities. In order to define the problems they encounter correctly and take appropriate action, they have to reconcile these two different realities. If combatants can't make the adjustment from civilian life to being in a war zone, then their chances of survival are decreased because they can't make use of the strategies needed to survive. However, men can also go too far and create their own "reality" in war as described in the book *Tiger Force* (Sallah & Weiss, 2006). Once a tour of duty is completed, the combatants, now veterans of war, have to put aside the experiences of Vietnam and adjust to civilian life once again. But, having undergone the experience of Vietnam they are different people from who they were when they first left home.

Moreover, veterans of Vietnam returned to a society different than the one they left behind a year or so ago. While they were changing as a result of being at war, society was also changing as a result of the war, developing new attitudes about war, country, and patriotism. I guess what I'm getting at is that my core concept needs to include the notion of physical, psychological and moral survival and be able to take the combatant from civilian life, through deployment in Vietnam, and then back again. It also needs to include the ability to reconcile all the different realities that come together to create the total experience of any Veteran.

I have several major themes to choose from–"the changing self," "shifting images of war," "culture of war," "homecoming"—none of which quite do it. However, in one of my early memos, a memo written during analysis of Participant #1, I had a concept, which I think fits what I am saying. The concept was "reconciling different realities." To me, this concept really puts emphasis on the active component in survival and the profound changes that combatants must undergo to survive. Other concepts such as "the changing self" and "shifting images of war" and the "war experience" can all be integrated under that concept. What is interesting to me is that "reconciling" necessitates by its very definition making changes in the self and shifting images of war. I might even dare to say that the "survival" has to do with how the different "realities—before, during, and after and war are reconciled." "Reality" being not the events of war per se but how those events are perceived and defined by the many different persons involved. The concept I am choosing as my core category is "survival: reconciling multiple realities." I think that this expanded concept explains the total experience focusing on the before, during, and after of going to Vietnam and being a combatant. It meets all of the criteria for a core category as spelled out in Strauss (1987, p. 36) and certainly has implications for other studies of persons who have to adjust to difficult situations in their lives. I think what makes war rather unique is that there are not only physical risks but psychological and moral risks as well.

---

### *Methodological Note*

It is important to make a comment here. A core category represents a phenomenon, the main theme of the research, though some researchers think only in terms of "basic social/psychological process." Thinking in these terms is fine for sociologists or psychologists, but does not fit for persons interested in educational, legal, business management, or architectural issues. So I hate to put emphasis on the core category being primarily a basic social process. One early reviewer of this text makes the point that the core category represents the main theme or phenomenon of the study, while the basic social process or whatever the process is can be found embedded in that main theme. There is the phenomenon like survival, then the process (whatever kind of process–legal, educational, psychosocial, etc.—that is, appropriate to the study). It in turn explains the phenomenon, and describes the how. Events, problems, and situations happen, and people respond to these. How a researcher defines the core category depends upon how he or she wants to place the emphasis.

Getting on with this study, how did I get to the point of integration? I must say that I put a lot of thought into the study and paid attention to my gut feelings. I wasn't satisfied with "surviving" alone to explain what I read in the Vietnam stories because it was too oriented to physical survival. I read the memos again and again. I sat and thought. I walked and thought. I kept coming back to the notion that physical survival didn't tell the whole story. Somehow I needed a way to bring out the physical, psychological, social, and moral problems inherent in war and the ways that persons respond to these both during and after returning from war. Once I hit upon the earlier concept of "survival: reconciling multiple realities," I knew that I had found the answer to that something. The core concept and other concepts come from data but "theory" doesn't just build itself; in the end, it is a construction built by the analyst from data provided by participants.

## Moving From Description to Conceptualization

Now that I'm satisfied that I have a core category that offers one plausible explanation about the survival experience of combatants in Vietnam, I am ready to see how the other categories can be linked up to it. I'll do that by retelling the Vietnam story, this time using the major categories and subcategories as the foundation for the theoretical structure.

---

## Integrative Memo 2

### July 24, 2005

### The Analytic Story

The "war experience" can be thought of as a "trajectory," or a course that extends over time. Entering into that, war experiences are "images of war" that begin long before an individual actually goes to war. Persons pick up "attitudes" and form "images" based on what they are told and see in their families, their communities, the media including movies, and from any contact they may have had with military personnel. Then (transitional hypothesis) when young men join the military they begin to formulate new but not quite "realistic" images of war based on their training in "boot camp." Though "boot camp" may be difficult and "war-like" it is not "war." It is not until combatants actually got to Vietnam and experienced actual combat that the "reality" of what war means set in. There is also the "self," the youth who entered the military. He came with all of the experiences of his past life, good and bad. Some came from middle-class patriotic and stable families. Others came from poor or difficult backgrounds and some were psychologically fragile before going to war. Most were inexperienced with life beyond home and family and ignorant about Vietnam and its people. Yet being able to make the transition to being in a "war zone" and adapting to an "inhospitable environment" is necessary if young men are going to have a chance at survival. It's not that they have to leave everything about themselves behind, but that once they arrive in Vietnam it is necessary to develop the skills, attitudes, experience, mental soundness, and moral strength to do what is necessary to survive. Surviving the "war experience" is about "reconciling the multiple realities" between a "civilian culture" and a "war culture," then adapting once again to a civilian culture upon "homecoming."

Reconciling is not a one-time event but something that goes on over and over again over time even into the present. "Reconciling" includes:

1. *Changing the self.* This encompasses going from "novice" to "seasoned combatant" by developing the technical military skills, emotional hardening, moral strength, and social resources necessary to overcome the "problems and risks associated with war." Just as there are changes in self that must occur in order to survive the war experience, there are consequences to the self as a result of having survived. One of the consequences to self from undergoing the war experience is the "growing older" than one's chronological years, learning about the self, learning new technical and social skills, maturing in

the sense of being able to take responsibility for self and others, and responsibility for any actions that one takes. Other possible consequences to the changes that occur to "the self" from going to war is coming home with "having unresolved anger" and/or "feeling alienated," from a society that has no idea what the combatant has been through.

2. *Shifting images of war.* Once in Vietnam, combatants had to let go of the "romantic" images of war derived from movies and replace the images with a more "realistic" view of what war is all about. Realistic images are defined as images that acknowledge "the enemy" as foe, and images that recognize and respect "leadership" and "rules of engagement" even if one disagrees with them. Other war images that must change include the notion of "sanctioned killing" but enough conscience must remain so that the combatant doesn't go overboard and kill indiscriminately. A combatant must accept that there are "moral contradictions" in war and that there is a thin line between right and wrong, yet also accept that one must not cross that line and let go of all morality. Having strong leadership is necessary to keep young men under control. Everyone's "reality" in Vietnam was different as "reality" depends upon who the person is, past experiences, psychological make-up, role in the war, and perception of events while there. Some of the consequences of having survived the war experience include bringing home the "residual anger" and "images of the horrors of war," including the "ghosts of war." Upon "homecoming" what seemed so "right" when in Vietnam often seemed so "wrong " when confronted with civilian protesters, and anger at those who were left behind and who failed to understand the "nature of war" and the things that can go wrong. The "reality of war" seen from the viewpoint of those who remained at home was completely different "reality" of those who fought in Vietnam. The clash of perspectives that many Veterans experienced on "homecoming" leads to "disillusionment" with war and government and with fellow members of society. To combatants, only those who have been there and who have experienced what they have could possibly understand.

3. *Culture of war.* This provided the context in which each individual soldier's, as well as the collective war experience, took place. The culture of war was derived from a combination of political, social, and historical conditions that combined to create the problems or obstacles to survival that combatants faced on the battlefield. In addition to the culture of war in Vietnam there was the "culture of war at home" that went from tacit support for the war in the beginning to a peace movement and a country divided about the war's morality. The collective move had considerable implications for returning soldiers. The culture of war was located between two other cultures, the "civilian culture" before going to war and the "civilian culture" after returning from war, the latter having changed over time also. The war experience was embedded in both of these, seen from the eyes of the novice upon arriving in Vietnam and seen from the perspective of the veteran after leaving Vietnam.

4. *Survival strategies.* These provided the active/interactive/emotional response to the obstacles that stood in the way of survival. But making use of

those strategies depended upon combatants' abilities to "reconcile their selves and images of war to the realities of war." The survival strategies used by combatants were matched to the problems at hand and can be classified into the "personal," "institutional," "collective or rescue," and "escape and evade" patterns. Though chance played a major role in who did and did not return, the active components of wanting to survive and a willingness to do what was necessary were essential to have a chance at survival and returning home. Sometimes the strategies to "rest" and "renew" themselves physically, psychologically, and morally didn't work and combatants began to "wear down" or become exceedingly fatigued and demoralized by the unending stresses and strains of war.

Here a researcher could write a hypothesis: As part of the experience of war, combatants who manage to survive physically, remain emotionally intact, and retain their moral integrity had to "reconcile themselves to the realities of war," that is, undergo profound physical, psychological, and moral changes in self and images of war going from novices to seasoned soldiers and utilizing strategies that would not only protect them from physical harm but enable them to renew themselves physically, psychologically and morally.

5. *Homecoming*. This meant leaving the war zone. Again, the combatants who are now veterans had to make drastic changes in self and images of war to fit with the "realities" of civilian life. "Reconciling" to civilian life was not easy for many veterans because they "carried back home the burdens of war" which manifested themselves in the form of "residual anger," "twinges at conscience," "ghosts that haunt," and "sense of alienation from society," along with "disillusionment" with government, society, and even in some cases with self, and in many cases considerable "anger." The burdens veterans carried upon returning home called for "healing." "Healing" requires letting go of anger, guilt, remorse, and fear with the ongoing support of friends and family and therapy groups. "Healing" after Vietnam was made more difficult by a society that had lost interest and support for the Vietnam War. Many veterans upon returning put up a "wall of silence" within their own minds and between themselves and others to protect themselves from the contradictions they experience coming from a war reality to a civilian one. They buried the war experience in the recesses of their minds. With time, and with a lot of support, some veterans, while never forgetting, were able to "reconcile" themselves in the present with the past and readjust to civilian life. They were able to go on with their lives and be successful, having matured by working through of the negative aspects of the war experience. Other persons had more difficulty with "healing." Some maintain their "wall of silence" even today. They remain angry, disillusioned, and alienated to various extents. Some have developed post-traumatic stress disorder, and some have turned to a life of alcohol and drugs to help them blot out the horrors they experienced.

In conclusion, the war experience can be described as an intense, powerful, life-changing "survival" event that encompassed many "different realities" that had to be "reconciled, if a combatant was to survive." The war didn't necessarily end when the combatant returned from Vietnam but continued to haunt some veterans for the remainder of their lives. Even many of the veterans who were not combatants per se but who served in supportive roles like nurses, doctors, and engineers, etc. were profoundly impacted in similar way as a result of going to war.

While the wall memorial (a wall of remembrance) was erected in memory of those persons who died, it also carries within it the pain of those who survived. I remember being so struck in those early interviews by the "wall of silence" that the guys put up around themselves. Now I understand. It serves a very important protective function in enabling them to live with the "reality" of their "war experience."

---

### *Methodological Note*

The memo above uses concepts derived from analysis (as denoted by quotation marks) to tell the story of survival and homecoming. Though perhaps there might be a better explanation, the conceptualization of the war experience as "Survival: Reconciling of Multiple Realities" seems to fit the data and offers one of several possible interpretations of what the research was all about. The major categories logically fit within our larger framework. While the framework does not account for why persons might not have survived (we have no actual data on these persons as they are dead) it does provide some insight into what the "war experience" is like for combatants, and how come some were able to survive.

When I started this study I had no idea where it was going. I let the data lead me. However, I was not a passive recipient of the data. During analysis I interacted with the data, and it was the questions that emerged from that interaction that led me forward. The interplay between analyst and data was constant.

## Refining the Theory

If this were a "real" study, it would now be time to take my general framework and (a) check for gaps in the logic and rework those areas where there seems to be gaps and (b) begin to use all of the memos that I had written and sorted to fill in the information under each major category.

### Checking for Gaps in Logic

I won't do this here. This is a good activity for individual readers to do at home and groups to do as part of a group session.

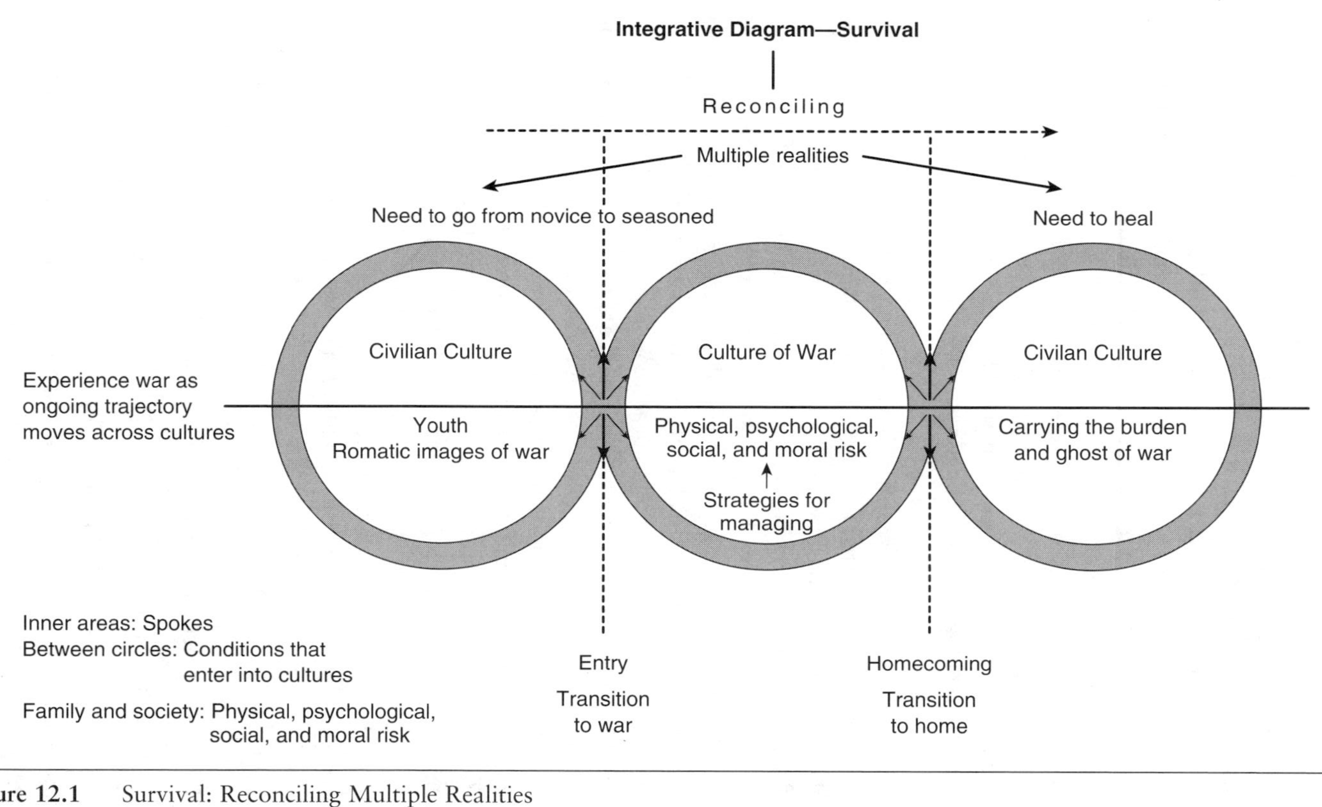


Figure 12.1 Survival: Reconciling Multiple Realities

## Filling In

The core category was "Survival: Reconciling Multiple Realities." Reconciling Multiple Realities represents the "the war experience" of combatants. It is what they had to do in order to survive—physically, mentally, and morally. Under the core category came several other major categories including: "the changing self," "changing images of war," the "culture of war," "survival strategies," and "homecoming." The changing self and changing images of war represent the changes necessary in order for reconciling to occur as combatants met head-on the challenges presented by the culture of war and upon homecoming. And "reconciling" was necessary in order for combatants to make use of the strategies that would enable them to survive physically, remain psychologically stable, and retain their moral integrity during the war and to heal once they got home.

All of the various memos that have been written about major and minor concepts would be sorted and used to fill in the details under each of these various statements of how the concepts related. By filling in, I would explain how reconciling takes place through changes in self and images of war. I would start with who the combatants were before going to war, how they changed from novices to seasoned soldiers, and how being seasoned enabled them to use survival strategies to solve the various problems they encountered while at war and to heal upon homecoming. I would also bring in variation by describing how some combatants were not able to reconcile as well as others and as a result became demoralized, fatigued, and wore down physically, psychologically, and/or morally. These are the soldiers who "wore down," and this explains why some did not heal as well as others upon "homecoming" and had more adjustment difficulties, with some going on to develop post-traumatic stress disorder.

I must add here that this research is not finished. To more fully understand the Vietnam War experience from the perspective of combatants it would be important to collect additional and more varied types of data. It was beyond the scope of this book to do this because the focus was on methodology and not on the research. The research was for teaching purposes only. However, should I continue with the study I would want to obtain more actual unstructured interviews with combatants and do the interviews with representatives of different military services, such as navy swift boat operators, a group not represented in the present data. It would also be important to obtain data from other participants in the war—doctors, engineers, military leaders, politicians—those who though not directly engaged in combat, certainly shared indirectly in the combat experience.

There were also many journalists who accompanied troops in Vietnam, and it would be interesting to have information on their perceptions and experiences with the war, and how they, through their experiences and the images they presented of those experiences to the public, shaped the outcome of the war. It would also be important to obtain more insight into the "war experience" from the perspective of the "enemy." Then there was the whole peace movement at home. As part of this study, I would want to have more information about how the images of civilians at home were shaped and reshaped over time, by whom or what, and the role the movement had in bringing about the end of the war. There is the whole political aspect of behind-the-scene activities that were going on in North Vietnam, South Vietnam, and the United States and the range of issues that impacted the ability to negotiate an end to the war. In order for any of this additional material to be relevant, it would have to be related back to the war experience of combatants because it was they who were the focus of this particular study. So there is a lot left to do.

Most research studies are not nearly as complicated or involved as the study of combatants in war. The important thing to remember is that regardless of a project's scope, one should include as many different perspectives on an issue or topic as feasible. Multiple perspectives add insight, richness, depth, and variation. It is also important to bring context into the discussion. It is not possible to arrive at every contextual factor that might impact upon the topic of study, but knowing how context enters into and helps define or create situations, and responses to those situations adds depth and validity to explanations. The moral here is to work within the limits of the time, energy, and money but do not rush through a project too quickly. The quality and contribution of one's work depends upon the depth and breadth of the investigation.

## Validating the Scheme

This is not a major issue here since this is not a "real" study. However, there are certain things that I did do to validate my scheme. I did send my early analysis to the three interviewees of the study for their comments. There were not a lot of comments and no criticisms. The interviewees thought that the process by which I arrived at the findings was interesting. Also, after coming up with the scheme I did return to some of the better memoirs and reread them. I felt that the scheme held up to that scrutiny. I also talked to other veterans of Vietnam whom I met casually and they seemed to think that the scheme worked. I didn't look for **negative cases** because the Vietnam study was not my focus for the book but a demonstration exercise. However, the topic remains interesting to me and I would like to do more with it.

## Summary of Important Points

Integration is the final step of analysis for researchers whose research aim is theory building. Integration is probably the most difficult part of doing analysis because it requires sifting and sorting through all the memos and looking for cues on how all the categories might fit together. Rereading memos, creating the story line, doing diagrams, and just plain thinking are all techniques that analysts can use to help them arrive at final integration. Just remember that doing qualitative analysis is an art as well as a science and that there is nowhere in the analysis where this becomes as apparent as in the final integration. The cues to integration are to be found in the data, as demonstrated in this chapter. Where the art comes in is in the ability to "make the scheme work" based on the data and the insight gained into the data in memos. The researcher must recognize when the scheme isn't working (there are missing links in the logic) and when this happens be willing to take the scheme apart and rework it again and again until the analytic story all falls into place and "feels right."

## Activities for Thinking, Writing, and Group Discussion

1. Look for gaps or breaks in the logic and explain how those gaps might be fixed.
2. Think of alternative core categories, ones that I might not have thought of, and write a summary memo showing how you would integrate the other categories around it. If you work with MAXQDA, make use of the Add-on MAXDictio as an explorative option: let MAXDictio give a list of the word frequencies and see if you get any additional ideas out of it. Discuss the result with your group.
3. Bring your summary memo to the group and present it for feedback.

# 13

# Writing Theses, Monographs, and Giving Talks About Your Research

*It is in the act of reading and writing that insights emerge. The [work of writing] involves textual material that possesses hermeneutic and interpretive significance. It is precisely in the process of writing that the data of the research are gained as well as interpreted and that the fundamental nature of the research question is perceived. (Van Manen, 2006, p. 715)*

## Introduction

After completing the analysis, it is time to present findings in papers, theses, monographs, and presentations. One of the interesting features about writing or doing presentations is that writing and presenting help clarify thoughts and elucidate breaks in logic. As one of our former students, Paul Alexander, stated in a memo dated September 19, 1996:

> Writing forced me to see the whole theory and highlighted those parts that didn't fit so well. . . . So I would go back to the data. . . . This kind of building and verifying of various aspects of the theory continued throughout the writing process especially in specifying the relationships between areas of the theory.

A few prefatory words should be said before beginning our discussion about writing and doing presentations. Why publish or give presentations? There are a variety of reasons. Without reviewing the many motivations (such as self-pride, career advancement, desire to contribute to reform, or to illuminate the lives of people studied), there is the paramount obligation to communicate with colleagues. Without writing and presenting, professional knowledge cannot be advanced, nor can implications for practice and theory be put into effect without their being made known through publications or presentations. Experienced researchers generally have this obligation built into their psyches. The less experienced, and especially graduate students doing research for the first time, may not only lack motivation to publish, they often undervalue their own research or are fearful of any criticisms it might generate.

The purpose of this chapter is to address those inevitable questions about presentations and writing that are associated with every research project. Questions such as: When should I begin writing? How do I know when the research is ready to put into print or ready to present? What should I write or talk about? What form(s) should the writing take—paper, monograph, or something else? What factors are different about writing papers from writing monographs, or giving presentations? Should I try to publish? Where should I publish? What audiences am I writing for or talking to (including when I am writing a thesis)? What should the writing look like? How do I get started on an actual outline for writing or presentation? How will I know when the writing is good enough to submit for publication?[1]

The chapter will be divided into three sections. The first addresses verbal presentations, the second monographs and theses, and the third various types of papers. For other suggestions on writing, consider two excellent texts devoted exclusively to writing qualitative research by Becker (1986b) and Wolcott (2001). See also the article by Wolcott (2002). Morse and Field (1995, pp. 171–194) and Silverman (2005, pp. 355–370) also have some good advice for would-be authors of qualitative research who want to get their studies published.

## Verbal Presentations

Often researchers present materials orally to see how a given audience will react before attempting to publish. Indeed, sometimes those who are being studied will directly or indirectly press a researcher, asking, "What are you finding? Can't you give us at least preliminary findings or interpretations?" Many investigators do oral presentations before publishing, either in an attempt to satisfy participant curiosity or to get feedback from colleagues. They even do this fairly early in their research projects. Qualitative research studies lend themselves to relatively early reporting because the analyses begin at the outset of the projects. It is not at all necessary to wait until analysis is completed to satisfy listeners with some of the fascinating stories told by participants.

Collegial audiences can absorb presentations at an abstract level, and even talks devoted to research strategies and experiences. Other audiences respond more to stories or discussions about interesting categories/themes spiced with sufficient descriptive narrative or case materials to make them interesting. Researchers also need to choose carefully the appropriate level of vocabulary for each audience. A bad choice of vocabulary, one that uses too much professional jargon, can turn off an audience. Too simplistic a presentation can bore colleagues. It is more important that researchers say something worthwhile to listeners. This means doing quality in-depth qualitative research (see Chapter 14).

All of this advice may sound rather general, although perhaps somewhat reassuring. What about the practical question of how one actually decides on the topic of a talk or speech? Keeping in mind that a talk's content should be matched as far as possible to the audience, I suggest the following answers. To begin with, generally it's preferable *not* to present the entire set of findings in a short presentation, especially if one has developed theory. There is the risk of overloading the audience. It takes a great deal of skill to present an entire theoretical scheme clearly enough in twenty minutes so that listeners can both understand and remember it after leaving the room. A researcher can, of course, sketch the main descriptive story before turning to an elaboration of one of the more interesting features of the research. However, I believe verbal presentations are more effective, and certainly better grasped and remembered, if they focus on one or two catchy categories and includes many descriptive examples, or tell the story of one or more research participants.

Returning to the study of Vietnam veterans, I might want to do a presentation focusing on two important subconcepts of survival: "becoming a seasoned soldier" and "wearing down" with lots of examples of each concept from the data. To prepare this presentation I would review all of the memos pertaining to those categories/themes. Using the memos as a guide I would first write a short summary statement about the main points I want to make in the presentation. That summary statement contains the logic of the presentation and might look something like this: Survival is the main focus of combatants. The young man going off to war has to physically, psychologically, and morally adjust to the realities of war if he is to survive physically, remain psychologically sound, and retain his moral integrity. With time and exposure a combatant becomes "seasoned," indicating that he has developed a set of strategies for handling the problems he encounters. However, with time, if strategies are insufficient or if the stress and strain becomes too great, combatants tend to "wear down" and their "realities become blurred and confusing," putting combatants' physical survival at risks and threatening their psychological sense of well-being and moral integrity.

With that guiding statement in front of me, next I would develop a clear outline including a few sentences about the main story of "Survival: reconciling multiple realities" so that I could put the concepts that I chose to talk about in context. I wouldn't make this too complicated or time consuming but I would describe what some of the multiple realities were and why the ability to reconcile was so important to survival. Then I would move on to the two main subconcepts, "becoming a season soldier" and "wearing down." First, I would describe what I mean by being seasoned, then explicate some of the conditions that foster "becoming a seasoned soldier," and finally explain how being seasoned contributed to survival, giving lots of descriptive data along the way. Next, I would discuss "wearing down": what it is, why it happens, and its possible impact on survival. The presentation would end with a few sentences about how to foster development of "being seasoned" and how to recognize "wearing down" and how it might be handled. It should be made clear to an audience that the researcher is presenting only one or more aspects of the total story.

## Writing Monographs or Theses

Over the course of a research project, the investigator develops a strong sense of what the research is all about. He or she also has learned a great deal substantively about the problem under investigation. Both of these will come into play during the writing. Of course, a researcher needs other skills also, such as a sense of how to construct sentences and how to present an idea clearly. Unfortunately a writer can be his or her own worst enemy. Aside from poor writing skills, a writer may have all the usual blocks described in books designed to help people write (see Becker, 1986b; also Lamott, 1994). Fortunately, by this time in a project, the researcher has a cache of memos and diagrams to provide the basis for writing. The writing requires:

1. A clear analytic story with the logic spelled out
2. A sense of what parts of the story the writer wishes to convey

3. A detailed outline

4. A stack of pertinent memos to fill in the detail of the outline

## Procedures

When beginning to think about writing up the results of a project, the investigator should review the last integrative diagrams and sort the memos until clear about the main analytic story. This review is followed by further sorting of memos until there is sufficient material to write a detailed outline. The sorting might even raise some doubts about the analytic story or point out some of the breaks in logic. If so, don't be discouraged. The worst that could happen is that the analytic story becomes qualified, and so improved. At any rate, the story must be translated into an overall outline. Some people do not work well with detailed outlines. Yet, I find from my own and from student's students' experiences, that it is advisable to at least sketch an overall logic outline, otherwise there may be gaps in the overall story that is presented.

There are additional procedures that can be of help in bridging the leap between analysis and outline. The first is to think intently about the logic that informs the story. Every research monograph, indeed every research paper, will have an internal logic. Each has a few key sentences or paragraphs that signal the author's underlying logic (Glaser, 1978, pp. 129–130), though sometimes the authors seem not to be aware of this. This signal of what is central to any given publication (or thesis for that matter) is often found in the first paragraph or pages and then again in the closing page or pages. As for a manuscript, even the first draft should have its essential analytic story presented clearly. When writing a thesis or monograph, unlike a presentation or even a paper, there should be an explication of the entire analytic story.

A second procedure for translating analysis into writing is to assemble a workable outline, then to write statements that link the sections together so that the writer remains clear about the progressive development of the theoretical story. Chapter outlines are detailed and ordered by thinking through what should be included in each section and subsection, keeping in mind the relation of the parts of the chapter to the entire book. Essential to these decisions, again, is the sorting of the memos that seem relevant. Even during writing, a researcher will frequently return to the memos for details and inspiration. The preface or opening chapter explains the purpose of the manuscript and perhaps even summarizes the analytic story; that is, what this thesis or monograph is all about. This statement as well as the outline itself can be revised if the investigator later deems it necessary.

A third procedure involves visualizing the architecture of the potential manuscript that is the conceptual form that the author wants the book or thesis to take. Visualizing the structure can be compared to creating a kind of spatial metaphor. For example, when writing *Unending Work and Care* (Corbin & Strauss, 1988), the authors carried in their minds the following metaphor. Imagine walking into a house: First a visitor would enter and pass through a porch, then the foyer, then enter a large room that had two prominent subsections, then leave the house through the back door. Then he or she would walk slowly around the entire house, looking into the main room through several different windows but now observing carefully the relationships of the various objects in the room. When the manuscript was finished, its form corresponded to this spatial metaphor: An introduction, a preliminary chapter, a large theoretical section composed of three chapters, then another long section consisting of several chapters that elaborated and drew implications from the theoretical formulations presented earlier.

If faced with writing a thesis, a researcher may find this third procedure (visualization) difficult to use. After all, dissertations in most university departments have fairly standard formats, even for qualitative research. These usually begin with an introductory chapter, followed by a review of the literature, the methodology chapter, then presentation of the findings (in two or three chapters), followed by the summary/conclusions/implications section. For all that, a dissertation writer may be able to think architecturally about the middle (content) chapters. At any rate, when constructing a dissertation based on the findings from a qualitative research study, a researcher should rely on the first two procedures touched on above: (a) developing a clear analytic story by sorting through the diagrams and memos, then (b) working out a main outline that will fully incorporate all important components of that story.

## What to Write?

Qualitative researchers often encounter a difficult problem when trying to decide what to write about their findings. The source of the problem is the fairly complex body of data generated during the entire research process. The big questions are: What of all this analysis should be included? How can I compress all of these findings into a couple of chapters? After all, the standard format for writing theses does not allow one to expound infinitum. In other words, how much depth does one go into when reporting the research? The answer is that first the writer must decide on what the main analytic message will be. Then he or she must give enough conceptual detail to convey this to readers. The actual form of the central chapters should be consonant with the analytic message and its components.

This answer nevertheless fails to specify, whether for writing a thesis or monograph, how much and which conceptual details to include and which can be excluded. It all goes back to answering the questions, "What was this research all about?" "What were the main issues and problems that these informants were grappling with?" Then there should be sufficient conceptual detail and descriptive quotations to give the reader a comprehensive understanding of these. Participants and those professionals familiar with the theoretical area should feel satisfied that the story has been told and understood.

## Example of an Outline for a Monograph

If I was writing a monograph about the survival experience of Vietnam War veterans, the outline might look something like the following. Notice how in putting together this outline my understanding of the findings has advanced even beyond that which can be found in the chapter on integration. By the time a researcher concludes the analysis and begins the writing, it's not uncommon for the analysis to become even more refined. After all, the researcher has had more time to think about it. Only a broad general outline is provided here. Understand that a researcher would use the memos to fill in the details.

### *Chapter 1: Introduction*

The war experience is a trajectory that can be broken down into periods, the prewar experience, during war experience or "the combat experience," and the postwar experience or "homecoming and beyond." Once the individual enters the combat experience, war can also be thought of as a "survival experience" because "surviving" the physical, psychological, and moral risks become the main issues or problems during combat and later during homecoming. There are two intertwining threads that run throughout the war trajectory, the self and images of war, both of which must undergo change if the individual is to survive. This change is accomplished through a process of "reconciliation" in which individuals come to grips with the situations of the present and manage the various risks through a series of physical, psychological, and moral strategies that help them to become "seasoned soldiers" and prevent "wearing down."

### *Chapter 2: The Prewar Period*

*The Self.* During the prewar years, the "self" can be characterized among other things as: youth, idealism, and inexperience with war. The prewar period includes the formative years that are influenced by family, cultural values and beliefs, education, and other experiences. Some of the youth were idealistic, patriotic, adventurous, and/or religious. Others were troubled. Some came from abusive families and others had trouble with the law. Some joined the military out of a sense of duty and honor, others to avoid a jail sentence or to get away from home. Others were drafted and didn't want to be there. These backgrounds, along with cultural beliefs and morals, are carried with the combatant into war and influence the ability to make the necessary adjustments and remain alive, psychologically sound, and retain moral integrity.

*Images of War.* Also formed during this prewar period are images of war. These images are derived mainly from war movies, television, and from stories carried over from World War II. In much of the media, war is or had been portrayed in a romantic sense. There is always a hero (or heroes) selflessly giving all for country and fellow soldiers. The United States is presented as invincible and while suffering setbacks it always emerges as a winner (a very important point that later impacted how veterans felt about losing the war). These images set the expectations that the youth brought with them to Vietnam, the belief that the war would be won, it would be short, and it would be adventurous. Once in the military, the youth were sent to boot camp where they supposedly were turned into soldiers with some time spent on survival in a jungle. Those with advanced education or special talent often made it into Officers Candidate School, a specialty school like sniper school, or flight school. During boot camp, the seeds of change in self and images of war are planted. Though boot camp is helpful in preparing youth for war, and promoting the human bonds necessary for survival, there still isn't the fear of facing the enemy and knowing he is there to kill you, or the smell of blood and death.

### *Chapter 3: The War Period or Combat Experience*

Upon arriving in Vietnam and experiencing the first battle, the self and images of war are shattered. The fear, the blood, and death become "real."

1. The idealistic and adventurous youth that came to war is now faced with a multitude of problems that threatened his very survival. Surviving becomes the main focus of the war experience.

2. The problems arose out of an ever-changing context. The context consisted of various aspects including historical, political, sociocultural conditions of nations, and the personal ideologies and beliefs of those who set the policies and conditions of war from those nations.

3. The problems arose out of situations and presented as a series of perceived risks that threatened survival. The risks were physical, psychological, and moral.

Each situation was different; it was an individual's perception of the risks and psychological makeup that determined response in the form of action/interaction/ emotional response.

4. In order to overcome the problems and reduce the risks, certain things had to happen. Soldiers had to change the self to become "seasoned" and avoid "wearing down." Becoming "seasoned" and avoiding "wearing down" necessitated a "reconciliation of past self and images of war" and adjusting to the "realities of the situational present." Combatants had to be able to define situations realistically in terms of risks and take protective action.

5. Survival action took the form of individual and institutional and collective strategies that were situational and aimed at preventing death and injury from wounds, maintaining health, supporting psychological well-being, and preserving moral integrity and most of all, have effective and experienced leaders.

6. But strategies had to be used or enacted in order to promote survival. Enabling their use were the "facilitators" of surviving. Blocking their use were the "obstacles" to surviving, including the element of fate.

7. Combatants who were successful reconcilers became "seasoned soldiers." They were able to make use of strategies to protect the self and avoid wearing down too far, and with a lot of luck survived. The unlucky ones didn't make it.

### *Chapter 4: Postwar Period or Homecoming*

Survival carried with it problems of its own.

1. First, there were the ghosts that haunt, guilt, nightmares, and the horrors of having gone to war and survived. The postwar experience often led to posttraumatic stress disorder. Some veterans suffered heavy wounds and were permanently disabled. Healing for them was more than psychological, for it was also intensely physical. Though some veterans did heal and recall the war experience as being "not too bad" and "maturing," even they carry with them the scars of war.

2. There was a changed society. A nation that once supported the war (at least tacitly) now openly turned against it. There were antiwar demonstrations, draft dodging, and flag burning. Veterans were derided and made to feel dishonorable for the actions of a few. Veterans couldn't understand how and why attitudes changed and why the youth that remained at home didn't understand the sacrifices they had made for their country, especially the death of 58,000 and an even larger number of wounded and those with permanent disabilities.

3. The social context that veterans found upon their return home called for another "reconciliation" of self and war. This reconciliation was necessary in order to "heal."

4. Some veterans were able to makes the adjustments in self and images of war and "heal" at least partially if not fully.

5. Other veterans were not able to reconcile or "heal." To survive, they put up a "wall of silence" to keep ghosts, fears, family, and society from breaking through and disrupting the fragile stability that keeps them afloat physically, psychologically, and morally. Drugs and alcohol are often used to blot out thoughts of war, fears, and nightmares and to keep ghosts away. It is the only way that some veterans can function in society. Even today, thirty plus years later, some veterans are still angry, and keep their thoughts about the war and their selves hidden behind the wall of silence. Some remain lost in a world of pain, alcohol, and drugs.

### *Chapter 5: Conclusions and Implications*

Today, there is a considerable knowledge about the impact of war on young men and women and programs established to help combatants make the transitions from war to home. Even the movies today about war present it in a more realistic and less romantic image. But there still is not enough counseling or support for returning combatants, either from the military or the general public. One of the most important things to come out of this study is an understanding of the war experience from the perspective of combatants. The experience is one of a daily struggle for survival, not only physical survival but psychological and moral survival as well.

1. The most effective solution would be to avoid war altogether.

2. That being impossible, it is imperative that the military and society give young men and women the support and skills that they need to be able to "reconcile" their selves and images of war and bring them more in line with the different realities or war. Then they must be provided with the support and counseling necessary upon homecoming so that they can take that next step in reconciliation necessary for "healing."

## The Issue of Self-Confidence

The increasing ease in accomplishing what is, after all, quite specialized writing is also related to the issue of a researcher's confidence in his or her own analytic and compositional abilities. This point is very important and valid, especially since understandings keep increasing and findings are evolving. In this regard, the following quotation expresses succinctly what an inexperienced researcher is likely to experience. The quotation refers more to analysis than to writing, but in writing itself the two skills are, as we have noted, tightly joined.

> This anxiety and anguish . . . can be further mitigated (also) by writing a paper or two before embarking, at least seriously, on the long and major writing task . . . getting a paper or two accepted for publication can give a considerable boost to flagging confidence or lingering doubts about one's ability at research. (Strauss, 1987, pp. 259–260)

Researchers may find themselves blocked when they begin to write, let alone later during the writing itself, if they lack confidence in their analysis. They may ask, "Do I really have it right? Have I left out something essential?" For those writing theory, there is the additional question of "Have I really identified the core category?" And if yes, still, "Do I have all of this in enough conceptual density?"

The answers may be yes, no, or maybe! But the issue here is not whether the analysis has been adequately and sufficiently done, but confidence that one really knows the answers to the above questions. Even experienced researchers are not always certain precisely where the breaks in logic are until they have chewed on their pencils for some time. Nor can they be certain that, after review, they know there are no important omissions in their analysis. Whether experienced or inexperienced, a common tactic for reducing uncertainty is to try out the story on other people, individuals, or groups, informally or formally.

Classroom seminars can give presenters confidence in their analyses, whether the theory be in preliminary or almost final form. Speeches given at conventions, if favorably received, can add further validation of an analysis and its effective reflection in readable prose. Nevertheless, when approaching or even during the writing period, there is almost invariably a considerable amount of anxiety about whether presentation of the research can be, or is being, accomplished effectively. After all, some people are perfectionists and cannot seem to settle for less than an ideal performance. That can mean, of course, no performance at all or a greatly delayed one. Others lack some measure of confidence in themselves generally, and this spills over into questions about ability to accomplish this particular kind of task.

## Letting Go

Having edited what probably should be the final draft, a researcher can have difficulty letting go of a manuscript. Letting go may not be due so much to a lack of self-confidence, though it can be that, but to a temporary failure of nerve. Have I really got the last details in? Got them right? These doubts are stimulated by the almost inevitable discovery of additional detail, both conceptual and editorial, and the relocation and rephrasing that occurs during each rewriting of a draft. Part of an increasing maturity as a researcher and writer is to understand that no manuscript is ever finished. If a writer is fortunate enough not to have a personal, departmental, or publisher's deadline, then he or she may profit from putting aside the final draft for some weeks or even months, in order to gain a bit of editorial and analytic distance from it. Also, a colleague or two might be pressed to read part or even all of the manuscript and to provide constructive feedback. Eventually a writer does have to let go of his or her work, convinced that the manuscript is as finished as it ever will be. A researcher can rest assured that once off to a publisher or committee there will always be feedback about improvements to make. The logic of letting go is that writing is only part of a cumulative stream of conveying ideas, which a writer may return to later to criticize in this or a later work. Incorporating one's own criticisms is no different from responding to other people's criticisms.

The psychology of letting go is, however, more complex. Basically, it comes down to avoiding the trap of dreaming of the perfect manuscript, and allowing oneself instead to be open to new projects, new ideas, and new data. It's important to strike a balance between profitable reworking of drafts and cutting loose from them. How to do this is difficult to convey. Of course, an experienced researcher who actually is familiar with the investigator's work can help with this problem, but in the end every writer must rely on his or her inner sense of rightness and completion.

If the researcher is writing a dissertation and is fortunate enough to study in a department that allows a certain degree of latitude in style, then he or she can write for audiences other than committee members and wider departmental faculty. Moreover, book publishers usually reject theses, as such, sent to them as possible publications, preferring a different format of presentation. So, if you are allowed to write a thesis or dissertation in a style that approximates a monograph, then the conversion to a potential publication is rendered that much easier.

## Audiences

There is also the question of a writer's conception of the audiences for his or her thesis. Perhaps this issue is less complicated than for other forms of publication (this will be discussed next) and for speeches, but it is one that plagues many students. After all, the immediate readers are the thesis advisor and other members of the doctoral committee. If they do not approve of the dissertation, then the entire enterprise will be a personal disaster. When doctoral committees consist of faculty who strongly disagree on their criteria for adequate work, students can be hurt by these methodological discrepancies. If fortunate or astute, students choose committee members who they know will agree among themselves about the desired standards and format acceptable for the dissertation, though perhaps with some revisions. There is no tried and true rule to suggest how this variable situation can be managed. My best counsel is to choose, if possible, a supportive yet critical advisor, and to write as good a manuscript as possible. If the student produces solid research, then the student is likely to earn a degree, unless some of the committee members are skeptical of qualitative studies. If that is a possibility, then the student should keep the number of such potentially adverse critics on the committee to a minimum.

There are some crucial differences between monographs and dissertations, though in the pages above we have tended to blur this distinction. Chief among their differences is that the discussion in a monograph should be conceptually fuller; that is, it should include greater depth and detail. Since there is more space and fewer page constraints, an author is freer to develop an analytic message. Moreover, the monograph can be more complex. In addition to a more extensive elaboration of categories and their relationships, it can present a much greater amount of substantive material. The latter may include case studies and even long quotations from interviews, field notes, and documents. The author may always choose to digress at times, bringing into the discussion minor and side issues, as long as these are consistent with the main thrust of the monograph. Also, a monograph can include some issues that were omitted from the more restricted dissertation or not fully worked out during the dissertation research. Inconsistencies that crept into the more hurried writing of a dissertation should be corrected in the monograph. Dissertation committees tend to emphasize findings, whereas the readers of monographs are more likely to appreciate or at least accept an analytically based argument, as well as a broader discussion of the research materials.

The author of a monograph has more latitude in choosing the style of presentation. In some part, the style should reflect the author's message, while taking into consideration the audience for whom the message is intended. Questions to consider are: Are the readers restricted to disciplinary or professional colleagues, or to some types of them; or does one hope to have readers from several fields, including perhaps those from practitioner fields? What about lay readership? For a monograph to be maximally effective, its author should ask, "What do I wish to say to each of these audiences?" Or if several audiences are intended, then, "What style can I use to reach each?" Usually, theory blended with sufficient descriptive detail to make it vivid and clear is the preferred combination. In short, the style and shape of presentation should be sensitive to and reflect the targeted audiences.

Suppose the author wants to address both disciplinary colleagues and laypersons. To reach both audiences requires giving considerable thought to the use of vocabulary, terminology, case materials, overall mood, and other aspects of writing style. Many monographs published by sociologists have both collegial and nonprofessional readers as targeted audiences. (Among the monographs that have been published are: Biernacki, 1986; Broadhead, 1983; Charmaz, 1991a; Davis, 1963; Denzin, 1987; Fagerhaugh & Strauss, 1977; Rosenbaum, 1981; Shibutani, 1966; Star, 1989; Whyte, 1955). Sometimes the targeted readers are the nonprofessionals, for instance, patients and their families, such as the book on epilepsy written by Schneider and Conrad (1983). Occasionally monographs are directed at lay audiences, colleagues, and professionals. Then they are published as trade books, as in a book on remarriage after divorce (e.g., Cauhape, 1983).

To write for multiple audiences is generally more complicated than writing for one's colleagues. Yet many researchers are eager or feel obligated by conscience to write for more than scientific or professional readers. Sometimes, too, they use their research as a platform for writing books that are not monographs. One possibility is to address policy issues, presenting an argument, though informed by one's research and perhaps also by professional knowledge (Strauss & Corbin, 1988). Or books can be written for practitioners, full of information based on research (e.g., Strauss, Schatzman, Bucher, Ehrlich, & Sabshin, 1964).

## Converting Dissertations to Monographs

How is a dissertation converted into a monograph? Guidelines bearing on how to do this were suggested implicitly in the preceding pages. However, the prior question that faces the author of a dissertation is if it should be written next in monograph form? Several questions pertaining to this decision should be carefully thought through, and preferably in the following order.

1. Are the substantive materials, findings, or theoretical formulations presented in the thesis sufficiently interesting to be worth my time and effort to write up for a wider audience or audiences? Some theses are natural candidates for such presentation. Other dissertations, no matter how important they may be to some colleagues, are not good candidates, though portions of their materials are likely to be published as articles and later may be widely cited.

2. If deemed sufficiently important, then how do I decide which are the most relevant topics and conceptualizations to include in a monograph?

3. Do I have sufficient time and energy to translate this thesis into a monograph? Am I really still interested in this subject matter? Am I saturated, bored, with it? Have I had it? Is it really my forte or should I move on to other, now more interesting, topics or areas? Of course sufficient interest in

doing it successfully can lead to very great personal satisfaction. Part of the commitment and resulting satisfaction may also derive from a sense of obligation to audiences, who ought to know about what one has discovered through the research.

4. There is still another question that many potential authors consider. Given a certain level of interest and sufficient time and energy, is it worth writing this monograph for career purposes? In some fields, writing a monograph (or other type of research-based book) is not especially important; papers published in refereed journals bring more prestige. However, colleagues in other fields, including the social sciences, especially when recruiting candidates for faculty or when they themselves are considered for promotion, know that monographs often weigh more heavily than do papers in the evaluation.

After considering each of these questions, as well as sometimes being impeded or confused by the counsel of faculty advisors, friends, sponsors, or other intimates, an investigator is still confronted with the additional question of how to translate a thesis into a monograph. In fact, trying to answer this question is very likely to affect the decision whether or not to write a monograph, since it includes weighing the time and effort involved. The actual conversion of the dissertation can be carefully guided by considerations touched on in preceding pages. The writer must think carefully about the targeted audiences. Plus, equally careful thinking must be done about the topics, or concepts, or theoretical formulations, that are likely to be of greatest interest or value to each audience. Those considerations lead to the issue of style. For instance, what format should be used? Should theoretical formulations be the major focus of the monograph and descriptive materials subordinated, or should these be kept in balance? Should a researcher argue the main thesis forthrightly, using existing theoretical formulations, or should he or she keep the argument low key or even implicit? Stylistic considerations of course also entail decisions about the kind and level of vocabulary to be used, modes of presenting selections from the data, the overall mood of the monograph, and so on.

As stated earlier, conceptual elaboration must be added to the original presentation in the dissertation. A researcher can do this by including theoretical materials already developed in the memos but omitted from the dissertation, and by thinking through aspects of theoretical formulations that were left unclear, ambiguous, incomplete, and even inconsistent. Also, in a monograph, the writer probably will wish to discuss at greater length certain implications of the research with reference to the theoretical literature, as well as implications for future research, and perhaps for practitioners or policy decision makers. Many researchers have found the experience of rewriting for a monograph tremendously rewarding. Others have translated theses into monographs primarily for career advancement and personal reputation, cashing in (literally) on that investment.

## Team Publications

When a project involves two or more researchers, then there is always a question of how publications are to be written. The answers depend, understandably, on the relationships between team members, their respective abilities and interests, their responsibilities, and the amount of time available to each. Some publications are written by the principal investigator of the project, with varying amounts of input by other team members. Other publications involve more truly collaborative writing, rather than just shared research. Presumably the possibilities are numerous. The same is true of papers based on the team's research.

## Writing Papers for Publication

This fourth class of research-based publications is scarcely a homogeneous one. The great variety of options for types of papers can be suggested graphically by a threefold breakdown of those possibilities.

1. For colleagues, a person might write papers with a major focus that is alternatively theoretical, substantive, argumentative, and/or methodological.

2. For practitioners, papers may provide theoretical frameworks for understanding and working better with clients, substantive findings, practical suggestions for better procedures, suggestions for reform of existing practices, and/or broad policy suggestions.

3. For lay readers, appropriate papers would include those describing substantive findings, suggestions for reform of current practices or policies, self-help guidelines or tactics for obtaining better services from practitioners or institutions, and those that provide assurance that others share their own experience (as in living through a divorce or adopting a child).

This variety of options for papers points to differences in purposes, emphases, styles, and of course different publication outlets. Nevertheless, research findings provide a firm basis for writing all of these types of papers. Qualitative research studies provide theoretical analyses, substantive content, and self-confidence. By completion of the research the investigator should have considerable sensitivity to issues, audiences, and the strengths and weaknesses of actors and organizations. The qualitative researcher will draw on this knowledge, too, when making decisions about what to write, for whom, and how. Decisions concerning those issues rest on reasoning and procedures not appreciably different from those discussed throughout this chapter. The few important differences can be stated briefly and are easily understood. Here are some conditions that may directly affect how and for whom and whether certain papers will be written:

1. As noted earlier, researchers may decide to publish papers even relatively early during the research process. They may do this for different reasons, for instance, to present preliminary findings, or to satisfy or impress sponsors, or because they have interesting materials bearing on side issues that can easily be written up now but might not get written at a later, more hectic time.

2. Sometimes researchers write papers either because they feel obligated to publish on a given topic or because they are pressured to do so. Of course this motivation will also affect what and how a researcher writes.

3. Researchers may also be invited to contribute papers to special issues of journals or edited volumes because they are known to be researching in given areas. They may also be urged or be tempted to convert verbal presentations into papers because listeners have responded well to them.

4. Another condition that can affect the writing of a paper is the existence of a deadline for getting the finished product to an editor. For some researchers, this can act as a stimulus, while others are daunted by any deadline.

5. The number of pages allowed by the editor also affects whether a paper will be written, at least for the particular publication, and what will be written and how.

6. Unless invited by an editor, there is the important decision to be made about which particular journal should be selected as a potential outlet for a given paper. Journals and papers have to be matched, otherwise it might be rejected and the time invested in writing it wasted. Or worse yet, the paper is accepted but for an inappropriate or insufficiently appreciative audience. Selecting an appropriate journal may be an easy task if the researcher knows that journal well, but otherwise issues of the journal should be carefully scrutinized. It helps to get the counsel, also, of people who are knowledgeable about specific journals. This is especially true when addressing audiences outside one's own field, as when a social scientist writes for a social work or medical journal.

Having noted these conditions, which are sometimes constraining but at other times stimulating, I can now discuss what else may be different about writing papers. The most important considerations are the interrelated ones of purpose and audience. Given the variety of purposes and audiences listed above, a reader can see that this is the central issue facing any researcher who writes a paper. (This is true even when invited to write one.) What should the writer say to an audience? Topics for some papers seem to emerge rather naturally during the research process. For instance, in their study of the chronically ill and their spouses (Corbin & Strauss, 1988), the authors were struck by the stylistically different approaches to management among couples. These ranged from highly collaborative relationships to ones characterized by considerable conflict. So, a paper was written on this topic relatively early in the research (Corbin & Strauss, 1984). Some papers may be conceived of early or midproject but do not get written until later, or the ideas become incorporated into the monograph.

Some ideas for papers take much longer to formulate than others, perhaps because they require deeper understanding of the phenomenon or more theoretical sophistication in order for the researcher to feel comfortable in writing about them. Writing papers that suggest reforms might be delayed because researchers are unable to commit themselves to a reform role until they become sufficiently disturbed at what they are observing; or perhaps because the directions in which reform alternatives can be specified are not yet clear to them. After the theoretical formulations are worked out clearly, there is the temptation to present the entire framework in one long paper. As I've stated, this is a very difficult task to do, since the framework will be very complex and dense with conceptualization. My advice is not to attempt this task in a paper. If the writer chooses to, then it is preferable to provide a frankly stripped-down version, referring readers to the forthcoming monograph. For example, Strauss and colleagues wrote a paper on "medical work" and its relationship to "safety work" and "comfort work" ((Strauss, Fagerhaugh, Suczek, & Wiener, 1985). In another paper, this research team wrote about "safety work," especially in relation to their continuous work with medical equipment that was so potentially hazardous (Wiener, Fagerhaugh, Strauss, & Suczek, 1979).

Other papers can be written around methodological issues or around policy issues. Then the theoretical materials will be kept subordinate but still give coloration to the main line of the discussion. A methodological focus may need both substantive and theoretical illustration to make sense to the reader. Policy arguments not only can be buttressed by data, they also can be explicitly or implicitly underpinned by a theoretical framework. Strauss and Corbin, for instance, gave an argument and suggestions for reform of the American health care system (Strauss & Corbin, 1988). These were based on criticism of the dominant acute care orientation of health professionals and institutions despite the prevalence of chronic illness today, a type of illness that has multiple phases, each of which requires a different type of care.

To return to our suggestion that a theoretically oriented article be restricted in the number of categories or ideas discussed, the question, as usual, is how is that discussion to be developed? The same general answer can be given as when writing chapters of a monograph, but modified for purposes of writing a paper. First, the researcher decides on a focus. What is the theoretical story that the writer wishes to tell? This decision may arise during the course of the research, or it may actually be prompted by thinking about the last integrative diagram or through sorting of memos. In regards to the Vietnam study, I might want to write a story about survival in war, detailing the types of risks to survival, strategies used to modulate those risks, and the conditions that facilitate or constrain the use of strategies. With the overall story to be told and the logic of the paper in mind, it is time to construct an outline of the paper. Just as with an outline for a monograph, once memos are gathered and read and the writer begins to piece together a story, ample illustrative description is brought into the narrative to make the paper interesting and to make the theoretical ideas more concrete and available to a wider audience. Though it might be interesting to write an article about "Survival: Reconciling Multiple Realities" as a general phenomenon, usually the first papers that are written from a project are less theoretical. They are written with the idea of educating colleagues and perhaps a lay audience about a topic. The notion of generating more formal theory can be left to a later date.

One danger when writing papers is permitting too much detail to flood thinking. Attempts to crowd too many findings into the short space of a paper may discourage the writer or at least impede the clarity of one's exposition. The working guideline here for what goes into a paper and what can be omitted, reluctantly or ruthlessly suppressed, can be expressed in the form of a dual question: Do I need this detail in order to maximize the clarity of the analytic discussion, and/or to achieve maximum substantive understanding? The first part of the question pertains to the analysis itself. The second pertains mainly to inclusion of data in the form of quotations and case materials.

As with monographs and theses, the drafts can be given a trial by sharing them with friends and colleagues, and even with accommodating practitioners or laypersons, if the materials pertain to them. Again, a writer might wish to have drafts scrutinized by a writing group or a student research group if one is so fortunate as to belong to one. A writer must also incorporate relevant literature. If it is a theoretical paper, the author might wish to think through its implications in order to make recommendations for changes in policies or practices. Then, when finally finished, and even more finally published, the researcher should already be on his or her way to thinking about, outlining, and beginning to write the next publication!

## Summary of Important Points

Making oral presentations and publishing written reports about findings of research introduces still another challenge for the researcher. With so much complex material available, how does a researcher make choices about what to present, to whom, and how? Generally, in a verbal presentation or an article, it is preferable to present only one or two concepts (categories/themes) in any depth with maybe one or two others woven in as related features. In writing monographs, the researcher has a wider range of possibilities, but even here he or she should carefully think through the logic and order of the material before doing a detailed outline. Dissertations present problems of their own for a standard format must be followed. Again, the writer must carefully think through how much detail to include and how to present the most relevant facets of the conceptual scheme while still retaining flow and continuity. The important point to remember is that researchers who have completed an in-depth analysis will have plenty of interesting material to write about for months to come and an appreciative audience for these writings.

## Activities for Thinking, Writing, and Group Discussion

1. Think about your present research project or one you might have completed in the past. What are some of the concepts that you might be able to write a paper or do a presentation on? Write an outline for a paper or presentation.

2. Take one or two of the concepts from the research on Vietnam. Write an outline for a paper. Consider the audience you want to write for and the journal you are targeting.

3. Bring your outline to the group and present it. Have them provide feedback. Where do you agree or disagree with their comments? How might your outline be revised, therefore improved?

4. Look through some of the journals that publish qualitative research projects. Take two papers, one that you consider to be a well-written one and one that you thought was too superficial and not very informative. Bring the papers to the group and compare each along the various properties that you believe make for good writing.

## Note

1. This chapter is a slightly revised version of the one published in the 2nd edition of this book and reflects the thoughts of both authors Strauss and Corbin on the topic. However to make it less confusing to readers who know that Strauss has been dead for some years, the pronoun "I" will be used.

# 14

# Criteria for Evaluation

> *Quality is elusive, hard to specify, but we often feel we know it when we see it. In this respect research is like art rather than science. (Seale, 2002, p. 102)*

## Introduction

I have to agree with Seale. Quality in qualitative research is something that we recognize when we see it; however, explaining what it is or how to achieve it is much more difficult. Though I have been concerned with quality for some time (Corbin, 2002, 2003), I find this chapter on evaluation difficult to write. I feel paralyzed, unsure of where to begin, or what to write. As I search the literature, I find that everyone agrees evaluation is necessary but there is little consensus about what that evaluation should consist of. Are we judging for "validity" or would it be better to use terms like "rigor" (Mays & Pope, 1995), "truthfulness," or "goodness" (Emden & Sandelowski, 1999), or something called "integrity" (Watson & Girad, 2004) when referring to qualitative evaluation? Then there is the question, can one set of criteria apply to all forms of qualitative research? The notion of judging the quality of research seemed so clear before postmodernist and constructionist thinking pointed out the fallacies of some of our ways. Now I wonder, if findings are constructions and truth a "mirage," aren't evaluative criteria also constructions and therefore subject to debate? As Flick (2002) states, the problem of how to assess qualitative research has not yet been solved (p. 218).

I enjoyed the irony in the title of Sparkes's (2001) article, "Myth 94: Qualitative Health Researchers Will Agree About Validity" (p. 538).

Despite all of the debate and the confusion from which I suffer, deep down, in the depths of my analytic soul, I still believe that qualitative research is both a "scientific" (Morse, 1999) as well as a "creative" and "artistic" endeavor, and that "quality" of the final product (findings) will reflect both these aspects, a point made by Seale (1999, 2002). As stated by Whittemore, Chase, and Mandle (2001), "Elegant and innovative thinking can be balanced with reasonable claims, presentation of evidence, and the critical application of methods" (p. 527). The purpose of this chapter is to explore the issue of "quality" in qualitative research, and to present some criteria for evaluating research based on the methodology presented in this book.

## Some Literature

I think that a good starting point for this chapter is a review of some of the relevant literature. One of the issues frequently mentioned is validity. Perhaps the person most widely quoted on the subject of "validity" as it applies to qualitative research is Hammersley (1987). He states that a research account may be considered valid if "it represents accurately those features of the phenomena, that it is intended to describe, explain, or theorize" (p. 67). Winter (2000) offers an explanation of validity based on Foucault's (1974) definition of multiplicity of truths and goes on to state that "[Validity] appears to reside within the appropriation of research methodologies to those systems of truth that their processes best represent" (p. 67). It seems clear that these "experts" associate "validity" with a kind of "truth," but truth in a more pluralistic than traditional sense.

In fact, Silverman (2005) states that "validity" "is another word for 'truth' (p. 224). He proposes five strategies for increasing the validity of findings. These five steps include: using the "refutability principle" or the refuting of assumptions against data as the researcher proceeds through the research; using the "constant comparative method," or the testing of provisional hypotheses against at least one other case; doing "comprehensive data treatment," or incorporating all cases into the analysis; "searching for deviant cases," that is including and discussing cases that don't fit the pattern; and "making appropriate tabulations," or using quantitative figures when these make sense as in mixed-method designs. Reliability according to Silverman (2005) can be achieved by tabulating categories if a researcher so chooses and also by being certain that when transcribing interviews all aspects of data are transcribed, even the most minute (pp. 209–226).

Morse, Barret, Mayan, Olson, and Spiers (2002) state that "it is time to reconsider the importance of verification strategies used by the researcher in the process of inquiry so that reliability and validity are actively attained, rather than proclaimed by external reviewers on the completion of the project" (p. 9). In other words, the researcher should take strategic action during the course of the research to ensure its validity and reliability. Morse et al. list several strategies for bringing "rigor" into the research. These include, "investigator responsiveness, methodological coherence, theoretical sampling and sampling adequacy, taking an active analytic stance and saturation" (p. 9). The use of these strategies is reasonable and they could be used with many types of qualitative research. However, though the strategies address the "scientific" aspect of doing qualitative research, they don't evaluate it for its creative aspects.

Creswell (1998) and (Creswell & Miller, 2000) propose eight different procedures for achieving what Lincoln and Guba (1985) call "credibility" and "trustworthiness" of findings. These include "prolonged engagement and persistent observation in the field," "triangulation," "using peer review or debriefing," "negative case analysis," "clarifying researcher bias," "in member checks," "rich think description," and "external audits" (Creswell, 1998, pp. 201–203). Again, these criteria are directed more at the validity aspects of doing qualitative research rather than the creative.

Chiovitti and Piran (2003) have delineated a list of criteria for achieving rigor in grounded theory research. The list includes: let the participants guide the process; check the theoretical construction generated against participants' meanings of the phenomenon; use participants' actual words in the theory; articulate the researcher's personal view and insights about the phenomenon explored; specify the criteria built into the researcher's thinking; and specify how and why participants in the study were selected, delineate the scope of the research, describe how the literature relates to each category which emerged in the theory. No doubt, following these procedures would lead to rigor, but there is nothing in the list about context, process, density, variation, or usefulness. Nor is there anything in the list about "vividness, creativity, thoroughness, congruence, or sensitivity," criteria for validity as described by Whittemore et al. (2001, p. 531).

Charmaz (2006, pp. 182–183) offers a list of criteria for evaluating constructionist grounded theory. Of all the criteria I've read, I find hers the most comprehensive because they address both the scientific and creative aspects of doing qualitative research. Charmaz breaks her criteria down into four categories. These are credibility, originality, resonance, and usefulness. The list is quite substantial and I don't want to reproduce it here in its entirety, so I will instead present a sample of the questions the researcher should be asking himself or herself. For example, under credibility she says, "Do the categories cover a wide range of empirical observations? Are there strong logical links between the gathered data and your argument and analysis?" Under originality she asks, "Are your categories fresh? Do they offer new insights?" Under resonance she says, "Do the categories portray the fullness of the studied experience?" And under usefulness she says, "Does your analysis offer interpretations that people can use in their everyday worlds?" There is only one problem (this is not a criticism but a comment) that I can see with the evaluative criteria proposed by Charmaz. They require self-evaluation during and after the research process and self-evaluation is tricky. It requires a certain degree of sophistication and experience to accurately evaluate one's own work and even then it's hard to remove the bias.

Before moving on, I want to review one more piece of literature that I think is important to this discussion. In their early book on grounded theory, Glaser and Strauss (1967) made the statement that, "By the close of the investigation, the researcher's conviction about his own theory will be hard to shake, as most field workers can attest. This conviction does not mean that his analysis is the only plausible one that could be based on his data, but only that he has high confidence in its credibility . . ." (p. 225). (Note that Glaser and Strauss use the term "credibility," meaning "believable," rather than "validity," "getting around the whole issue of "truth.") The reason these authors give for this confidence is the fact that the researcher has been so immersed in the social world under investigation that he or she can offer a "plausible" explanation about it. Glaser and Strauss offer the following criteria for judging the "credibility" of a study. Though written for theory building research, the criteria also have significance for more descriptive forms of research.

The first is that there be sufficient detail and description so that readers feel that they were vicariously in the field (thus able to judge for themselves). Second, there should be sufficient evidence on how the data were gathered and how the analysis was conducted (so that readers can assess how the researcher came to his or her findings or conclusions). Glaser and Strauss state that multiple comparison groups make the credibility of the theory greater (because the findings are based on more than one group). Finally, the researcher should specify the kinds of data upon which his or her interpretation rests (pp. 223–235). In addition to "credibility," Glaser and Strauss (1967) take up the notion of "applicability," for which they also provide criteria. From my standpoint, any research findings with claims to quality should be "applicable," therefore I also want to mention the criteria of Glaser and Strauss (1967). The criteria for applicability include a theory should "fit" the area from which it was derived and in which it will be used, a theory should be readily "understandable" by laymen as well as professionals, a theory should be sufficiently "general" to be applicable to diverse situations and populations, and finally, a theory should provide the user with sufficient control to bring about change in situations (pp. 237–250).

As I look back, I realize that even though Glaser and Strauss were talking primarily about theory building research they were on to something. Furthermore, what they had to say pertains to descriptive as well as theory building research. If the research findings are "credible"; that is, believable or plausible and "applicable" in the sense that findings can be readily used because the findings provide insight, understanding, and work with diverse populations and situations to bring about desired change, then it seems to me all this philosophic debate about "truth," "validity," and "reliability" is superfluous. In other words, the "proof is in the pudding," so to speak. If it "fits" and it is "useful" because it explains or describes things, then what is all the concern about rigor and everything else? Rigor must have been built into the research process, or the findings would not hold up to scrutiny, would not fit similar situations, and would be invalidated in practice.

## Some General Thoughts About Quality

I have come to the conclusion that, for me, quality and validity are not synonymous. Quality findings also have an innovative, thoughtful, and creative component. Here I want to quote something from Agar (1991). Though in the article he is talking about how the use of computer programs can stifle as well as help with analysis, the quote brings out how thinking and creativity are built into the analytic process and in many ways it is these aspects of analysis that bring out the richness and authenticity of qualitative research, something that computers cannot do. He says,

> The problem is that software presupposes a way of seeing the problem and the situation, the pinnacle of the top-down framework that guides one into gathering of material, development of research categories and perceptions of relationships among them. The critical way of seeing, in my experience at least, comes out of numerous cycles through a little bit of data, massive amounts of thinking about that data and slippery things like intuition and serendipity. . . . (Agar, 1991, p. 193)

Also I (Corbin) don't feel comfortable using the terms "validity" and "reliability" when discussing qualitative research. These terms carry with them too many quantitative implications (a personal bias). Somehow the word "truth" also bothers me in the sense that it seems that no matter how you define "truth," the term carries with it a certain degree of dogmatism. I would rather use the term "credibility" (Glaser & Strauss, 1967; Lincoln & Guba, 1985) when talking about qualitative research. To me, the term "credibility" indicates that findings are trustworthy and believable in that they reflect participants', researchers', and readers' experiences with a phenomenon but at the same time the explanation is only one of many possible "plausible" interpretations possible from data. Finally, I am in agreement with Rolfe (2006). I don't believe that the same judgment criteria can be applied across qualitative methodologies. Each method deserves its own set of judgment criteria.

## What Is "Quality," or, What Are Its Properties?

How would I describe "quality" as it applies to qualitative research? Quality qualitative research is research that makes the reader, or listener, stand up and say things like "wow," "I'm touched," "now I understand," "that has power," "I feel like I've walked in those participants' shoes," "there is so much depth in the study that it covers detail that I never knew about this subject," "this is something I can use in my practice, in my life." In other words, quality qualitative research resonates with readers' and participants' life experiences. It is research that is interesting, clear, logical, and makes the reader think and want to read more. It is research that has substance, gives insight, shows sensitivity, and is not just a repeat of the "same old stuff" or something that might be read in a newspaper. It is research that blends conceptualization with sufficient descriptive detail to allow the reader to reach his or her own conclusions about the data and to judge the credibility of the researcher's data and analysis. It is research that is creative in its conceptualizations but grounded in data. It is research that stimulates discussion and further research on a topic.

## What Are the Conditions That Foster the Construction of "Quality" Research?

Quality doesn't just happen. It is something that has to be worked toward, and there are certain research situations or conditions that foster it.

The first condition is "methodological consistency" (Flick, 2002, p. 219; Morse et al., 2002). If a researcher says he or she is going to use a particular method, then he or she follows through, using all of the relevant procedures as designed. This is not to say that researchers can't be creative in their use of analytic strategies or how they carry out particular procedures. But if researchers do what Baker, Wuest, and Stern (1992) call "method slurring" or combining philosophically different qualitative methods, and if researchers use only some but not all of the major procedures that are part of a method, then they are likely to lose some of the credibility associated with that method.

Methodologies are designed to do certain things and, with usage over time, have attained a certain degree of "credibility" when used in a manner consistent with the design. To mix up different methodologies, or use only certain procedures and not others, erodes at that credibility. For example, though there are many versions of "grounded theory," the procedures that are consistent with the different versions are the "constant comparative" method of analysis, the use of concepts and their development, theoretical sampling, and saturation. And, for those researchers who want to build theory, there should be the construction of a well-delineated theory. Therefore, if a researcher uses only one or two of these procedures or doesn't build theory, he or she can't claim to be doing grounded theory. It would be much more accurate to say that he or she is using some of the procedures associated with grounded theory to do a "such and such" study.

A second condition is that the researcher has clarity of purpose. A researcher should be very clear at the onset of a study whether the aim is description or theory building. It is difficult to do quality research if a researcher is unsure about the difference between description and theory. Findings are likely to appear muddled and fail to live up to either good description or theory. The quality and value of a piece of research lies not in whether or not it is theory or description. As Sandelowski (2000) explains, "qualitative description" has its place in nursing research and I might add, in other disciplines also. The value of any research study lies in the substance, depth, and innovation of the product that is generated.

A third condition is having self-awareness (Hall & Callery, 2001). Since the researcher (as interpreter) is such an integral part of both the research process and the findings, it is important that a researcher remain aware of his or her biases and assumptions. Keeping a journal and/or writing frequent memos about the researcher's reactions and feelings during the data collection and analysis can help the research recognize the influence that the researcher is having upon the research, and just as important, that the research is having upon the researcher.

A fourth condition is that a researcher should be trained in doing qualitative research. What the researcher brings to the analysis in terms of qualifications, experience, perspective, as well as underlying philosophical orientation will make a major difference in the quality of findings. Everyone wants to get in on the qualitative research act, or so it seems today. As a consequence, there is much variation in the quality of the work that is being produced, much of it looking more like poor journalism than research. One of the problems, I suspect, is that many researchers think that doing qualitative research is easy, and that anyone can do it. That is not true. Anyone can pull a few themes from data, but not everyone knows how to build well-developed themes, arrive at thick rich description, or know how to construct an in-depth narrative. Nor does everyone know how to build theory. Doing quality qualitative research requires a sound educational foundation in methods, data gathering, and analysis—just as quantitative research does.

A fifth condition is that a researcher has "feeling" and sensitivity for the topic, for the participants, and for the research. To do good analysis you have to be able to "step into the shoes of participants" and feel at a "gut level," otherwise you lose some of the richness and depth of the data. A cold and distant researcher may serve to enhance the "validity" of qualitative research but can erode the "credibility" of findings by preventing the researcher from developing the sensitivity, empathy, carefulness, respect, and honesty (Davies & Dodd, 2002) needed to accurately capture the viewpoint of participants.

A sixth condition is that the researcher must be willing to work hard. Doing qualitative research is like any worthwhile endeavor. It takes time, thought, and a willingness to work and rework it until you get it right. It takes time to write memos and to do all of the other tasks associated with qualitative research. There are no shortcuts to doing "quality" qualitative research, whether one is doing description or building theory.

A sixth condition is a willingness to relax and get into touch with the creative self. Getting out of your conceptual ruts (Wicker, 1985) means being willing to brainstorm, turn things upside down, make theoretical comparisons, and think about things in new ways. The qualitative researcher has to be open to new ideas and use strategies flexibly and creatively in order to get at the essence or meaning of what participants are telling us. In fact, Hunter, Lusardi, Zucker, Jacelon, and Chandler, in their article "Making Meaning: The Creative Component in Qualitative Research" (2002), demonstrate how important creativity is to qualitative research.

A seventh condition is one borrowed from Seale (2002, p. 108), what he calls "methodological awareness," indicating that the researcher should be aware of the implications of decisions he or she makes throughout the process. Methodological awareness calls for anticipation of potential criticisms, and carrying out data collection and analysis in ways that contribute to credibility while attending to methodological problems as they arise.

An eighth condition is a desire to do research for its own sake. I believe that the reason that we see so much variation in the quality of qualitative research is because there is so much push in master's level programs and in academia to do research. There is this mystique that "doing research" is somehow the "end all and be all" or mark of an educated person. But some persons are better teachers or practitioners than others. And some persons are better researchers than others. I believe that quality in qualitative research would be less of an issue if, as professionals, we could give credit to what persons do best—practice, teaching, or research—and not make everyone feel that they have to do research.

## Criteria for Judging the Quality of Research Using This Method

I do not mean to imply that meeting the conditions above would necessarily guarantee quality. In the end, it is the quality of the findings that matter and it is this quality that will be judged by others. However, my tendency in this section would be to "let the research findings speak for themselves," or better yet refer readers to Charmaz (2006), Creswell (1998), Guba and Lincoln (1985), Hammersley (1987), Morse et al. (2002), Silverman (2005), or Wolcott (1994), or any number of other researchers, depending upon the methodology used. I realize, too, that I can't abdicate my responsibility to provide a list of criteria for judging the quality of research findings based on the methodology explicated in this book. I do not believe that these criteria must be applied to all qualitative research methods or even to other grounded theory methods. Making judgments about the quality of qualitative research is difficult because so much depends upon who is doing the research, its purpose, and the method that is used. Below is a list of general criteria that can be used to evaluate the "quality" of research findings. These criteria are drawn from multiple sources, including the literature mentioned earlier in this chapter. Many are similar to those proposed by other qualitative researchers.

The first criterion is *fit*. Do the findings resonate/fit with the experience of both the professionals for whom the research was intended and the participants who took part in the study? Can participants see themselves in the story even if not every detail applies to them? Does it ring "true" to them (Lomberg & Kirkevold, 2003)? Do they react emotionally as well as professionally to the findings?

The second criterion is *applicability* or usefulness of findings. Do the findings offer new explanations or insights? Can they be used to develop policy, change practice, and add to the knowledge base of a profession?

A third criterion is *concepts*. Concepts are necessary for developing common understandings and for professionals to talk among themselves, therefore one would expect that findings would be organized around concepts/themes. How the findings are presented is not what is relevant. What is important is that findings have substance, or that they must be something more than a mass of uninterpreted data that leave the reader trying to figure out what to make of it. And of course, concepts should be developed in terms of their properties and dimensions so that there is density and variation.

A fourth criterion is the *contextualization of concepts.* Findings devoid of context are like jelly donuts devoid of jelly, to use a rather graphic simile. I mean to imply that findings devoid of context are incomplete. Without context, the reader of research cannot fully understand why events occurred, why certain meanings and not others are ascribed to events, or why experiences were one way and not another. Imagine trying to understand the experience of combatants in Vietnam without knowing something about the historical factors that led up to the war or the political decisions that determined how the war would be fought. The reader would feel that something essential was missing from the story.

A fifth criterion is *logic.* Is there a logical flow of ideas? Do the findings "make sense"? Or are there gaps or missing links in the logic that leave the reader confused and with a sense that something is not quite right? Are methodological decisions made clear so that the reader can judge their appropriateness for gathering data and doing analysis?

A sixth criterion is *depth.* While concepts provide a common language for discussion and give organizational structure to the findings, it is the descriptive details that add the richness and variation and lift the findings out of the realm of the ordinary. It is depth of substance that makes the difference between thin, uninteresting findings and findings that have the potential to make a difference in policy and practice.

A seventh criterion is *variation.* Has variation been built into the findings, meaning are there examples of cases that don't fit the pattern or that show differences along certain dimensions or properties? By including variation, the researcher is demonstrating the complexity of human life. Anselm Strauss was very adamant about the fact that life is very complex and that any research that is done should capture as much of that complexity as possible by building in variation.

An eight criterion is *creativity.* Are the findings presented in a creative and innovative manner? Does the research say something new, or put old ideas together in new ways? No one wants to hear the same old things. It is not that the topic needs to be new, but that new understandings of that topic are brought forth. To do that, procedures must be used consistently, creatively, and flexibly rather than in a dogmatic fashion.

A ninth criterion is *sensitivity.* Did the researcher demonstrate sensitivity to the participants and to the data? Were the questions driving the data collection arrived at through analysis, or were concepts and questions generated before the data were collected? In other words, did the analysis drive the research or was the research driven by some preconceived ideas or assumptions that were imposed on the data? The latter may or may not be okay, depending upon how careful the researcher was to put aside bias and honestly seek to find contradictions in the data to his or her assumptions.

A tenth criterion is *evidence of memos*. Finally, but certainly ranking up there among the most important criteria, is evidence of memos. Since a researcher can't possibly recall all of the insights, questions, and depth of thinking that goes on during analysis, memos are among the most necessary of all procedures. Memos should grow in depth and degree of abstraction as the research moves along. Thus, there should be some evidence or discussion of memos in the final report.

## Additional Criteria for Evaluating the Quality of Research

Additional criteria are necessary for evaluating the "credibility" of descriptive findings or theory constructed using the research process described in this book. In judging the credibility of research, both the "doer" and the "reviewer" of research should be able to make judgments about some of the components of the research process. Even in a monograph, which tends to be longer than an article, if there are no indications of the research process, there is no way that readers can accurately judge how the data were collected or the analysis was carried out. To remedy this, it would be useful if researchers provided the kinds of information bearing on the criteria given below. The detail need not be great, even in a monograph, but sufficient to give some reasonable grounds for judging the adequacy of the research process as such. The criteria are presented below in question form. These criteria were originally published in Corbin and Strauss (1990).

*Criterion #1.* How was the original sample selected? How did later sampling occur?

*Criterion #2.* What major categories emerged?

*Criterion #3.* What were some of the events, incidents, and/or actions (indicators) that pointed to some of these major categories?

*Criterion #4.* On the basis of what categories did theoretical sampling proceed? That is, how did theoretical formulations guide the data collection? After the theoretical sampling was done, how representative did the categories prove to be of the data?

*Criterion #5.* What were some of the statements of relationships made during the analysis and on what grounds were they formulated and validated?

*Criterion #6.* Were there instances when statements of relationships did not explain what was happening in the data (negative cases)? How were these discrepancies accounted for? Were statements of relationships modified?

*Criterion #7.* How and why was the core category (if applicable) selected? On what grounds is the final analytic decisions made?

*Criterion #8.* Are the concepts systematically related? To have theory, there must be systematic development of concepts and linkages of those concepts to form a theoretical explanation about some phenomenon. The key word is that theory *explains.* It doesn't just describe. You have to answer the questions of "why" or "how come" and "when" this happens, then "is this likely" to happen? Linkages may be presented as a list of hypotheses or propositions. Or, they may be woven throughout the text in more subtle forms. It may be helpful to the reader if the researcher makes this clear to the reader.

*Criterion #9.* Is variation built into the theory? Variation is important because it signifies that a concept has been examined under a series of different conditions and developed across a range of dimensions. Some qualitative studies report only a single phenomenon and establish only a few conditions under which it appears; also they specify only a few actions/interactions that characterize it, and a limited number or range of consequences. By contrast, when using this methodology there should be considerable variation built into the theory. In a published paper, the range of variations touched upon may be more limited, but the author should at least suggest that other writings include their specification.

*Criterion #10.* Are the conditions and consequences built into the study and explained? Any explanation of variation should include the conditions under which it can be found and some of the consequences of action/interaction/emotional responses. Conditions should not be listed merely as background information in a separate chapter, but woven into the actual analysis with explanations of how they impact the events and actions in the data. These include, but are not limited to, economic factors, organizational policies, rules and regulations, social movements, trends, culture, societal values, language, and professional values and standards.

*Criterion #11.* Has process been taken into account? Identifying process in research is important because it enables theory users to explain action under changing conditions. The conceptual scheme used to explain process is less important than attempts to bring it into the analysis.

*Criterion #12.* Do the theoretical findings seem significant, and to what extent? It is entirely possible to complete a theory generating study, or any research investigation, yet not produce findings that are significant. If a researcher simply goes through the motions of doing research without drawing upon creativity or developing insight into what the data are reflecting, then this researcher risks the possibility of arriving at findings that are less than significant. By this, I mean that the research fails to deliver new information or to offer new insights or explanations. Remember there is an interplay between the researcher and the data, and no method can ensure that the interplay will be creative. This depends on four characteristics of the researcher: analytic ability, theoretical sensitivity, ability to think about data in different ways, and sufficient writing ability to convey the findings.

Of course, a creative interplay also depends upon the quality of data collected or utilized. An unimaginative analysis may, in a technical sense, be adequately grounded in the data, but be limited for theoretical purposes. This is because the researcher either doesn't draw on the fuller resources of data or fails to push data collection far enough.

*Criterion #13.* Do the findings become part of the discussions and ideas exchanged among relevant social and professional groups? Findings are time and place specific, however, major concepts often have continued usefulness. Take concepts such as stigma, division of labor, uncertainty, stress, and negotiations. These concepts have proven their usefulness throughout the years, though the specific findings associated with them may have been modified and changed with time

## A Final Note

Readers should keep in mind three additional comments about evaluative criteria. First, these criteria should not be read as hard and fast evaluative rules, either for the researcher or for readers who are judging others' research publications. The criteria are meant as guidelines. Certain investigations may require that the research procedures and evaluative criteria be modified to fit the circumstances of the research. Imaginative researchers who are wrestling with unusual or creative topics and analytic materials might depart somewhat from standard ways of doing and analyzing research. In such unusual cases, the researcher should know precisely how and why he or she departed from conventional ways of doing things, say so in the writing, and leave it up to readers to judge the credibility of the findings.

Second, researchers should provide a brief overview of what their research procedures were, especially in longer publications. This would help readers to judge the analytic logic and overall adequacy or credibility of the research process. It would also make readers more aware of how a particular research investigation differs from those using other modes of qualitative research. In specifying this information, readers are apprised precisely about what operations were used, and their possible inadequacies. In other words, the researcher should identify and convey the strengths and inevitable limitations of the study.

Finally, it might be useful in certain publications for a researcher to include a short explanation of his or her own research perspectives and responses to the research process. This enables readers to judge how personal reactions might have influenced the investigation and interpretations placed on data. Writing reflective memos during the research process is one way of ensuring that the researcher will be able to do this at the end of the study (Rodgers & Cowles, 1993).

## Evaluation and Computer Programs

I have to admit that when it comes to using computers for analysis, like Agar (1991), I am a "concerned" advocate of their use. By that I mean I worry that somehow using computer programs for analysis will stifle creativity, mechanize the analytic process, or worse yet, do what Agar (1991) and Fielding and Lee (1998) fear, which is to lead users down a path in which they succumb to "technological determinism" or letting the computer program rather than the analyst structure the analysis. Yet, when I look around and see the computer "savvy" of most ten-year-olds I know the time has come to recognize the place of computers in the future of qualitative research. Not only are computer programs helpful for organizing, storing, and shuffling data and memos, they play an important role in evaluation. Computers certainly increase "methodological awareness" (Seale, 2002, p. 108) because the researcher has an indisputable record of his or her decisions. The image in front of the researcher is like a mirror indicating flaws in the logic, undeveloped categories, and insufficient conceptualization. With the "data" right out there, researchers can't pretend to have achieved "saturation" or "conceptual density" when they have not. Computer programs are perfect for creating an "audit trail" (to use the term of Guba & Lincoln, 1985) because the record of the researcher's work is accessible and can be reconstructed without a great deal of effort.

Computer programs don't do the thinking, and they can't write the memos (only store them), but despite their limitations computer programs can enhance the creativity of analysis because they enable researchers to try things first one way and then another, thus seeking alternative explanations. Dey (1993) states that, "Computers can help us confront data more effectively, by making it easy to analyze data in different ways" (p. 227). I still worry about distraction, about the production of sterile findings and all the possible pitfalls of using computer programs for analysis. However, I realize that I came to the use of computers relatively late in life. Those researchers who come of age into the world of computers today no doubt have the facility to use computer programs creatively and in a manner that will enhance the quality and credibility of their research. And, who knows what the future of computer programs will hold in store. For a discussion on the use of computer programs in research and alternative views on these points, see Bong (2002), Bazeley and Richards (2000), Fielding and Lee (1991), Flick (2002, pp. 250–261), Kelle (1997), Roberts and Wilson (2002), and Weitzman and Miles (1995).

## Concluding Remarks

A reviewer of this manuscript suggested that it might be worthwhile to continue with the Vietnam study by applying evaluative criteria to that study. But that would be unfair, as the study was meant as an example only and not as a finished product. As mentioned in Chapter 12, there are much more data that need to be collected before the study could be considered finished. Categories could be more fully developed, more variation could be built in, and there could be greater exploration of alternative conceptualizations. At this point I am not satisfied that the "quality" is what I expect of myself as a researcher, though I think that the findings that I have arrived at are "credible." I think it would be interesting for students and readers of this text to do an evaluation of the study and to point out the flaws. Therefore, I will suggest that in one of the activities.

## Summary of Important Points

The quality is what is often missing in qualitative research. When this happens, it is usually because researchers are not well trained, are in too much of a hurry, or are not certain how to judge the quality of their own and others' work. This chapter explores the notion of "quality," examining it in light of its properties and conditions for achieving it. It then suggests some criteria for judging the "credibility" or "plausibility" of findings. Though I strongly believe that the notion of "quality" as applied to qualitative research should be taken seriously, the criteria presented here are meant as guidelines only. I still think that the findings "speak" for themselves and when we see quality we will know it. I also recognize that there are special research circumstances requiring different approaches to doing research and standards of judgment. In these situations, it is important for a researcher to explain the specifics of why and what was done, leaving it up to readers to judge the results.

## Activities for Thinking, Writing, and Group Discussion

1. Think about research studies that you have read. Pick out one that you thought of as having "quality." What were the characteristics of the study that made you think so?

2. Would you describe the study above as also being "credible," "plausible," or "believable"? Why or why not?

3. As an individual or as a group, take the criteria for quality and credibility outlined in this chapter and apply them to the Vietnam study presented in this book. Discuss the strengths and limitations of the study and what more needs to be done to provide greater quality and credibility.

# 15

# Student Questions and Answers

*Artists and writers, however, may present the single case, ignore the scientific (and even the popular) literature, and be unconcerned with "truth" or "reality," however that might be conceived. Researchers do not have such freedoms and cannot dodge their responsibilities to their participants while still expecting to be taken seriously and be considered to "do" research. Again, there are no shortcuts. (Morse, 2004, p. 888)*

## Introduction

In writing the revision of this text, I jumped ahead to this chapter, thinking that maybe it was outdated. What I found was that the questions posed in this chapter are still relevant and that the chapter has withstood the passage of time. It was originally written in response to questions raised by students in class, during consultations, and following presentations. The questions arise from various concerns. Students are puzzled because certain procedures or techniques seem unclear, are ambiguous, or run counter to those used in more conventional research methods. Other students want to know how to respond to criticism from mentors, thesis committee members, and friends. Below are a few of the most frequently asked questions with responses. This chapter has been placed at the end of the book, rather than at the beginning, because it summarizes many of the major points made throughout the book.

*Question 1.* "I've heard that there are some very good computer programs available that can help with analysis. Do you know anything about these and how they are used?"

*Answer.* In 1962 a man named Douglas C. Engelbart was asked to write a summary report for the Air Force Office of Scientific Research, Stanford Research Institute in Menlo Park, California. Recently, while surfing the Web, I came upon a copy of that report. Considering the year that the report was written, I found it to be quite fascinating because the words were so prophetic. Consider the following paragraph taken from that report:

> You can integrate your new ideas more easily, and thus harness your creativity more continuously, if you can quickly and flexibly change your working record. If it is easier to update any part of your working record to accommodate new developments in thought or circumstance, you will find it easier to incorporate more complex procedures in your way of doing things. This will probably allow you to accommodate the extra burden associated with, for instance, keeping and using special files whose contents are both contributed to and utilized by any current work in a flexible manner—which in turn enables you to devise and use even more complex procedures to better harness your talents in your particular working situation. (Engelbart, 1962, p. 5)

What Engelbart (1962) was getting at in his report was the ability of the computer to augment the human mind by doing a lot of the detailed and tedious work involved in many endeavors, thus freeing up the user to be creative and thoughtful. And this really is what computer programs do for qualitative analysis. They make many of the chores—like sifting and sorting through data—a lot easier, leaving the researcher freer to do the thinking necessary to do "quality" analysis. Not that using a computer program is necessary for doing "good" qualitative analysis. Most of the classic qualitative studies that we are all so familiar with in our respective fields were not done using computer programs. Anselm Strauss played around with various computer programs, especially ATLAS/ti and Nvivo (Nudist in his day) because he found them fascinating. However, when it came down to doing the actual research, he used his computer to write and maybe sort memos, but never actually as part of the analysis. I do believe that if Anselm were coming of age today, things would be different. He would take advantage of what computer programs for data analysis have to offer and use them to augment his already prolific and creative mind.

Notice that Engelbart does not say that computers do the thinking for the user. Unfortunately, computers have not reached that stage of accomplishment yet. Even the most sophisticated software programs for qualitative analysis do only what their users tell them to do. But, what they can do is plenty. Before

I get into what computer software programs can do for the researcher, though, let me provide a word of caution. Novice researchers tend to be very careful in their approach to analysis. They don't trust themselves or the research process and therefore tend to hold on to methodological procedures or software programs as lifelines. I think this is one of the reasons why some qualitative researchers are reluctant to use, or have their students use, computer programs. They worry that the students will be so focused on the software programs that they won't be free to think for themselves.

However, I am convinced that computer assisted qualitative data analysis will become, in the future, standard in the field of qualitative data analysis. Why? Because many of the present, and certainly the next generation of qualitative researchers, will or already are using computer programs to augment their mental and physical abilities. Software programs for analysis will be just one more extension of their selves. I can only dream about the future capabilities of future computer software programs for data analysis. But for right now, let me review for you some of how they can be used.

From my perspective, some of the most important contributions of software programs to qualitative data analysis (QDA) are as follows. They contribute to creativity in the sense that the researcher is able to try out different axial views of data, looking at relationships first "this way" and then "that way" without having to spend a lot of time retrieving and organizing data. The computer does the retrieval and layout work while the researcher does the mind work. Since with the use of a computer program the researcher can always retrace his or her analytic steps, the research process can be made transparent to self and others, leaving the audit trail at one's finger tips. Then too, the researcher doesn't have to guess at what he or she was thinking or wrote in memos months ago. These can be pulled out of the data bank in moments, making the analysis more consistent and the findings more reliable. The researcher doesn't have to ask, did I already use this code? If so, how? All the researcher has to do is look to the list of codes, turn back to the raw data and any memos written on the subject. Want to do a diagram? It's easy, just turn to the graphic part of the program. Most important from my standpoint is the value computer programs have when it comes to writing the research. It is so easy to access codes, to return to the raw data to use as examples or for quotes, to retrieve memos, do diagrams, correct mistakes, find gaps in logic, and most of all rewrite.

Some programs are more sophisticated than others. Some are more difficult to use than others. My advice is as follows. Try out different programs. Many can be downloaded over the Internet for limited use. Find the program that works best for you. But remember, you don't want to be so concerned about learning and using the software program that the research becomes secondary or lost somewhere in the process. The challenge should be doing the research, not learning the program. Nor do you want the computer program to direct the research. By that I mean it's easy to follow the boxes or windows outlined by the program, thus letting the program guide the research process, rather than the researcher using the program flexibly as an extension of self to store, retrieve, organize, and reorganize ideas about the data. Analysis is about thinking, and thinking is the one thing the computer can't do.[1]

*Question 2.* Stephan, an anthropologist who is surrounded at work by psychologists, says, "They are continually asking, 'Where are the numbers?'" This is also a frequent question asked by thesis committee members, and more quantitative researchers.

*Answer.* Though there are some qualitative researchers who do quantify their data, as a rule qualitative researchers are not as much concerned with numbers as they are with identification of process and social mechanisms. Qualitative researchers seek to identify significant concepts and to explore their relationships. They are more interested in understanding what is going on than they are in testing hypotheses. If committee members insist on numbers, students can add a quantitative component to the study by including some relevant measuring instruments. These satisfy committee members and often provide additional findings of interest.

*Question 3.* "What is the focus of analysis, if not numbers?"

*Answer.* This question is a variant of "where are your numbers?" The skeptic is assuming you can't arrive at conclusions unless you use statistical modes of sampling and analysis. For us, the unit of analysis is the *concept.* As explained in Chapter 7, the sampling procedures are designed to look at how concepts vary along a dimensional range, rather than measuring the distribution of persons along some dimension of a concept. Therefore, researchers collect data from places and/or persons and/or on things where they expect potential variations in that concept will be maximized. For example, in the Vietnam study presented in this book, after the analysis with Participant #1, a "noncombatant" who described his experience in Vietnam as "not so bad," I followed up on that dimensional description of "experience" ("not so bad"). I went out and looked for a sample of "combatants" and even other nurse "noncombatants" in the same war to see how they would describe the experience. It was variation in the "experience" that I was looking for, combatants and noncombatants just being the sources of those data. The ability to sample on the basis of concepts is very important because it provides researchers with the flexibility to follow the analytic leads, and in doing so build variation and density into their findings.

Later, if researchers want to test some aspect of their findings by doing cluster analysis, correlations, or some other type of sophisticated statistical analysis, they may do so. Remember, the primary purpose of doing qualitative research is discovery, not hypothesis testing. At the beginning of the research, the analyst doesn't know which variables are important or what their properties are, or how these vary dimensionally. Therefore, sampling is guided by concepts and what they tell us about phenomena rather than on numbers that tend to quantify phenomena.

*Question 4.* "Can we use data that has already been collected? Must we code all our data? Should we sample randomly? Are there other ways of sampling?" These questions often are raised because students (and other researchers) have already collected their data before coming to the seminar or before they begin their analysis. Sometimes, their concern is, "Do I have to start data collection all over in order not to violate the procedure of theoretical sampling that states data collection and analysis are interwoven procedures?" Other times, their concern is, "How do I manage so much material, especially since I don't have unlimited time?"

*Answer.* The response to the first question is as follows. Essentially, working with already collected data is no different from doing secondary analysis on one's own or someone else's—perhaps long since collected. Also, the problems associated with already collected data are very similar to those confronted by anyone who discovers a large cache of archival materials or memoirs and wishes to analyze them. The major difference, perhaps, is that with personally collected materials a researcher has some familiarity with the materials.

Researchers should approach already collected data, secondary or archival materials, or memoirs exactly as they would their own data. To handle these kinds of data, researchers characteristically begin as usual, examining the earliest interviews, field notes, or documents for significant happenings and events. At first, they might scan the data and find a passage that interests them, then begin careful initial coding. Likewise, since sampling is done on the basis of concepts, a researcher can sample theoretically already existing data, sorting through interviews, observations, or videos, to look for examples of relevant concepts and analyze these. Analytic problems do sometimes arise with already collected or secondary data, for example when researchers attempt to saturate categories, or find variations in properties and dimensions, only to discover to their dismay that there are insufficient or incomplete data. When this situation arises, the analyst must either return to the field to collect additional data or live with gaps in the theory.

In response to the second question, the answer is, "No, not every single bit of data has to be analyzed 'microscopically.'" However, as stated earlier in this book, close inspection of data during the early phases of the research process is necessary to build rich dense description and tightly integrated theory. Usually, microscopic analysis is reserved for early interviews or observations in order to delineate categories. Categories must be filled in, linked, extended, and validated through more data gathering and analysis. There is no substitute for intensive coding during the early phases of the research. Once the foundation of the analytic story is established, the researcher can be more relaxed in his or her approach.

Random sampling is more appropriate to quantitative studies than to qualitative ones for all the reasons listed in Chapter 7. As stated, qualitative researchers are not trying to control variables, but to discover them. They want to identify, define, and explain how and why concepts vary dimensionally along their properties. So, while random sampling is possible, it could be detrimental because it prevents the analyst from following analytic leads and discovering the answers he or she is seeking.

As for other types of sampling, in almost any qualitative research, the first data are gathered through a variety of procedures—cashing in on lucky observations, using "snowball sampling," networking, and so on. The lucky researchers are those who have unlimited access to sites, and who know where and at what times they might find the comparable situations that will enable them to extend and elaborate their concepts. Sometimes, researchers don't know which persons or places to go to in order to find examples of how concepts vary. Instead, they sample by "sensible logic" or by "convenience," hoping to come upon that variation. Variations almost invariably exist because no two departments, situations, or happenings are quite the same. Each situation has the potential to present different features of phenomena. The more interviews or observations that a researcher conducts, the more likely it is that conceptual variations will be found in data.

*Question 5.* Valerie and Stephen (a psychologist and an anthropologist) say, "Psychologists are taught to think up 'mini-theories' out of their heads to see if they work. That's just the opposite of your way of doing research."

*Answer.* These "mini-theories" are essentially hypotheses, perhaps grounded a bit in a psychological researcher's experience and reading. However, these hypotheses are not derived through systematic analysis of data and validated during the research process. From a practical standpoint, the mini-theories have merit, especially for practitioners who need knowledge to handle problematic situations. Much depends, of course, on how those mini-theories were derived. If not grounded, they can be misleading.

*Question 6.* "Do qualitative researchers do much describing or descriptive quoting from interviews and their field notes? Some research reports feature more quotes than analysis."

*Answer.* There seems to be a tendency among some researchers and some disciplines to do less analysis and more quoting, leaving the interpretation up to the reader. It all depends upon the method that is being used and the researcher's philosophic orientation. The position taken in this book is that while quoting makes fascinating reading, it doesn't provide the reader with any framework for making sense out of those readings. Nor does it explain why certain quotes were chosen over others or the underlying conceptual message or understanding the researcher is trying to convey. That said, quotes do add interest and provide evidence for skeptics, therefore a good sprinkling of them throughout a research report is important. For example Strauss, Schatzman, Bucher, Ehrlich, and Sabshin, in their book titled *Psychiatric Ideologies and Institutions* (1964, pp. 228–261), presented materials bearing on the beliefs of psychiatric aides working in a psychiatric hospital. One of the points made in the materials was that the aides, who though uneducated in psychiatric principles, nevertheless considered themselves to be "doing good" for the patients. The aides recognized the professional work of the nurses and physicians, but sometimes thought that they did more good for specific patients than did the professionals with all their psychiatric ideologies. To convince potentially skeptical readers, long quotations from interviews with the aides were given. However, the more usual practice for these authors is to balance description and quotes with conceptual explanations. Individual qualitative researchers handle the matter of quotations differently, and it is suggested that students read a number of monographs and papers to get an idea of the variation.

*Question 7*. Krystof asks, "I did an organizational study of one factory in Japan. A colleague asks, 'How can you generalize from studying just this one factory to all other Japanese factories?'"

*Answer*. The answer to this question is quite complicated. True, you can't generalize from one factory to all factories. But then generalization is not the purpose of qualitative research. The idea behind qualitative research is to gain understanding about some phenomenon, and a researcher can learn a lot about a phenomenon from the study of one factory or organization. Remember, as researchers, we are analyzing data for concepts and their relationships. Manifestations of concepts might be found a hundred or more times in this one case. For example, in their study of work in hospitals, Corbin and Strauss identified the concept of "workflow" as being relevant. They asked the question, "What enables the work to 'flow,' or keeps it going on a daily basis, and what happens when it is disrupted and why?" There is much to be learned about workflow through the study of one organization. However, it is impossible to learn everything there is to know about workflow from one case (person, family, factory, organization, community, nation). Explanations based on a single case will be somewhat limited and require further study in other organizations in order to elaborate upon the concept. By specifying the contexts (set of conditions in which specific phenomena [concepts] are located), all we can say is that this is why the work keeps going here. If similar conditions exist in your organization, then perhaps much of what we've learned about workflow may help you to understand what is going on in your organization, or at least stimulate you to think about the concept of workflow as it pertains to the organizations you are studying.

Therefore, if a person asks a researcher, "Is this one case representative of all cases?" The answer is probably "no," in the traditional meaning of the word "representative." However, if the same person were to ask, "Is there something we can learn from this case that will give us insight and understanding about how and why work flows, or keeps going, in other organizations?" Then, the answer can be "yes" because the concept of workflow should apply to many organizations, though the specifics might differ.

*Question 8*. "If I am collecting data in a foreign country should I translate my interviews into English in order to code them, or should I code them in the original language (providing of course I speak that language)?" The usual reason given is that translating takes "so much" time. This question is one that is often asked by doctoral students from outside the country who are pressed by their thesis committees to translate their interviews into English.

*Answer*. There are several reasons for doing only minimal translating. A main reason for doing some translating is so that English-speaking readers can get at least some degree of feeling/insight into what the interviewees are saying and thinking, as well as a sense of what the coding looks like. On the other hand, there are considerable difficulties with capturing the nuance of meaning in translation. Few of us are specially trained or natively skilled at overcoming those difficulties, especially for extended passages. Foreign students report additional difficulties in trying to code in English. One such difficulty is that often there is no equivalent word in English capable of capturing the subtle nuances of the word in the original language. "Meanings" to quote Eva Hoffman (1989) become "lost in translation." For presentations or publications in a country other than the one where the data were collected (if the language is different), key passages and their codes can be translated, approximating the original as close as possible. However, as a general rule, too much valuable time and meaning can be lost in trying to translate all the research materials. Also, many of the original subtleties of meaning can be lost in translation.

When working with students in foreign countries, I do ask them to translate some passages, otherwise I can't work on these data. However, the presenting student is asked whether a given translated word or phrase really approximates what the interviewee intended. Usually, after some discussion, an understanding of what is meant by a word is reached. In other words, in seminar or teamwork sessions, there are additional opportunities to explore the parameters of translated meanings and to avoid imposing outsider interpretations on the data. Naturally, when it come times to write up the research, some of the quotes will have to be translated into the language of the publication. See, for example, Saiki-Craighill (2001a, 2001b).

*Question 9.* "Are there special problems with doing qualitative studies in non-industrialized societies, or in industrialized non-European cultures? After all, so much emphasis is placed on close linguistic analysis in this methodology."

*Answer.* This question raises a thorny issue, which surely deserves serious consideration. In a general sense, qualitative analysts face precisely the same difficulties when trying to comprehend the meanings of acts, events, and objects when these are profoundly "cultural" in nature. It is all too easy for people living in Western countries to misinterpret foreigners or persons only partly assimilated, when comparing their acts and words to American ones. As the anthropologists have taught us, to avoid such misinterpretations, a researcher must spend a fair amount (some say a great deal) of time at the foreign site, and engage in a lot of observation and conversations (informal interviews). Also researchers must understand at least some of the language, besides examining their own often culture-based assumptions. Even with this counsel, anthropologists can't guarantee that misinterpretations (sometimes gross ones) have not occurred.

However, if a foreign student is studying here but wishes to collect data in his or her own country, then most certainly he or she can use this or other qualitative methods. It is important not to "borrow" theories derived from other cultures, but to develop theory specific to reflecting a society's time and place. Alas, a mistake frequently made is to superimpose theories developed in one society upon another society. Even if cultural differences are very subtle, they are there. The imposed theories may sound good, but if not carefully evaluated in terms of fit, they can be misguiding.

As for the use of procedures, there is no reason why the procedures described in this book can't be used to analyze data collected in any country. After all, the procedures work when studying ethnic Americans, other "subcultural" groups like "punks" and "junkies," whose cultural meanings and behavior often differ from the usual. As an illustration, one of the American students studied conceptions of health among the Sioux Indians, living among them on a reservation and previously working there as a public health worker for several years. She concluded that anthropologists who had studied these people did not accurately grasp how Sioux philosophy of the world affected their conceptions of health and medicine—ideas very different than the usual Western ones.

*Question 10.* Krystof, a visiting sociologist from Poland, has a mass of data already collected. He asks, "How does a researcher handle this much data?"

*Answer.* The answer to this question is similar to the response given above about previously collected data. Suppose a student is studying a business organization that is flourishing despite a bad recession, and wants to know how the organization has managed this feat; that is, the basis for the decisions they've made, the visions that guide their executive's actions, the incentives they provide, and so forth. The data might consist of organizational documents only, but masses of them. To begin, the analyst would chose some documents and familiarize himself or herself with their contents, just as if they were interviews. Then, once the analyst has a sense of the types of information the documents contain, he or she can begin intensive coding. With a beginning list of concepts, the analyst could turn to successive documents, analyzing each as if it were incoming data. Not every bit of data has to be analyzed. Once categories are saturated, the researcher can skim the remaining materials to see what new ideas they contribute to the findings.

*Question 11.* "Can the analytic process be hastened or shortened? Also, many practitioners and professionals don't have the time required for theory development but want to do good research; what should they do?"

*Answer.* As stated, there are no shortcuts to doing qualitative analysis. A researcher must go through the process (as it pertains to the method of his or her choice) if he or she wants to do thick rich description or develop a dense well-integrated theory. Researchers can choose not to saturate categories or look for context. Researchers can choose not to do memos and so on. It's all a matter of time, money, and training. However, findings will reflect this and the researcher should be prepared to accept the limitations.

Though many publications purport to use a method such as grounded theory, in fact what some researchers do is pick and choose among the procedures using those that most suit their purposes. They might make use of constant comparisons but not adopt the method in its entirety. Sometimes they use certain analytic procedures but use them in conjunction with other qualitative methods. This issue was addressed in Chapter 14 and it is suggested that the reader return to that chapter for further elaboration on what constitutes quality qualitative research. If a researcher identifies categories (themes), but doesn't want to take the time to develop the categories elaborately in terms of their properties, dimensions, variations, or relationships, then the findings will be "thin" and perhaps not very informative and probably won't add much new knowledge to the profession. If a researcher chooses to use bits and pieces of a method, then he or she can't claim to be using that method and most certainly will lose some of the credibility that a consistent follow through with a method will provide.

*Question 12.* "Can you say something about the work involved in doing qualitative analysis: the amount, kinds, and so on?"

*Answer.* First, there is the work of collecting data. Data may be collected by the primary investigator or by research associates or paid data collectors. Then there is the work of doing analysis, though we don't like to think of it as work in a negative sense. Analysis does take effort and it is time consuming, but it also is very interesting and rewarding. A researcher using this method should allow himself or herself a considerable amount of time, especially if the researcher is doing the transcription of interviews, in addition to doing the interviewing itself and the data analysis. If the transcription is done by someone else, then perhaps the workload can be reduced somewhat. If the researcher encounters difficulties in data collection or in analysis, understandably there is more work.

Unquestionably, a most important issue bearing on the amount and kinds of work is the ultimate aim of the researcher. If the researcher is aiming for densely conceptualized theory, then there will be more analytic work than in studies aiming at description. Yet doing think rich description can also be complicated and time consuming.

Another issue deals with what kinds of work are involved. If this book has been read carefully, the researcher is aware of the many forms of work involved in data collection and analysis. There is the work of data collection (with all the potential difficulties); also of recording, perhaps transcribing (even translating), and then the coding. When all of this is finally finished, there is still the work of writing papers or books and making presentations. Before the study begins, there is the work of grant writing, of obtaining human subject committee consent, and so on. In short, the only major difference between doing theory building research and other forms of qualitative analysis, or even other forms of research, is the amount of work that goes into the coding process. A computer can help with the work, but it still requires effort on the part of the analyst.

There is also the issue of what kinds of resources are needed for this kind of work, in addition to the requisite skills. Really, nothing is needed in addition to notepads, telephone, tape recorder and tapes, computer or typewriter, the usual paraphernalia of qualitative research. Sometimes money is necessary for travel and occasionally for paying interviewees. A good research library can also be very helpful and even a necessity, but with computers even these are becoming less important. Important also are one or more consultants or helpful friends. Included in your resource pool, if you are lucky, are an indispensable supportive spouse or significant other.

*Question 13.* "What's the relationship of everyday life explanations to our theoretical explanations?"

*Answer.* As we said elsewhere, you must listen very carefully to what the various actors are saying. Their words and expressions provide in-vivo concepts, give meaning, and provide explanations. Also, everyday explanations are usually revealing of the actors' perceptions, ideologies, and unwitting assumptions. So note these, be respectful of them, and integrate actors' explanations into your own interpretations.

*Question 14.* "If you've been trained in psychoanalytic theory or some other disciplinary approach, how would you integrate this into qualitative analysis?"

*Answer.* The actual techniques and procedures for qualitative analysis can and have been used by people trained in different disciplines and with their respective theoretical approaches. What these disciplinary theories do is tend to focus the analysis on certain problems, and at the same time provide a perspective for interpretation of data. For example, a person coming from a Freudian perspective might be more concerned with hidden motives and deep psychological meaning than an organizational sociologist, who is more interested in social organizational process and structure. The important thing is to be aware that perspectives influence interpretation. A researcher who uses qualitative methods is interested in making new discoveries and uncovering new understandings. Arriving at new understandings can be blocked if a researcher fails to think "outside the box." Procedures presented in this book have been designed to help researchers think differently about data but cannot guarantee it. It is up to the researcher to use them and to use them wisely. Once the analysis is complete, findings can be related back to the literature, provided the researcher explains where the findings are the same and where they differ from the literature.

More specifically, there is a basic tenet of the methodology that is relevant to the question. All assumptions of preexisting theories are subject to potential skepticism, and therefore must be scrutinized in light of your own data. The latter allow you to question and qualify, as well as give assent to your received theories. Concepts must "earn their way" into a study and not be blindly accepted and imposed on data. ("Received" theories may work brilliantly for some data, but not so well on yours.) Therefore, to summarize, psychoanalytic theory or any other theory must pass the "grounding" test.

*Question 15.* "How many interviews or observations are enough? When do I stop gathering data?"

*Answer.* These are perennial research questions asked by all researchers using qualitative methods. For most theory-building researchers and for achieving thick rich description with data collection, it is safe to state that the researcher continues to collect and analyze data "until theoretical saturation takes place."

However, there are always constraints of time, energy, availability of subjects, and other conditions that affect data collection. These can impose limits on how much and what types of data are collected. The researcher must keep in mind, however, that if data gathering stops before theoretical saturation occurs, the findings may be thin and the story line not very well developed. Sometimes a researcher has no choice, but must settle for a theoretical scheme that is less developed than desired.

*Question 16.* "How is this methodology similar to and different from case analysis?"

*Answer.* This is another one of those complicated questions, because in some part the answer depends on what you mean by a "case" and its analysis. The book *What Is a Case?* (Ragin & Becker, 1992) reflects upon this problem. Two sociological authors asked a number of respected colleagues to discuss how they used cases in their research. There was a wide disparity both in the nature of these cases and in how they were analyzed. Frequently, when one speaks of cases, persons interpret that to be an in-depth study of a single person or group. Often these take the form of a narrative life story, a career, or the handling of a personal crisis. But a moment's reflection tells us that a case can also be a study of a business organization, an African village, or a public celebration. Whether the researcher is analyzing a single organization or several, the process of analysis remains the same if using this methodology. The researcher would want to sample theoretically, and continue sampling, until categories are saturated.

*Question 17.* "Is using a 'basic social/psychological process' the only way to integrate a study? I notice that some researchers seem to assume this."

*Answer.* Usually, when persons say this, they mean that the findings are integrated around a concept and explained in terms of how the concept evolves in steps or phases. No, the use of a basic social/psychological process isn't the *only* way to integrate the data to construct theory. This assumption (certainly not made in Barney Glaser's [1978] discussion of basic social processes) represents a grave underestimation of the complexity of the phenomena that are likely to be encountered in any given study. It also hampers the potential flexibility of this methodology, restricting the strategies for integrating analyses. In every study, one finds process, but process should not be limited to steps and phases. Nor should it be restricted to basic social or psychological processes, unless the term "social process" also includes family, organizational, arena, political, educational, legal, community processes, and/or whatever processes might be relevant to a study. In summary, one can usefully code for a basic social or psychological process, but to organize every study around the idea of steps or phases or social/psychological processes limits what can be done with this method.

*Question 18.* "You emphasize that your method is both inductive and deductive, yet I often see it referred to in the literature as wholly or primarily inductive. Sometimes the reference is appreciative and sometimes critical. Can you comment?"

*Answer.* Again, this is a misunderstanding. In some part, it stems from a misreading of *The Discovery of Grounded Theory* (Glaser & Strauss, 1967). Glaser and Strauss emphasized induction because of their attack on "ungrounded" or speculative theories. The desire was to focus readers' attention on the inestimable value of grounding theories in systematic analyses of data. However, that book also emphasized the interplay of data and researcher. Since no researcher enters into the process with a completely blank and empty mind, interpretations are deductions or researcher's abstraction of what the data are indicating. This method is inductive in the sense that findings are derived from data. It's deductive in the sense the concepts and the linking statements are interpretative; that is, constructed by the analyst from data.

*Question 20.* "I am absolutely flooded with interviews. Unfortunately I haven't been able to prevent the flood. I never dreamed I would get caught up in this situation and not be able to stop the stream of interviews. I am so sated with the interviewing and information that I can't even think of asking new interview questions. Worse yet, I haven't followed the rules and done analysis while interviewing. What should I do?"

*Answer.* Your plight puts you in exactly the same position as most interviewers who put off analyzing data until most of the data are collected. This situation is precisely why data collections should be guided by analysis. Therefore, the best thing to do at this point is stop interviewing and start analyzing, get phone numbers, and make a later date with respondents. You will need these people later to fill in categories and validate the evolving theory.

*Question 21.* "Can you tell me something about the differences between the research method described in this book and, say, auto-ethnography?"

*Answer.* I can't tell you much because I must admit that I am not an expert in other methods. What I think is most important for the novice qualitative researcher to know is that each method has its own theoretical foundation, purpose, and procedures for collecting and analyzing data. Quality qualitative research can be done using all of the different methods as long as the user is true to the method. My suggestion is that the novice researcher wishing to do qualitative research explore all of the different methods before embarking on a study. Different methods appeal to different researchers. I probably would not be comfortable doing auto-ethnography for two reasons. I doubt I could be unbiased in analyzing data about myself. And second, I would feel uncomfortable revealing that much about myself. I could talk about myself in an anonymous interview, but not reveal myself in a monograph. I would be too self-conscious. That is just my personal bias; each researcher has to determine what he or she feels comfortable with.

## Summary of Important Points

This concludes our chapter on questions and answers. There are, no doubt, a great many more that could be asked. The chapter also concludes this book, thus there are a few words of wisdom that we wish to convey to our readers before closing. Readers are advised not to worry needlessly about every little facet of analysis. Sometimes a researcher has to use common sense and not get caught up worrying about what is the right or wrong way. The important thing is to trust oneself and the process. Stay within the general guidelines outlined in this book and use the procedures and techniques flexibly according to your abilities and the realities of the study.

## Activities for Thinking, Writing, and Group Discussion

1. Think about the questions that you have after reading this book.
2. Write them down and bring them to the group.
3. As a group, come up with answers to the questions. No doubt, there will be some disagreements as to the appropriate answers, and this is okay because probably there is more than one answer that is possible. Be creative, be flexible, and most of all be willing to put your ideas out there so that others can react to them.

# Note

1. I want to thank Anne Kuckartz from VERBI Software for her help and support with answering Question 1 and also, again, for her assistance in adding a computer component to this book.

# Appendix A

## Exercises for Chapter 4 and 6

### Field Notes

Biography Study

These interview notes represent just a few pages of a much longer interview, and are intended to accompany the activities presented at the end of Chapters 4 and 6. The study topic was "the biographical impact of a life-threatening cardiac event."

This person went into the emergency room with chest pain that was radiating down her arm.

The "event" happened while she was outside pruning her roses.

R = Researcher

P = Participant

Researcher: J.C.

R: Now getting back, when you were undergoing this procedure [placement of a stent into two blocked blood vessels] at any time were you frightened of dying or having something go wrong, or did you just trust that the health care system would take care of it all?

P: When going through the procedures, I was afraid I was going to die on the table. I remember thinking that. I better not die here. But the thought that I was going to die has never really entered into it for some reason. I have to internalize that. I'm having nightmares and things like that. But it's very interesting. The denial is incredible. I won't accept the fact. It's like my sister. She doesn't really ever think she had a heart attack. She thinks she really had—she refers to it yet as her event. She doesn't say she had a heart attack. And as I look at this thing that happened to me, it wasn't really a heart attack, it was just a little narrowing and they opened it up before they [narrowed blood vessels] did anything.

Now, the fact of the matter is everybody dies. The only reason I went into the hospital [in] the first place was because I knew my mother had three silent MIs, my sister only had pain in her elbows, my aunt—everybody in the family has these and they don't even know they're having them. And so I thought, well, I don't want to be one of those that doesn't know. At least I want to know what's going on, if there is some change coming about. So I knew that the threat was there, but it didn't feel like I was going to die. I keep internalizing that, internalizing and trying to make it click for me. Because if I don't make it click for me, I will die, that's it.

It's like doctor X said, she said, well, if you don't change anything, the same thing will happen again. It's as simple as that. And she's right. It will. I've got a bunch of other vessels that are waiting to close down, or one of them could drop off a piece of plaque. But [if] I would die early I would be one of those in the family that died early. I'm trying to get hold of that, and even if it doesn't internalize, it doesn't seem that way. I know that I have to take steps to make my life different or I will die.

R: One of the interesting things about cardiac disease is that you can't see anything. And so it's very hard to incorporate that into your being. Nothing shows on the outside.

P: Nothing shows different. You know, it's interesting. Now, the nightmares. I'm thinking that—and I know what it is. I know there's this little stent in there, this little wire cage. And I don't know whether I was half-thinking it or half-dreaming it, that it came loose. They don't do that. But it came loose, or it went sideways and it blocked the artery and I woke up terrified that it would shut off the artery. It isn't something you can look at, you're right. It's the history. Even my complaints about the whole cardiac experience in the hospital are not that bad. When I came home, I was having, you know, you keep thinking what is this, what is this, is there something going on? And I think, maybe that's a pain, maybe that's it. And by the time I go to find my nitroglycerine, the pain is gone. And I got, and I thought, when I came home, I had trouble with asthma. Now I don't know whether that was asthma, as I look at it, or whether I was having some kind of an anxiety attack. Because I never had asthma. I was given albuterol for coughing. When I got home I was coughing, and so I took some albuterol and was having premature ventricular contractions. Now that scared the bejeebies out of me. So I called the doctor and they told me to stop the albuterol and they put me on cardizam which slows down the heart and takes the sensitivity away, so

I did that then I went off that and went on, Flovent, which is a cortisone type of medication. They didn't want me to take albuterol again. Finally the funny stuff stopped. I took the Flovent for a week or so, and I didn't want to take it. I know, it's such a miniscule amount I was told to take. I mean it's topical, it doesn't get into your system at all. But somewhere in my head I was thinking you don't heal well when you're on cortisone preparations and I wanted those stents to heal because I was only going to be on blood thinners for a month and I wanted to make sure the healing took place in that period of time.

So anyway I tried to avoid it, but I had to take it, I couldn't breathe. I was coughing, coughing, coughing. I took it for about a week, two puffs a day, and the coughing stopped. I took the medication down to one puff a day, then I stopped it entirely. I've been off it every since. But I don't know whether it was asthma or whether I was having an anxiety attack because the worse coughing attack I had was when I came home. I sat there and started coughing, I could hardly breathe and I didn't know whether it was the cat or what it was. I noticed when I sat here for the whole week you think I would have spent most of my time in the garden. Usually, I'm out there fussing around in the garden in my free time. But I didn't do it. I didn't finish pruning the roses on the fence because I was home alone and I was afraid. I was going to tell you I called the doctor to change the prescription to get the Flovent instead of the other medication and I said to him, he said have you been having any chest pain. I said, I get this funny little twinge but I'm not sure what it is. By the time I find the nitroglycerine. . . . He said, wait a minute. I don't care what you get. He was one of those guys and I thought, oh, all right. He was serious enough about it, he said no matter what it is take a nitroclycerine and take yourself over to the emergency room. So I thought maybe there is more here. Or maybe I'm not hearing what they say. Because I thought it was a done deal, you had the stents put in and that was the end of it.

So I didn't do much that week. Actually it was like being on a retreat. I read my books and I looked out the window and I fussed a little in the backyard. I just love to be alone. I could be a hermit very nicely. But I didn't do much of anything. Usually while my partner and son are gone I do some huge project, paint this or do that, but this time I didn't. And shortly after they came back, we went to X and of course I wasn't going to go. And what I'm finding is that I'm treating myself like I'm frail. I got the flu while I was there and that was worse yet. Everybody was waiting on me, making my bed and bringing me ginger ale. And you know, I'm locking into this sick role. I'm thinking why am I doing this? Usually I hate that. But I didn't go places. We usually go to the same places, but I didn't walk this time. I didn't go down hill and I didn't walk along the ocean this time. I stayed on the top and I watched them do it. So I'm thinking to myself, I've got to stop this crap because somewhere I'm incorporating frailty into this. And I don't know whether I'm scared because I'm listening to them or if it is something else.

I guess it's because everybody's watching. You see that's it. I didn't want to get down the hill and not be able to get back. I signed up for cardiac rehab to get over this. I've got to make sure my insurance covers it. The doc wanted me to go to rehab. She said some people have good luck with it. I think it will be good because I haven't gotten on my bicycle yet either. I'm afraid to be out there without somebody to be with me, although I have my little telephone. I don't even know how far I can go. It's very interesting. . . . I think part of my reaction is at an unconscious level.

# Appendix B

## Participant #1: Veteran's Study

### Face-to-Face Interview

R = Researcher
P = Participant #1
Researcher: A.S.

P: Basically, I come from a middle class family, very patriotic, God fearing and religious. We were a very loving family and continue to be. I have three brothers and one sister. May father is dead. My mother died in her eighties. We all [get] together for a family reunion at least one time a year.

I left home at sixteen. I worked a couple of years at menial jobs, well not necessarily menial but low paying. I worked as an orderly in a hospital and that's how I became exposed to the nursing profession and decided to pursue that. I was twenty-one-years-old when I was first licensed as a nurse. Now that I'm fifty I have a long history of nursing in there. This was back in the 60s. I worked one year at a veteran's hospital in the city of X, where I was exposed for the first time to veterans, people who had been to wars. Primarily, there were elderly World War I people, some middle-aged World War II people, and a few Korean veterans thrown in. And I was pretty much interested in listening to them talk about their experiences and all that, so in 1966 when the government finally made a commitment to Vietnam, sending lots of men and women and materials, I volunteered to go. Well, kind of volunteered. I was one step ahead of the draft. So I volunteered to go. I did basic training at Fort Sam Houston in Texas, a six-week wonder. I came out as a second lieutenant and was immediately sent to Vietnam. I . . . most of the time I was there I worked in transport and an evacuation hospital. We went out in helicopters and picked up people from aide stations, which were pretty much . . . it's hard to say because there were really no defined lines. The lines could change every day, two to three times a day but the aide stations were in the areas of conflict. We would transport the most seriously wounded back to Saigon, which was about 75 miles away and the less seriously injured back to the evacuation hospital, which was about 25 to 30 miles away. Let's see . . . I was pretty young, twenty-one-years-old, very patriotic and gung ho, and thought that we had every right to be there and doing what we were doing.

I was very much anti-Vietnamese like most of the soldiers always feel about their enemies. I guess during the time I was there I started to become aware at little nips at my conscience, inconsistencies, but don't think that I paid much attention to them. There was too much going on to have really given a lot of thought to that. And I'm not sure that it's not some sort of unconscious mechanism that keeps you from looking at what you're doing and evaluating it. I don't know if it's because you don't want to or you choose not to. I'm not sure. It's pretty hard when you're in the middle of something to be evaluative while you're doing it. I actually can't say that my experience there was all that bad. I was young and kind of enjoyed that experience. I think it's the most maturing thing I've ever done in my life to be there and realize that people would want to kill me! As far as I know I never killed anybody else even though we had to carry weapons at times. I never shot at anyone. Not on purpose anyway. It was a strange time in my development.

A lot of things that I hold sacrosanct such as the value of human life I guess I saw that diminish I was there in '66 to '67 during the Tet Offensive when the North Vietnamese fought back and really won a great victory. I can remember in this one village, the village was called "Cu Chi," after they had been routed, there were dead Vietnamese, these were South Vietnamese, killed by the Viet Cong, and they were stacked along the road like racks of firewood and I can remember not having any emotion about that. It was just like "Hey this is war!" This is what kind of happens. So that kind of confused me because before that the thought of someone dying would send me into some sort of scurrying behavior. Working in a hospital, if someone is dying you really get concerned and upset about that. And I just really didn't feel anything about that. Like this was all well and good, that's the way thing[s] should be in war. It was a strange feeling. And if I remember correctly, most of the people around me didn't show any emotion about that either. In fact, there was a lot of jocularity. "Well that is one less 'gook' we have to worry about." That was a common name for the Vietnamese, "gooks". . . so let's see . . .

For a while then I worked in an evacuation hospital. They kind of rotated you from job to job. The strange thing is these were Quonset huts set up like hospital units and there were . . . we would have three kinds of people in there at one time, which was strange. We would have wounded American soldiers, we would have wounded South Vietnamese soldiers, and we we'd have wounded Viet Cong or North Vietnamese. So we kind of depersonalized those people. I remember when we would give report to an oncoming shift we would talk about our soldiers, use their names and stuff like that. I remember when giving report on a North Vietnamese or a South Vietnamese we would say bed #12 or the "gook" in room such and such. It was a way of depersonalizing that person so you didn't have to feel for them. You couldn't communicate with them because you couldn't speak the language. You very seldom had a translator or interpreter around. What I do remember about these men was how stoic they were. I can't remember them asking for something to ease their pain, which as I think back they must have been in. At the same time, unfortunately, I don't remember myself, or any of the other nurses or doctors ever taking the initiative to find out if they were in discomfort.

The wounds of war can be terrible. I don't know. I never thought about that at the time. I don't remember ever giving a Vietnamese anything for pain. They were very stoic. I do remember one incident where I felt sorry for this Vietnamese person and I don't remember if he was an enemy Vietnamese or a friendly Vietnamese, it's when he woke up after surgery and looked under the covers and saw that one of his legs was missing and he was crying. Being unable . . . I don't remember anyone, myself included, being able to comfort this person in any way. Hmm. . . . Then again this would be abnormal behavior on the part of a medical person outside a war zone. We wouldn't let people suffer emotionally or physically the way we let these people suffer. At times there would be conflicts in the units because we would have these three groups of people. Some American soldiers or South Vietnamese would see that their enemy was in there, the North Vietnamese or Viet Cong and there would be conflict. We would always protect them from the other people. We would never allow our soldiers to physically abuse them, although I do remember a lot of verbal behaviors, threats and all but I never saw any physical violence.

There was never a question about who would get care, or who would get supplies as they were needed. Always, the Americans or the Australians came first. There was an Australian division next to ours and they would wind up in our hospital. Ah . . . they always got priority of care and supplies. Generally there was enough to go around. So ah . . . I recall one incident where I didn't make the choice, but a choice was made to take a North Vietnamese off a ventilator and use it for an American solider because it was the only one available. That is the only time I remember that kind of decision being made. Most of the time was more of a case of benign neglect of their needs, to see if they really did want or need something. Sometimes I can remember the South Vietnamese interrogation team came into the hospital to interrogate the Viet Cong and I can remember at times they took the people out of the hospital. I can only imagine what happened to them. They would take them out. They said they were going to take them to another hospital but I'm sure they were taken and interrogated or even killed. But again, at the time, in all reality that didn't bother me. It was war and they were just faceless people. They were just another North Vietnamese to me. [Pause]

Like I said there were times when it would slip into my consciousness, I would think about the inconsistencies. It was not only the treatment of the Vietnamese that bothered me but there was a hierarchal system within the American army system. I was an officer so I had a lot more privileges than did the basic soldier. They would have to work a twelve to eighteen hour shift at a stretch whereas officers did not. They were the "grunts," but that's the military. That's consistent worldwide with military everywhere. I'm trying to think about my peers, to think back to see if we had any discussions about what was going on. I don't recall any. I really don't know anything about how other people were feeling while they were there, if they were having any problems with what they were seeing or not. It amazes me how comfortable you can get in that situation. You get up and go to work and it just doesn't seem to bother you a great deal. I guess that's part of the whole human adaptation that goes on. You just adapt to the surroundings. But life took on an almost normal feel at the time. You had parties. At times the big concern was where are we going to get enough beer. Or can we trade some penicillin to another group for some whiskey or something like that. We never thought that maybe some other group needed that medicine.

R: Were you ever attacked? Did you ever feel in any danger when you were there?

P: Do you mean the compound or the hospital itself? The hospital itself came under fire very often and there were people killed in the encampment. When fire did come we had to move patients out of their beds on to the floor on their mattresses. The buildings, the Quonset huts were made out of tin and when a shell would hit there would be shrapnel flying around. But we never moved the North Vietnamese. They stayed in their beds. Americans went on the floor on their mattresses out of the line of fire. [Pause] Some of the other inconsistencies were that during the day we allowed Vietnamese to come into the encampment to work, clean up the place and that kind of thing. You don't know if at night they went out and put on their black pajamas and became Viet Cong. It's like in the daytime you are okay. We can see you. We don't know who you are at night, that kind of thing.

I stayed there for a year. In retrospect it was not a terrible year. It went very fast. It was very maturing for me. Um . . . it was in '67 that I came back. That was when the peace movement was starting to be heard very vocally. I remember my first stop after Saigon was the San Francisco airport. They made us take off our uniforms and change into civilian clothes because people in the airport were throwing things at the soldiers coming back from Vietnam and calling them murderers and things like that. That made me really mad. I thought I had gone over there and taken part in something all well and good and how could they treat us like that. Over the years my feelings about that have changed. It was senseless for us to have been there. It's hard to lose your patriotism. It's hard to give that up. What I think that the experience did to me is give me the motivation to do something. I was maybe twenty-two or twenty-three by then. I don't remember which but by then I had formulated plans of what I wanted to do when I was discharged. I came back to X to finish my time [military] out there. I applied to the university and received a bachelor's and master's in nursing. I was very busy. I worked part-time and went to school. I was really too busy to think about that whole experience. I just put it on the back burner and went on with my life. I really, at this point can say that there weren't any major negative affects of the war on my life. It's hard to know over the years and my feelings about war and killing have changed. It's hard to say what cause the change, whether it's a maturation process or whether it was just becoming aware of all the inconsistencies and feeling the futility of war. I normally have avoided situations where I would bring this stuff back into consciousness. I have never gone, never went to watch a movie about Vietnam. Those never had any appeal to me at all. I don't know why they don't appeal. I never tried to maintain any friendships with any of the people that I knew in Vietnam. I got out of the military. I knew I never wanted any more of that, so I got out. When I got out of the military I severed that relationship completely. It's almost like that it was a part of me that I find almost difficult to recall. It's like that experience was part of me, it's over with, and it's gone. It's something that I seldom ever think about and less ever talk about.

When I think about the impact of the war on me it was a positive one. It seems strange to say that war can have a positive impact. I met some people in Vietnam, motivated people and it kind of motivated me to go on to school. [Pause] I would say if I had to put any kind of weight on it, it was probably more positive than negative. It was a maturational process. I probably would have matured anyway but this was kind of instant maturity. I was still angry when I got out of the military. This was 1967 and the peace movement was big. I was in college and I would get angry with the student marchers, groups, and stuff like that. There were still soldiers over there and I know that it hurt them to watch that, to see the news and all of that. Now looking back as I said before I admire the marchers. At the time I was seeing them from my viewpoint, a patriot and they were seeing the war from their viewpoint, "this is all wrong." So looking back now I admire those people who at the time had more insight into that situation than I did at the time. It was wrong.

R: Let's talk about that a little bit. There are two things I'm interested in. One is that war is a maturing experience, certainly understandable. Can you say more about that? Then I'd like to know more about the looking back and the change in perspective about war that has occurred with time.

P: I guess the maturity came from learning how to set priorities. Ah, being very self-reliant, learning to speak up for myself. [Pause] Along with a maturing experience it was also a hardening experience. I think I learned during that situation to not be so sensitive about things, people suffering, the human condition because if you allow yourself to be that way when you are in that kind of situation I don't think you could function very well. Maybe it did harden my sensitivity to people suffering, to pain, death those kinds of things. Ah. . . .

R: You went into the war, like other friends of mine, with a pro military background and totally accepting of the American government. Okay, what happened to that in this?

P: If I follow your train of thought, I was able to separate myself being an American from the government imposition of war on the people. It changed me as an American. It now and was the beginning of a process . . . back then I felt that the government would do the right thing, that our leaders would always do what was best for our country and at that time what was best for our country was supposedly good for the world. We were riding high then. I guess that I lost that naivety . . . well you should turn over all that personal power to the government, that those people up there in Washington would always to the right thing. So that was part of the maturational process. My two older brothers were also in Vietnam. I was there at the same time as one of them. It is interesting that over the years that Vietnam has never been a topic of discussion. Ah . . . they've gotten on with their lives and have been successful. It is not something that we reminisce about at all. Again, I'm not sure what that means. I'm not sure if that means that it was something that we are not proud of or something that is history and not worth bringing up.

R: You got on with your life in one sense?

P: That's the phrase I like to use, "just get on with life." It's one stepping stone and you go on. It's really hard for me to say how it impacted my life. It's been almost thirty years. And things happen along that continuum of life that make me who I am now and so it is hard for me to directly relate who I am now to that experience, I really can't.

R: Say more about that. I'm not trying to pin an impact of the war upon you. As you say it is only one set of events. But as you look back now, where does it fit in to the additional steps that you took.

P: I think it was a stepping stone for me, a motivator to maybe try to fulfill my life as well as I could. Maybe I saw the futility that life can lend. I'm not sure but I think of that experience as a springboard.

R: Would you have sprung into the university otherwise?

P: You know, I doubt it.

R: Why?

P: I don't know. I was quite content with my level of education. I don't know if I would have stayed that way. I could have moved on. I think it was some of the people that I met in the military.

R: Nurses?

P: Yes, I admired them. I thought that they were very competent, the higher-ranking officers, the older people who had been in the Army Nurse Corps much longer than me. At the same time I didn't want to emulate them. There was something about being a career officer that didn't appeal to me. There was something about the way they approached life, their attitudes about life that did not appeal to me.

R: And what was that?

P: I think that they gave over the decision making for their lives to someone else. I'd always hear them talking about, "I don't know where the army will send me next." And I thought I know where the army will not send me next because I will make the decisions about where I go and where I live. There was a certain hardness about them. That's not categorical. There were some really good people but in general the career people were more interested in what this experience in Vietnam would do for their careers more than anything else. It was a great opportunity to get promoted. I went from 2nd lieutenant to captain in two years, which in time of peace would take ten years. Now it would probably take fifteen years to get anywhere. And for many of the career officers who had been majors or colonels since the Korean War, this was an opportunity to get their long-awaited promotions. They did all kinds of things. I remember a couple of men, they were physicians, the men were all in the tents they slept in and we were having a discussion one night. One man cut his foot somehow and he was wondering if he could get a purple heart for this because everyone was out to get as many medals and accommodations as they could get and I thought, I remember thinking there are people who are getting their legs blown off and their eyes blinded and who will get a purple heart and you're thinking about a purple heart for a cut on your foot that you probably did out of carelessness. But again they were looking at how the war would help out their careers.

R: Did that kind of thing shake you up in terms of the military?

P: Absolutely, absolutely. Hm . . . a lot of wheeling and dealing went on including a lot of black marketing, especially in medical supplies. That used to bother me. Like I said, they used to trade a case of antibiotics for something else, a case of beer or something like that.

R: So despite the good care they were giving they were doing other things too.

P: Well you say good care, that's relative also. I can remember times when the doctors and nurses would be so drunk that they didn't know what they were doing. However that was the exception and not the rule.

R: So why did you stay in nursing when you got out?

P: I was a nurse already.

R: But you could have shifted into something else when you went back to school.

P: I could have. That part I didn't have any problem with. I thought that there was room in nursing for anything you wanted to do. That experience didn't shake that part of it. Again because that was just one experience. Vietnam was one experience, one year of my life. It really didn't change my professional focus.

R: One of the things that J. spoke about with one of her students who studied Vietnam nurses was that people were upset in their professional hearts because they were saving people to go back to battle. The severely wounded, the very hurt, were allowed to die, which is the reversal of the usual medical way of treating the worse off first.

P: Again, that is the military way. The goal of military medicine is to return people back to the position they came from, be it a foot soldier, a pilot, whatever. So, there would be situations after a bad battle or attack where our hospital would be inundated with 150, 250, 300 people and there was a definite triage that went on in that people were shunted to different treatment areas. People were kept comfortable they were given narcotics to ease their pain. I think there were six operating rooms and there might be a backlog of 100 people. And those people who were severely injured never went into the operating room. They were allowed to die. I wasn't part of that triaging. I'm trying to think how I would have reacted. I think that I would have been okay with that. Again because it was the "military way."

R: Let's go on. After you were discharged you say you took the next step with your education and went back for a nursing degree. Can you recapture some of the things that went on in the university around '67, '68, and '69?

P: I was pretty busy most of the time. I went to school full-time and I worked part-time. I never took part in any of those demonstrations, if that is what you mean. At the same time I don't remember feeling. . . . After a time I began to feel that they were really right. I never supported the demonstrations. At the same time I was never negative about them.

R: Why did you think they were right?

P: Umm. . . . In the late 60s early 70s it became apparent not just to me but the whole nation that we had been caught up into something that was . . . unavailable and that was peace. We would have all these false stops and starts, treaties, stop firing and start up again. Then the political situation in Washington, we were committing billions of dollars and yet our social system was breaking down here in the States. I think I thought more about it being wrong in those terms as opposed to the wrongness of people dying. I thought the war was causing social unrest and upheaval, an impact upon our country. And I think at the time to me that was more wrong than what we were doing to those people over there because I still kind of depersonalized them, the Vietnamese.

R: And what were the specifics of what you saw that made you feel that way?

P: I can't remember the specifics. I'm trying to think about what was going on here economically, but I think that is was more that needed social reform was not going on here. I remember that the age of students was going down and inflation was going up. I remember that I was starting to have a hard time living on the money I was earning. I can't think of anything specific. Mostly I led an insulated life. I was living my own little life and really wasn't aware of the whole big picture.

R: Why did you say the social order was breaking down?

P: That was the rioting. I remember the 1970s and Kent State because I was still in college at the time and I remember a lot of demonstration[s] on our campus and thinking how could we turn that way on our own people and shoot them. I had some sympathy for those that were caught up in that situation. That may have been one of the turning points in my attitude about government. I'm not sure. I was losing more and more confidence in the government.

R: Did you have contact with other vets?

P: No I really had no contact. I remember that there was a veteran's organization on campus but I didn't have any desire or time to be part of hat. It wasn't cool to be a vet at that time. I can remember in some of the classes, sociology classes, that the topic of Vietnam would come up and I never volunteered and I never spoke up. Absolutely, I never would, I never wanted to be identified in any way as a Vietnam vet. I was a little older than most and I was taking Soc. 101 with kids eighteen and they were all worried about the draft and the unfairness of it all. I never opened up myself to any of that.

R: And yet your attitudes about the government were changing.

P: It was a gradual shift. I started losing confidence in the government. [Pause] Again I can't see where any governmental policies were having any great affect on my life because I was really focused on what I was doing and I was doing okay. But remember at the time that the Head Start program was disbanded because there wasn't enough money to fund it, everybody was talking about all the money going to pay for the war and that things were not being taken care of at home. I remember in one class that there seemed to be general disapproval of the government in the classroom. People were negative about the government. It was starting to get to me. I hadn't yet lost all my confidence. You might say it was slowly eroding. It's hard to say because where I lived was home to one of our major presidents and it is hard to be negative when you respect these people. It's hard to let go of that respect.

R: The war went on and on and you went on with life.

P: Yes, the war went on until 1975 but you can draw a curtain on a part of your life. I did not spend a lot of time thinking about it. I can remember how excited I would get at times when they said they had reached a truce and the fighting was going to stop. The next day it would start over again. I don't recall having any negative feelings about the Vietnamese per se though at this time because I was losing confidence in the government. I thought we were just as much the blame as they were. And then I started thinking that maybe we should get out.

R: Was this early or later?

P: It was later. One thing I can remember doing, they would publish the name in the paper of soldiers who had been killed. I always read the names to see if there was anyone on the list that I knew. [Pause] But as far as trying to keep up with the day-to-day happenings of what was going on with the war I did not. As for the peace talks in Paris, I remember them going on but I don't remember being that interested in them. I think that like most Americans I felt that we got out with our tails between our legs and that I think that was when I really made the decision for myself that war is futile and nobody wins. And I think there was some anger towards the government because they never really committed themselves to the war. And I remember how the government would never call it a war. It was the Vietnam Conflict. They would never come out and call it a war because Congress never declared war. And so it was a play on words war vs. conflict.

R: What about Cambodia?

P: That was going on all the way through because the place where I was stationed Cu Chi, was only 75 miles from the Cambodian border. It was not unusual to go into Cambodia because in those countries there is not a well-defined border. It was more like behind this tree is Cambodia, behind that one is . . . it was not news to us that they had been in Cambodia.

R: Tell me more about the time after leaving the military.

P: I graduated with a master's degree in the early 70s and took a teaching position in a school of nursing. I taught for twenty years after that. That must have been pretty much the right decision at the time because the career lasted. I stayed at the university and went on and got my doctorate mostly because it was being required for teaching and tenure. By then I was ready for a move and left the state I was teaching in and came here.

R: Have you been to the Vietnam Wall? And did it have any impact on you?

P: The wall itself did. I went to Washington just to see it and I remember becoming very overcome emotionally with the wall. I went specifically to look for someone's name, someone I had known who was killed, and when I found the name I remember a real rush of emotion. At the same time, I think that it was probably for me a cleansing experience. After I had been there, seen it, expressed my emotion, that was the end of it. I didn't have any lingering problems with it.

R: And so the whole wall experience so to speak centered around the person you were looking for?

P: It seemed to be that way. If I were to conceptualize that and say one incident that characterized that whole situation that name would have been it. It was finding that name on the wall.

R: Describe that day for me.

P: I remember it was a cold day in Washington that morning. It had been raining earlier that morning. I was sloshing around that part of Arlington. I couldn't quite find the wall and so I had to ask someone. I remember that the person I asked didn't have any idea where the wall was and I remember thinking that is strange. Then when I finally found the wall it was completely deserted. I thought there would hordes of people around but there wasn't. I was the only person at the wall at that particular time. I remember looking at the wall and trying to make some sort of sense out of it. I read what it was supposed to represent, which I forgot already and I remember looking at it from different angles and just sitting there. There are some benches and a table with something like a history book. It kind of helps you find the name of the person you are looking for, what part of the wall their name is on. I remember flipping through that. I don't remember any special emotions at that moment. Then I found out where the name of the person was supposed to be. I went over to the wall and when I got a little closer there was evidence of people having been there. There were mementos and flowers. That was a little more encouraging because I think that I was disappointed that there weren't more people around that there wasn't more an expression of grief because in a cemetery there are all sorts of monuments and stuff. To me I thought it should be "the monument!" In retrospect it was probably no more important than the monuments to the World War persons that died. But to my mind it should have been outstanding. There should have been bands playing, people there all that kind of stuff. But there wasn't. It was actually lonely. It was lonely. I don't remember any specific emotions after I left. It was one of those things that I wanted to do, and that was that.

R: What about all the attention Vietnam now is getting?

P: I find it very interesting that Vietnam is opening up a tourist trade. They want our people to come there. And people are going back to see where they were at. It's kind of like that if you can't fight them, join them kind of thing. To be quite honest I would like to go back. I think mostly out of curiosity. I don't think that I am looking for anything specific or trying to solve any leftover problems. I think it is curiosity that drives me. I think it would give me a picture of how futile it all was because nothing has really changed. They are still there and we are still here. Nothing much has changed. Um . . . I don't have any animosity towards the Vietnamese whatsoever. I think the war was something imposed on the people and they had no choice in fighting. Of course they had a history of occupation for many years. And their loyalties are to whoever is in power. That's how they adapt and survive. I have no problem with that.

R: How do you feel towards the Vietnamese living here?

P: Not a great deal. I've taught a great number of Vietnamese students. I did go to their Tet festival one year. What I found so interesting was that the Vietnamese children who were born here, the children of the immigrants were about six feet tall and their parents about this tall. We always think about Vietnamese people being so small. They are small there for one reason, and it's the diet. I just found it funny to see all those tall Vietnamese kids walking around. I have a lot of admiration for the Vietnamese who have come here and made successes of their lives. They've gone on with their lives.

R: Did you even have a close friend that died in the war? And have you read any books or novels about this war or other wars?

P: No I didn't. The person who died there died after I came back and he was the brother of a friend. So that was the closest that I came to having someone who actually died. As for books, I've read a couple of funny lighthearted ones. There are two books, one is called *The Tunnels of Cu Chi*. It was a story about how the Vietnamese dug tunnels. That's why we could never get them out. They lived in those tunnels. Since I was in Cu Chi that was interesting to read. I read one that was almost a farce on Vietnam. It was called *The Book That Picks Up Bullets*. It was basically biographical. It's pretty funny, about the absurdities of war. I know there are many books out there but I have no desire to read them. There are no heroes out of that war.

R: Did you carry any images with you when you went off to war?

P: I don't know if I did or not. My family did not have a military background. Patriotic yes, but not military. And my brothers went after me, so I didn't have any preconceived ideas about military life.

R: What you are saying in summary then is that the war hit a boy in his early twenties and that it was a maturing experience. It made you grow up fast in certain ways. It doesn't seem to have crippled you because you've had a good career since. On the other hand you've sealed off certain things.

P: I'd say that I sealed them off, yes. I don't think, personally, I don't think that the war has been a negative in my life.

R: I'd like to return for a moment to the life in the evacuation hospital. In your mind, or structurally, was there any distinction made between the North Vietnamese and the South Vietnamese in how they were treated?

P: They were treated differently. I remember that there would always be a contingent of the South Vietnamese Army that would come to the hospital and talk to their soldiers and bring them little gifts, things like that. Of course, there was no one to visit the Viet Cong except the interrogators from the South Vietnamese. But, we did give more attention to the South Vietnamese, the friendly Vietnamese as we called them. Again, we had difficulty being able to communicate with either group because we couldn't speak their language. So sometimes what looked liked neglect . . . I think in my own mind it is just that we weren't able to do as much as we would have wanted to. But it was a hospital and we gave care as best we could. There was not much of a military feel inside the place because there were no guns in there.

R: When you went to the wall why did you look for that person?

P: Well he was the brother of someone I was close to. I draw a corollary on that. As I told you before, I'm going to Washington in October because the AIDS quilt is going to be displayed there and I see a lot of corollaries between the quilt and the monument. And I'm going for the same reason. My lover died last year of AIDS and he has a quilt panel we did and I think a lot of it is the same type of cleansing experience that maybe I was looking for when I went to the wall.

R: During the war and your time in the hospital did you begin to have a kind of distancing not only from the army, but from what was going on, some doubt about what was going on. At age twenty-one or twenty-two to be able to differentiate yourself, how were you able to do that?

P: I think mostly in my spare time, my free time I started giving a lot of thought to things that I wanted to do when I left Vietnam, like places that I wanted to visit, where I wanted to live, where I wanted to go to school, things like that. I think that this futuristic orientation kind of helps you separate from the reality of the situation that you're in. I thought more about the future than the present.

R: Is it also a reflection of maturation even if it is somewhat defensive?

P: I think so. I think anyone that did not deal with anything beyond that day . . . I just think that they would have more difficulty dealing with that . . . I could see. . . . And probably by formulating my plans about the future also subconsciously did tell me that I had a future, that I was not going to die, that I was going to get out.

R: Were there any individuals that you would be willing to mention that played a role in the shift in your life or was maturation all an internal process for you?

P: Well there were individuals both on a personal and professional level. At the university when I returned I certainly met some exciting people, teaching and those kinds of things. One of the things I had forgotten to mention, left out . . . I don't know if it is important. . . . When I was in Vietnam I came to grips with the fact that I was gay. And I met someone when I was there and so it is kind of interesting. It was an exciting time for me. I came to grips with whom I was. This same person who was there in Vietnam with me, we ended up moving to X together and lived together for six years. So there is that both on a personal and professional level that helped mold me. I've often wondered why it happened there. Maybe it was the freedom there and maybe it was, there may not be a tomorrow. You better experience today. My lover was drafted into the army, whereas I joined. He stayed out until they drafted him. He was not there by choice. But he was more professionally oriented. I can remember he was more concerned about conditions in the hospital than I was. Things that he saw that could be done better, how people behaved. I think he was somewhat of a role model for me.

R: Did he share the curtain?

P: Again you'd think that two people who had been through that experience and lived together for six years would talk. But I don't remember us ever [having] discussion [about] that part of our lives. We just kind of moved on and that was it.

R: I have another question. Given AIDS and all that it has stirred up, do you see any relationship between this and the war or are these separate events?

P: No, I think . . . I consider the fight against AIDS a war. The people who are most affected are mostly young men. So you can draw that kind of corollary between a war zone and the people who are dying around the world from this disease. Also, I think that what we've seen we have these dichotomies very severe very distinct those people who are very pro as far as winning this war against disease. You have other people who don't really care. And the same thing about the war. You have those people who want to get in there and do all the right things and you had people who didn't care. Socially I think that there are a lot of relationships and similarities between the two. And I don't have any more confidence in the government committing themselves this way than to their commitment to winning the Vietnam War. I've been so touched by AIDS that I can't separate that. I guess I'm more antigovernment because of AIDS than I've ever been.

R: A lot of veterans that came back from Vietnam eventually became converted by the demonstrations and are still angry and upset because their own efforts were discounted.

P: You know I think that any soldier can say that about any war because it has not made that much difference. That about all the countless wars, all the people who died. Is the world any better because of their deaths? I think not. Look at what's going on now. I would like to think that everyone who died, that made the effort, who gave up something, made a difference. I'm no longer able to say those kinds of things are worthwhile. I don't think that they have any lasting value for society. Apparently we don't learn from them.

# Appendix C

## Participant #2

### Part 1: Electronic Correspondence/Questionnaire

Dear Participant #2,

I was so happy to receive a reply from you and I can see from the dates that you gave me that you were there in the thick of things.

If we were doing a face-to-face interview, I would ask you to tell me your story about Vietnam and sit back and listen. But since we are not face to face, I will give you some topic areas and you can take it from there, adding or deleting as you see fit. I may ask you after you respond (if you continue to choose to do so) some follow-up questions based on what you said for clarification.

1. First it would be good to get a couple of lines of background information on you—when you went to Vietnam, such as your age, something about your family relationships, if you have siblings and did they serve, were they patriotic and supportive.

I was 21 when I went to Vietnam. I came from an average southern family—my father being a schoolteacher, coach and athletic director. My mother was a homemaker and I had one sister 19 months younger than me. I wasn't married or engaged. My father was a WWII combat veteran flying 50 combat missions on a B24 out of Toretta, Italy. My family was supportive of my choices, not necessarily the war in Vietnam.

2. Did you volunteer or were you drafted?

I was a volunteer as all marines were when I entered service in 1964. I did not serve with any draftees in Vietnam.

3. What was your role, combattant, noncombatant?

I was a Combat Marine Rifleman also certified in 3.5 inch rocket launchers.

4. Describe something of what it was like for you to be there, to be engaged in battle (if you were), to be fighting, and how were the enemy. (This is really the heart of it.)

The Viet Cong were a very well-trained and disciplined military force who gained foot holes in local villages by terror, killing, and torture. Marines like myself were extensively trained to follow orders, no question why or the politics of the situation. I could and would kill without hesitation as that was my job and I was trained to do just that. It doesn't take long for one to get into the grove seeing his friends wounded and killed. The killing becomes a habit and self-defense as time goes on and you survive. Marines fight for other marines and the corps, not necessarily the cause.

5. Did you feel supported while you were there and how did it feel to come home to all the antiwar movement?

I was always supported when I served. There were a few of us that did not want to be there but no one wants to be in a life or death situation in combat if they have a choice.

As far as the antiwar movement was concerned, that's one of the reasons GIs fight. The right of free speech, right to protest and right to live free. However, when that movement attacks GIs due to their choice to serve, call them baby killers just to mention one name and to have never served this country in anyway with the exception of running their mouths about things they know not or never will know anything about, I detest to this day and to my grave. These groups will be the downfall of the United States as we know it. The anti-war movement did nothing but gain a dishonorable peace and disrespected 58,000 Americans who paid the ultimate price for the rights of its citizens. The GIs of the Vietnam War were treated like traitors to the student and activist antiwar movement of that era. That should never again happen to an American GI.

6. Would you describe the experience as a maturing experience, a bad experience?

Maturing? I considered it a surviving experience.

7. Did having been in the war impact the rest of your life in any way?

Every combat veteran and some who were not are affected for a lifetime by the killing, carnage, loss of friends and family. Some carry the burdens easier than others. Outwardly anyway.

Like I said, basically I want your story as you are looking back on it today.

In closing, I joined the Marine Corps by choice out of State University. At that time we only had advisors in Vietnam. Myself as well as my entire unit did not join the corps especially for the Vietnam cause. I joined as John Kennedy said, "Ask not what your country can do for you. Ask what you can do for your country." I wanted to give something back to the country and people I so love. Myself and tens of thousands of others were in the same boat when the leaders of this country who were elected by the people took us into the Vietnam cause. I'm a true American patriot and believe that those who choose to serve or are required to serve should do just that in an honorable way. Those who choose to attack us for our service, those who ran to other countries are not the foundation this country was built on. These attitudes carry to this day with many and never should have been tolerated or excused by the American people. The difference

> with Vietnam compared to WWII or WWI, we weren't attacked by a foreign force. The GIs of all those eras are no different in their service to the United States. Just the cause.

I do thank you so much for taking the time to do this. If you wish I can let you see what I am doing with these materials. Also, when the book is finally finished I can send you a copy. As I said, it is a methodology book but I do need materials in order to demonstrate to students how you work with qualitative data.

## Part 2: Electronic Correspondence/Questionnaire Follow-Up

Dear Participant #2,

In the first interview, which by the way was done with a good friend of mine, several themes came out and I wonder if you could respond to them. I think in some way you have but wonder if you might say more. One is about the "culture of war" and how that conflicts with standard behavior. Because of that conflict, at times my friend had pangs of conscience at what he was seeing and doing but the only way to survive that was to push those thoughts aside, see the enemy as the "enemy," one who would kill you if given the chance, call them "gooks" to distance oneself from them being human, and just not talk about it. In fact he had never talked about the war with anyone during or after the war up until the time of the interview. He just blended into the college campus when he got out, avoiding all antiwar activities and discussions on campus.

1. Did any of that haunt you then or afterwards and how did you deal with it?

It has haunted me everyday of my life. Not a day passes that I don't remember something about that era. I never mentioned or talked about Vietnam to anyone including my wife of thirty-seven years until the late 90s.

I guess what I'm getting at is that you say that you thought of it as a survival experience, but what were those strategies that enabled you to survive?

Surviving the war was a matter of pure luck. You happened not to be in the wrong place at the right time. That was merely luck. You could not survive the war by being careful, a coward or trying to stay in the rear with the gear. I know guys who served an entire combat tour without even a briar scratch and then I knew others who were there less than thirty days and nearly blown in half.

2. How did you deal with the death that was happening all around you?

Death and mutilation is all around you in war and it becomes a matter of acceptance and habit. You mentally try to remove yourself from all the carnage and put your mind in another place and another time. Your mind spends hours upon hours at home in a warm, dry, clean, safe bed with family and loved ones. It's my opinion that marines were better trained than some of the other services to deal with the carnage. Not better GIs, just better trained and much closer to each other.

3. How do you turn that off?

I was able to mentally remove myself from the carnage. I always felt if I dwelled on it and allowed it to consume me I would be the next one hit.

Then and now?

Since Nam and now I put it completely out of my mind with friends, family and loved ones. I avoided drinking completely as booze would bring on the most vivid mental attacks of rage, anger and depression. I would not be talking about it today unless a great friend of mine through boot camp and Nam found me after forty years and all the memories flooded back into my mind. Talking with a brother you served with is easy but not the general public. This guy was a machine gunner in my weapons platoon and now we see each other regularly which allows us to dump all the memories on each other which is like taking a drug. I've been so lucky to have a woman in my life who never pushed the issue, never asked questions, held me quietly when the nightmares came and gave me her unyielding support.

4. Just the name of your Web site intrigues me "n.g.a."

N.g.a. as you have guessed has to do with the ghost of war and Vietnam. The name popped into my head in 1996, thirty-one years after my Nam tour. Several dozen Nam vets used to gather at a Web site put up by a lady and Vietnam vet supporter who was never associated with a veteran or Vietnam in anyway. It became too much for her to deal with over the years so I put up a chat room and Web site to honor my unit and maintain contact with many Vietnam veterans I've met over the years. Mostly marine combat vets but we have a few others from other services including the air force, army and navy who join us weekly. We're a very tight knit group and stay to ourselves for the most part. During our gatherings online we try to avoid the ghost of Vietnam. Therefore the name, n.g.a.

5. Another theme has to do with "the enemy," who they are, how one thinks of them. Did you ever have any direct contact with the enemy, such as prisoners, and if so, what was that like?

The contact I had with the enemy were dead or dying. I watched several last breaths and can see each one today as I did then. We had intimate contact with ARVN (Army of the Republic of Vietnam) which in some cases I'm convinced were VC, the enemy. There was no difference in the Vietnamese friend or foe as far as the people were concerned. They were of a different culture and religion but human. I never view friend or foe as nonhuman or villains.

6. My friend was a medic and so at times had to "treat" the enemy and this was difficult because they supposedly were "the enemy." Also, there was the fact that during the day Vietnamese were allowed into the base to do work and all the while you knew that these same people probably put on pajamas at night and took shots at you. So there was always this internal conflict and sense of distrust when dealing with Vietnamese people even those from the south.

Like your medic friend I did not trust any of the Vietnamese, friend or foe. You never knew what they were from one day to the next. Under the right pressure of being killed or tortured, your friend on Monday was your foe on Tuesday. They were still human, just the enemy. You depended on your GIs who came from the same land as you.

7. Would you say that the war hardened you, made you more sensitive and feeling, disillusioned you about war?

Unfortunately, war has become a necessary evil of the world as there are cultures who want to murder us, each and every one. I'm not against war under the right circumstances and Vietnam for sure did not make me a pacifist. I viewed myself as hard nosed before Vietnam, owned my first gun when I was seven. Hunted alone before I was nine. Things that your parents would go to jail for today. Not then. Vietnam showed me how many Americans really are in their attitudes about God and Country. I learned they are all about themselves and will kill Americans to have their own way or force their views on society. What ever one wants to call these people need to give this old GI a wide berth in life. If you want to brand that hardened, yes, I'm hardened. It's my feeling the elected leaders of this country should put GIs in harms way only as a last resort. WWII was a last resort. I'll have to say, I'm not sure about Vietnam, Korea or Iraq. The average American does not have the information at hand as our elected leaders have to make the determination of war. History will prove whether these other wars made a difference in the world or the well being of the USA. I wish I would be here for those answers. I detest seeing humans abused, tortured and killed now and before Vietnam. I think we are blessed as a country as well as a people which puts us in a mindset to help others. Is this a justification of war? I'm just not sure and don't have all those answers.

8. Have you been to the war memorial and how did that affect you?

Yes, I've been my one and only time. No way I can explain how seeing those 58,000 names many being GIs I served with as well as friends from high school and college affect me. I will say, I never want that feeling again.

Thank you again. I do appreciate your willingness to share some of that experience with me.

Ms. Corbin, in closing I just want to warn you if you don't already know, asking these questions of some Vietnam vets will bring on aggressive responses and sometimes verbal attacks including guys who patronize my Web site. I would say most of them as matter of fact. I choose and never have edited the message board and the guys know it. We offered our lives for freedom of speech as well as all other GIs who have served. Who am I to censor free speech? I've tried to accommodate teachers and students like yourself over the years with basic input to enable those who were not involved to see the views of many, especially the views of veterans in a feeble attempt to create an understanding of their views. Just don't take it personal if some tell you to take a hike.

# Appendix D

## Participant #3

### Part 1: Electronic Correspondence

Hey J.,

I read your post at N.G. I am a Panama, Saudi, Bosnia, veteran. I served with the U.S. Marines. What can I do for you?

Dear Participant #3,

I thank you for your response. I am interested in your war experiences. My interest in this topic started as I was looking through materials that I had at home to demonstrate to students for a text on qualitative research (3rd edition, Sage Publications) that I'm writing and I found an interview with a Vietnam War veteran. I had it but had never really read it. You know you read something but don't really read it. After reading it I became very interested in the war experience from the perspective of those who have to serve in those wars, the front-line soldiers. The subject now goes beyond the book because I think it is a story that can't be told enough. I've read some of the memoirs from Vietnam and frankly am astounded at what soldiers face and how little we know or understand what it is that they go through. So basically, I am asking any marine who will tell me, to tell me your war story, things like your background, then why you enlisted, what you did in the service, did you see battle, what was it like, how you lived through it, and how you now live with those memories, anything that you want to tell me or want others to know. I always remove any identifying information from my database. If you are still interested let me know.

Thank you,

J.C.

## Part 2: Electronic Correspondence/Journaling

Hey Julie,

To start here is an excerpt from my personal writings. Therapeutic in nature, no plans for them. It was a way to start the healing process. I am still working on that. It's more like an evolution. I will write more and send it as I do. Do you have a deadline? Ask me anything you would like and I will answer them as emotionally honest as I can.

One perspective of mine that helps me is that if you take for granted the freedoms you have and demand more, is that we defended the freedom so well that you do not have to lose sleep over it, or have to constantly think about it as others do in their countries. That in itself is a nice payment. Ask me anything you want, part of healing is having to remember these things and process them as we didn't have time for it when we were there. Don't try to protect me and don't treat me as a child. If I can stand a post with an M-16, I can handle what you would like to know. I am on my 3rd marriage, I am a fire fighter/Paramedic now, and I am starting to enjoy life a lot. Even with all my quirks and even when the neighbours think that I am losing it by digging up the front lawn to build a series of ponds and waterfalls. I look forward to this, as no one has ever asked to hear my story. Thank you for taking an interest. Please leave my name and other things out as I do not want to have any unfounded attention. I am just one of millions of men who have done this. We all did it as a team and we all deal with it in our own ways.

I wasn't really shot rather I was hit with a hand grenade just one tiny sliver that went through my right armpit area and collapsed my lung. Yes it did hurt one of the surprising aspects is that it felt very hot. After I healed I went home to my pregnant wife and started to drink. In March 1990 my wife was in a motor vehicle accident and we lost [our] son. In [A]ugust 1990 I went to Saudi. I had been promoted and this time I was playing Dad to my troops. Seasoned is a good word for it. I wasn't scared now it was a job. I could pick out the bad guys with out hesitating. Keep the moral[e] up, and keep my guys together. You don't get used to it, you act then later it all comes back and you wonder why anyone would want to do it. You understand the big picture but the one on one with a guy that is no different then yourself. Raising a family, paying bills. They have pictures of [their] family in their wallet like you do. You are amazed at the amount of lead that is in the air flying all over that more people aren't hit. Being wounded actually made me less vulnerable, as you experience things in life the less strange they seem to you or scary. I hope that makes sense. It makes me realize that life goes by faster [than] I initially thought when I believed that I had eighty years. I think it motivated me to live more of it. However I did lose about ten years with drinking a lot. Something that I am actually pretty ashamed of because deep down that wasn't me. I was aggressive and belligerent. Not really in my nature so to speak. Bosnia I am still not sure why we were there at all. There was no really defined mission. I miss my buddies who didn't come home and even those who did. That is what NG is all about. Vets talking to [v]ets. My memorial to them are the waterfalls, to my buddies who didn't come home, and 343 firefighters that didn't come home on 9/11. My spirit took a long time to come home, physically I was home in twenty-four hours, emotionally and spiritually most of me is here. That I owe to my wife. Great girl teaching me how to live again that's harder [than] anything I have ever done, Dying is the easy part.

## Part 3: Electronic Correspondence—Follow-Up

Dear Participant #3,

Why the alcohol? What did it do for you? What did you carry back with you that was so painful? Would you have used alcohol in that way if you had not been a marine and gone to war?

J.C.

> Why the alcohol, it was socially [acceptable]. My platoon would get together for some "beers" to forget and unwind, then you wanted to forget faster so you started drinking Jack Daniels. It was easy. In hindsight it did nothing for me but make it worse. Behavior problems, nightmares, and you could never quite drink enough to forget although you tried. Everything was painful, what you did and to whom, what you saw was burned into your brain. Your buddies who did not come back you missed you were constantly grieving and [angry]. [Extreme] anger you hated the world and wanted to kill it. But since you couldn't it created internal stress like a steam boiler just about to blow up. I wouldn't have drank if I had not seen what I did nor would I have drank if it wasn't the only "socially [acceptable]" form of a theorized relief. I didn't drink in high school and never had an interest in it. At twenty-one I had been in two wars, divorced, had a son killed and was hundreds of miles away from home. I was too young to have a support network of friends, family and I didn't have any idea how to process this emotionally. Physically I was on U.S. soil, spiritually and emotionally I never came home. Even though my body was twenty-one mentally I was about fifty to sixty in regards to experience of life. Everyone around me only saw a twenty-one-year-old in this society if you're just a kid you aren't trusted with anything. I didn't come home until 2002.

## Part 4: Electronic Correspondence—Follow-Up

Tell me more.

What I have found to be true is that a veteran goes through a grieving process, denial, bargaining, anger, and acceptance. After the "imprint of horror," a video is imbedded into the memory of a soldier. The video often replays continually until the coping skills are exercised and the imprint is reduced. Anger stays as the primary emotion because this is where everything is stuck, anger at loss of life, loss of innocence, loss of the "fun years," loss of power, loss of any number of things. The average age of a service man is eighteen to twenty-five. What do you remember of those years and why do you remember it? College, spring break, friends, all nighters, etc., these are fond memories in contrast of war for the veteran. The secret is to get the veteran to enable them to use coping skills they don't know they have because they were never taught to use them as you were with "critical thinking" in college. Emotionally 'til the veteran uses coping skills they can't advance in emotional or cognitive age. They are stuck with thoughts, hormone imbalance, etc. Some need not only counseling but medication also to help maintain psychological homeostasis. I learned how to use coping skills with meds, counseling, support network of other veterans, and my wife. That's why I am finally back in college going after what I wanted to be fifteen years after the normal age of doing that. Regret is another hang up. Have you ever done anything that you regret because you didn't think it was you really doing it? Regret turns into confusion emotionally and it in turn creates anger. It's a cycle that continues until you break it.

## Part 5: Electronic Correspondence—Follow-Up

### Why so much anger?

The anger comes from several avenues. It starts in boot camp. They are training you to protect, defend, and to kill if it comes to that.

They frustrate you, intimidate you and irritate you because any sane person would not make you do the things that they do. Then if you do go to war and experience it, our anger splits, like an atom does and creates heat. Anger is volatile. You are sent someplace to defend your way of life, to protect your country, her women and children and her divine right to exist. You get mad because you don't understand why the other guy [enemy] hates you because you're an American. It builds and builds because everything you were told as a child you have to protect. You are afraid it will be taken away. This adds to the anger. You do your job. You win and you get to go home and everyone has been protected. No one loses sleep while I'm protecting you.

You come home and no one cares that you fought for them. They didn't feel the pinch, the lead flying around, the bullets, smell the death, smell diesel fuel, the napalm, the gun powder, the smells that are burned into the soldiers brain. Because they didn't experience it nor did they actually feel that their liberties were in jeopardy, they don't think that you did anything for them. So their freedom was never really challenged in their eyes so quit overreacting you didn't do anything for me. This reaction from an ungrateful person adds to the anger, it continually compounds. Now remember you are still young and you do not have the coping skills because so much happened to you so fast that the coping skills are short circuited in the process.

Now you begin to think it was a waste and your buddies died for nothing and you got shot for what? More anger you're like an atomic bomb with its atoms splitting. It is a continual reaction. Add alcohol to this already explosive mixture. You are in hell, you don't understand. You did it right. You were a Marine and defended America. You did what you were supposed to do. Why does life hurt so bad and why do I not want to be here anymore? You can't think it through there is no logical thought pattern that will help you put this together. Now add the hormones, the dopamine, the epinephrine, because you were in a constant state of excitement and fear your hormones that flow in the brain to maintain emotional stability are all screwed up and stuck high. You can't process it now if you wanted to.

The anger is actually a chain of events, then it goes to a chemical reaction in the brain, then add the Jack Daniels to this, the anger does not go away till one of these chains are broken. That's why it takes years to "come home."

I hope this answers the question for you. Take this info and help more guys to be able to "come home." You will help me by bringing all of us home.

# References

Agar, M. (1991). The right brain strikes back. In N. G. Fielding & R. M. Lee (Eds.), *Using computers in qualitative research* (pp. 181–194). London: Sage.

Alvarez, E., Jr., & Pitch, A. S. (1989). *Chained eagle*. New York: Donald I. Fine.

Anderson, D. (1981). Doc. In A. Santoli (Ed.), *Everything we had* (pp. 66–75). New York: Ballantine Books.

Baker, C., Wuest, J., & Stern, P. N. (1992). Method slurring: The grounded theory/phenomenology example. *Journal of Advanced Nursing, 17*(11), 1355–1360.

Baszanger, I. (1998). *Inventing pain medicine: From the laboratory to the clinic.* New Brunswick, NJ: Rutgers University Press.

Bazeley, P., & Richards, L. (2000). *The NVivo qualitative project book.* Thousand Oaks, CA: Sage.

Becker, H. S. (1982). *Art worlds.* Berkeley: University of California Press.

Becker, H. S. (1986a). *Doing things together.* Evanston, IL: Northwestern University Press.

Becker, H. S. (1986b). *Writing for social scientists.* Chicago: University of Chicago Press.

Becker, H. S. (1998). *Tricks of the trade: How to think about your research while doing it.* Chicago: University of Chicago Press.

Bell, K. (1993). *100 missions north: A fighter pilot's story of the Vietnam War.* Washington, DC: Brassey.

Berg, B. C. (2006). *Qualitative research methods for the social sciences* (6th ed.). Boston: Allyn & Bacon.

Beveridge, W. I. (1963). Chance. In S. Rapport & H. Wright (Eds.), *Science: Method and meaning* (pp. 131–147). New York: Washington Square Press.

Biernacki, P. (1986). *Pathways from heroin addiction.* Philadelphia: Temple University Press.

Bird, T. (1981). Ia Drang. In A. Santoli (Ed.), *Everything we had* (pp. 34–43). New York: Ballantine Books.

Blumer, H. (1969). *Symbolic interactionism.* Englewood Cliffs, NJ: Prentice Hall.

Bong, S. A. (2002, May). Debunking myths in qualitative data analysis. *Qualitative Social Research,* 3(2). Retrieved December 6, 2005 from http://www.qualitative-research.net/fqs-eng.htm

Broadhead, R. (1983). *Private lives and professional identity of medical students.* New Brunswick, NJ: Transaction Publishers.

Cannaerts, N., Dierckx de Casterlé, B., & Grypdonck, M. (2004). Palliative care, care for life: A study of the specificity of residential palliative care. *Qualitative Health Research, 14*(6), 816–835.

Caputo, P. (1977). *A rumor of war.* New York: Ballantine Books.

Cauhape, E. (1983). *Fresh starts: Men and women after divorce.* New York: Basic Books.

Charmaz, K. (1983). Loss of self: A fundamental form of suffering in the chronically ill. *Sociology of Health and Illness, 5,* 168–195.

Charmaz, K. (1991a). *Good days, bad days: The self in chronic illness and time.* New Brunswick, NJ: Rutgers University Press.

Charmaz, K. (1991b). Struggling for a self: Identity levels of the chronically ill. In J. Roth & P. Conrad (Eds.), *Research in the sociology of health care. Vol. 6. The experience and management of chronic illness* (pp. 203–321). Greenwich, CT.: JAI Press.

Charmaz, K. (2006). *Constructing grounded theory.* Thousand Oaks, CA: Sage.

Chesney, M. (2001). Dilemmas of self in the method. *Qualitative Health Research, 11*(1), 127–135.

Chiovitti, R. F., & Piran, N. (2003). Rigour and grounded theory research. *Journal of Advanced Nursing, 44*(4), 427–435.

Clarke, A. E. (1991). Social worlds/arena theory as organizational theory. In D. R. Maines (Ed.), *Social organization and social process* (pp. 119–158). New York: Aldine de Gruyter.

Clarke, A. E. (2005). *Situational analysis.* Thousand Oaks, CA: Sage.

Corbin, J. (1993). Controlling the risks of a medically complicated pregnancy. *Journal of Perinatal Nursing, 7*(3), 1–6.

Corbin, J. (2002, August). *Taking the work seriously: Putting the quality in qualitative research.* Keynote address presented at the 2nd Thinking Qualitative Workshop series, International Institute for Qualitative Methodology, University of Alberta, Alberta, Canada.

Corbin, J. (2003, November 28). *Taking the work of qualitative analysis seriously.* Paper presented at a meeting of qualitative researchers sponsored by the Japanese Red Cross, Tokyo, Japan.

Corbin, J., & Cherry, J. (1997). Caring for the aged in the community. In E. A. Swanson & T. Tripp-Reimer (Eds.), *Chronic illness and the older adult* (pp. 62–81). New York: Springer.

Corbin, J., & Morse, J. (2003). The unstructured interview: Issues of reciprocity and risks when dealing with sensitive topics. *Qualitative Inquiry, 9*(3), 335–354.

Corbin, J., & Strauss, A. (1984). Collaboration: Couples working together to manage chronic illness. *Image, 16*(4), 109–115.

Corbin, J., & Strauss, A. (1988). *Unending work and care.* San Francisco: Jossey-Bass.

Corbin, J., & Strauss, A. (1990). Grounded theory method: Procedures, canons, and evaluative procedures. *Qualitative Sociology, 13*(1), 3–21.

Corbin, J., & Strauss A. (1991a). Comeback: The process of overcoming disability. In G. L. Albrecht & J. A. Levy (Eds.), *Advances in medical sociology* (Vol. 2, pp. 137–159). Greenwich, CT: JAI Press.

Corbin, J., & Strauss, A. (1991b). A nursing model for chronic illness management based upon the trajectory framework. *Scholarly Inquiry for Nursing Practice, 5*(3), 155–174.

Creswell, J. W. (1998). *Qualitative inquiry and research design: Choosing among five traditions.* Thousand Oaks, CA: Sage.

Creswell, J. W. (2003). *Research design: Qualitative, quantitative, and mixed methods approaches* (2nd ed.). Thousand Oaks, CA: Sage.

Creswell, J. W., & Miller, D. L. (2000). Determining validity in qualitative research. *Theory Into Practice, 39*(3), 124–130.

Cutcliffe, J. R. (2003). Reconsidering reflexivity: Introducing the case for intellectual entrepreneurship. *Qualitative Health Research, 13*(1), 136–148.

Davies, D., & Dodd, J. (2002). Qualitative research and the question of rigor. *Qualitative Health Research, 12*(2), 279–289.

Davis, F. (1963). *Passage through crisis.* Indianapolis, IN: Bobbs-Merrill.

Davis, F. (1991). Identity ambivalence in clothing: The dialectic of the erotic and the chaste. In D. R. Maines (Ed.), *Social organization and social process* (pp. 105–116). New York: Aldine de Gruyter.

Denzin, N. K. (1987). *The alcoholic self.* Newbury Park, CA: Sage.

Denzin, N. K. (1989). *Interpretive biography* (Qualitative Research Methods, Vol. 17). Newbury Park, CA: Sage.

Denzin, N. K. (1994). The art and politics of interpretation. In N. K. Denzin & Y. Lincoln (Eds.), *Handbook of qualitative research* (pp. 500–515). Thousand Oaks, CA: Sage.

Denzin, N. K. (1998). The art and politics of interpretation. In N. K. Denzin & Y. Lincoln (Eds.), *Handbook of qualitative research* (pp. 313–371). Thousand Oaks, CA: Sage.

Denzin, N. K., & Lincoln, Y. S. (1998). The art of interpretation, evaluation, and presentation. In N. K. Denzin & Y. S. Lincoln (Eds.), *Collecting and interpreting qualitative materials* (pp. 275–281). Thousand Oaks, CA: Sage.

Denzin, N. K., & Lincoln, Y. S. (2005). *The Sage handbook of qualitative research.* Thousand Oaks, CA: Sage.

Dewey, J. (1917). *Creative intelligence: Essays on the pragmatic attitude.* New York: Henry Holt.

Dewey, J. (1922). *Human nature in conduct.* New York: Henry Holt.

Dewey, J. (1929). *The quest for certainty.* New York: G. P. Putnam.

Dewey, J. (1934). *Art as experience.* New York: Minton Blach.

Dewey, J. (1938). *Logic: The theory of inquiry.* New York: Henry Holt.

Dey, I. (1993). *Qualitative data analysis.* London: Routledge.

Dey, I. (1999). *Grounding grounded theory.* San Diego, CA: Academic Press.

Downs, F. (1993). *The killing zone: My life in the Vietnam War.* New York: W. W. Norton.

Ellsberg, D. (2003). *Secrets: A memoir of Vietnam and the Pentagon Papers.* New York: Penguin.

Emden, C., & Sandelowski, M. (1999). The good, the bad and the relative. Part Two: Goodness and the criterion problem in qualitative research. *International Journal of Nursing Practice, 5*(1), 2–7.

Engelbart, D. C. (1962, October). *Augmenting human intellect: A conceptual framework* (Summary Report AFOSR-3223 under Contract AF 49(638)-1024, SRI Project 3578). Menlo Park, CA: Air Force Office of Scientific Research, Stanford Research Institute. Retrieved March 1, 2007, from http://www.bootstrap.org/augdocs/friedewald030402/augmentinghumanintellect/ahi62index.html

Fagerhaugh, S., & Strauss, A. (1977). *The politics of pain management.* Menlo Park, CA: Addison-Wesley.

Fielding, N. G., & Lee, R. M. (Eds.). (1991). *Using computers in qualitative analysis.* London: Sage.

Fielding, N. G., & Lee, R. M (Eds.). (1998). *Computer analysis in qualitative research.* London: Sage.

Finlay, L. (2002). "Outing" the researcher: The provenance, process, and practice of reflexivity. *Qualitative Health Research, 12*(4), 531–545.

Fisher, B. (1991). Affirming social value: Women without children. In D. R. Maines (Ed.), *Social organization and social process* (pp. 87–104). New York: Aldine de Gruyter.

Fisher, B., & Strauss, A. (1978). The Chicago tradition: Thomas, Park, and their successors. *Symbolic Interaction, 1*(2), 5–23.

Fisher, B., & Strauss, A. (1979a). George Herbert Mead and the Chicago tradition of sociology. *Symbolic Interaction (Part 1), 2*(1), 9–26.

Fisher, B., & Strauss, A. (1979b). George Herbert Mead and the Chicago tradition of sociology. *Symbolic Interaction (Part 2), 2*(2), 9–20.

Flick, U. (2002). *An Introduction to qualitative research* (2nd ed.). Thousand Oaks, CA: Sage.

Flicker, S., Haans, D., & Skinner, H. (2004). Ethical dilemmas in research on Internet communities. *Qualitative Health Research, 14*(1), 124–134.

Foster, W. F. (1992). *Captain Hook: A pilot's tragedy and triumph in the Vietnam War.* Annapolis, MD: Naval Institute Press.

Foucault, M. (1972). *The archaeology of knowledge* (A. M. Sheridan Smith, Trans.). New York: Harper & Row.

Foucault, M. (1974). *Power/knowledge: Selected interviews and other writing.* New York: Pantheon Books.

Fujimora, J. (1987). Constructing doable problems in cancer research: Articulating alignment. *Social Studies of Science, 17*, 257–293.

Fujimora, J. (1991). On methods, ontologies, and representation in the sociology of science, where do we stand? In D. R. Maines (Ed.), *Social organization and social process* (pp. 207–248). New York: Aldine de Gruyter.

Gerson, E. M. (1991). Supplementing grounded theory. In D. R. Maines (Ed.), *Social organization and social process* (pp. 285–301). New York: Aldine de Gruyter.

Geertz, C. (1973). *The interpretation of cultures: Selected essays.* New York: Basic Books.

Gilgun, J. F., Daly, K., & Handel, G. (1992). *Qualitative methods in family research.* Newbury Park, CA: Sage.

Glaser, B. (1978). *Theoretical sensitivity.* Mill Valley, CA: Sociology Press.

Glaser, B. (1992). *Basics of grounded theory analysis.* Mill Valley, CA: Sociology Press.

Glaser, B., & Strauss, A. (1965). *Awareness of dying.* Chicago: Aldine.

Glaser, B., & Strauss, A. (1967). *The discovery of grounded theory.* Chicago: Aldine.

Goulding, C. (2002). *Grounded theory: A practical guide for management, business, and marketing.* Thousand Oaks, CA: Sage.

Green, D., Creswell, J. W., Shope, R. J., & Plano Clark, V. L. (2007). Grounded theory and racial/ethnic diversity. In A. Bryant & K. Charmaz (Eds.), *The Sage handbook of grounded theory.* London: Sage.

Greene, J. C., Kreider, H., & Mayer, F. (2005). Combining qualitative and quantitative methods in social inquiry. In B. Somekh & C. Lewin (Eds.), *Research methods in the social sciences* (pp. 274–281). London: Sage.

Guba, E. G., & Lincoln Y. S. (1998). Competing paradigms in qualitative research. In N. K. Denzin & Y. S. Lincoln (Eds.), *The landscape of qualitative research* (pp. 195–220). Thousand Oaks, CA: Sage.

Gubrium, J., & Holstein, J. A. (2001*). Handbook of interview research: Context and method.* Thousand Oaks, CA: Sage.

Guessing, J. C. (1995). *Negotiating global teaming in a turbulent environment* (Microform 9613463). Unpublished doctoral dissertation, University of Michigan.

Hage, J. (1972). *Techniques and problems of theory construction in sociology.* New York: John Wiley.

Hall, W. A., & Callery, P. (2001). Enhancing the rigor of grounded theory: Incorporating reflexivity and relationality. *Qualitative Health Research, 11*(2), 257–272.

Hamberg, K., & Johansson, E. (1999). Practitioner, researcher, and gender conflict in a qualitative study. *Qualitative Health Research, 9*(4), 455–467.

Hamilton, R. J., & Bowers, B. J. (2006). Internet recruitment and e-mail interviews in qualitative studies. *Qualitative Health, Research, 16*(6), 821–835.

Hammersley, M. (1987). Some notes on the terms "validity" and "reliability." *British Educational Research Journal, 13*(1), 73–81.

Hammersley, M., & Atkinson, P. (1983). *Ethnography: Principles in practice.* New York: Tavistock.

Herr, M. (1991). *Dispatches.* New York: Vintage International.

Hildebrand, B. (2007). Mediating structure and interaction in grounded theory. In T. Bryant & K. Charmaz (Eds.), *The Sage handbook of grounded theory.* London: Sage.

Hildreth, R., & Sasser, C. W. (2003). *Hill 488.* New York: Simon & Schuster.

Hoffman, E. (1989). *Lost in translation: Life in a new language.* New York: Penguin.

Holroyd, E. E. (2003). Chinese family obligations toward chronically ill elderly members: Comparing caregivers in Beijing and Hong Kong. *Qualitative Health Research, 13*(3), 302–318.

Hughes, E. C. (1971). *The sociological eye: Selected papers.* Chicago: Aldine-Atherton.

Hunter, A., Lusardi, P., Zucker, D., Jacelon, C., & Chandler, G. (2002). Making meaning: The creative component in qualitative research. *Qualitative Health Research, 12*(3), 388–398.

Isaacs, A. (1997). *Vietnam shadows: The war, its ghosts, and its legacy*. Baltimore, MD: Johns Hopkins University Press.

Kelle, U. (Ed.). (1995). *Computer-aided qualitative analysis: Theory, methods and practice*. London: Sage.

Kelle, U. (1997). Theory building in qualitative research and computer programs for the management of textual data. *Sociological Research Online, 2*(2). Retrieved December 9, 2005, from http://www.socresonline.org.uk/socresonline/2/2/1.html

Kuckartz, U. (1988/2007 [latest version]). MAXQDA—Professional Software for Qualitative Data Analysis [Computer software]. Berlin, Germany: Verbi Software.

Lakoff, G., & Johnson, M. (1981). *Metaphors we live by*. Chicago: University of Chicago Press.

Lamott, A. (1994). *Bird by bird: Some instructions on writing and life*. New York: Doubleday Anchor.

Langguth, A. J. (2002). *Our Vietnam: The war 1954–1975*. New York: Simon & Schuster.

Lincoln, Y. S., & Guba, E. G. (1985). *Naturalistic inquiry*. Beverly Hills, CA: Sage.

Lofland, L. H. (1991). The urban milieu: Locales, public sociability, and moral concerns. In D. R. Maines (Ed.), *Social organization and social process* (pp. 189–205). New York: Aldine de Gruyter.

Lofland, J., Snow, D., Anderson, L., & Lofland, L. (2006). *Analyzing social setting: A guide to qualitative observation and analysis*. Belmont, CA: Wadsworth Thomson.

Lomberg, K., & Kirkevold, M. (2003). Truth and validity in grounded theory—A reconsidered realist interpretation of the criteria: Fit, work, relevance and modifiability. *Nursing Philosophy*, *4*(3), 189–200.

Lonkila, M. (1995). Grounded theory as an emerging paradigm for computer-assisted qualitative data analysis. In U. Kelle (Ed.), *Computer-aided qualitative analysis: Theory, methods and practice* (pp. 41–51). London: Sage.

Long, T., & Johnson, M. (Eds.). (2007). *Research ethics in the real world: Issues and solutions for health and social care*. Edinburg, UK: Churchill Livingstone, Elsevier Health Sciences.

Márquez, G. G. (1993). *Strange pilgrims* (E. Grossman, Trans.). New York: Knopf.

Marrett, J. G. (2003). *Cheating death: Combat air rescues in Vietnam and Laos*. Washington, DC: Smithsonian Books.

Marshall, C., & Rossman, B. (2006). *Designing qualitative research*. Thousand Oaks, CA: Sage.

Mays, N., & Pope, C. (1995). Qualitative research: Rigour and qualitative research [Electronic version]. *BMJ, 311*, 109–112. Retrieved on December 6, 2005, from http://bmj.bmjjournals.com/cgi/content/full/311/6997/109

McMaster, H. R. (1997). *Dereliction of duty*. New York: HarperCollins.

McNamara, R. S., Blight, J. G., Brigham, R. K., Bierstaker, T. J., & Schandler, H. Y. (1999). *Agreement without end*. New York: Perseus.

Mead, G. H. (1917). Scientific method and the individual thinker. In J. Dewey (Ed.), *Creative intelligence: Essays in the pragmatic* (pp. 176–227). New York: Henry Holt.

Mead, G. H. (1956). *On social psychology: Selected papers* (A. Strauss, Ed.). Chicago: University of Chicago Press.

Mead, G. H. (1959). *Philosophy of the present* (Arthur E. Murphy, Ed.). Chicago: Open Court.

Mead, G. H. (1967). *Mind, self, and society* (Works of George Herbert Mead, Vol. 1, C. W. Morris, Ed.). Chicago: University of Chicago Press. (Original work published 1934)

Mead, G. H. (1972). *Movements of thought in the nineteenth century* (Works of George Herbert Mead, Vol. 2, M. E. Moore, Ed.). Chicago: University of Chicago Press. (Original work published 1936)

Mead, G. H. (1972). *The philosophy of the act* (Works of George Herbert Mead, Vol. 3, C. W. Morris, Ed.). Chicago: University of Chicago Press. (Original work published 1938)

Miles, M. B., & Huberman, A. M. (1994). *Qualitative data analysis* (2nd ed.). Thousand Oaks, CA: Sage.

Mishler, E. G. (1986). *Research interviewing*. Cambridge, MA: Harvard University Press.

Moore, C. J. (1992). *Oral history reflections of army nurse Vietnam veterans: Managing the demanding experience of war*. Unpublished master's thesis, Department of Nursing, San Jose State University, San Jose, CA.

Moore, H. G., & Galloway, J. L. (1992). *We were soldiers once and young*. New York: HarperPerennial.

Morgan, D. L. (1996). *Focus groups as qualitative research*. Thousand Oaks, CA: Sage.

Morse, J. M. (1994). *Critical issues in qualitative research*. Thousand Oaks, CA: Sage.

Morse, J. M. (1999). Editorial myth #93: Reliability and validity are not relevant to qualitative inquiry. *Qualitative Health Research, 9*(6), 717–718.

Morse, J. M. (2001). Toward a praxis theory of suffering. *Advances in Nursing Science, 24*(1), 47–59.

Morse, J. M. (2004). Alternative methods of representation: There are no shortcuts. *Qualitative Health Research, 14*(7), 887–888.

Morse, J. M. (2005). Creating a qualitatively derived theory of suffering. In U. Zutler (Ed.), *Clinical practice and development in nursing* (pp. 83–91). Aarhus, Denmark: Center for Innovation in Nurse Training.

Morse, J. M., Barret, M., Mayan, M., Olson, K., & Spiers, J. (2002, Spring). Verification strategies for establishing reliability and validity in qualitative research [Electronic version]. *International Journal of Qualitative Methodology 1*(2). Retrieved July 22, 2006, from http://www.ualberta.ca/~ijqm

Morse, J. M., & Field, P. A. (1995). *Qualitative research methods for health professionals* (2nd ed.). Thousand Oaks, CA: Sage.

Nhu Tang, T. (with Chanoff, D., & Van Toai Doan). (1986). *A Viet Cong memoir: An inside account of the Vietnam War and its aftermath*. New York: Vintage.

Ninh, B. (1993). *The sorrow of war: A novel of North Vietnam* (Phan Thanh Hao, Trans.). New York: Berkeley Publishers.

Oleson, V. (1998). Feminism and models of qualitative research. In N. K. Denzin & Y. S. Lincoln (Eds.), *The landscape of qualitative research theories and issues* (pp. 300–332). Thousand Oaks, CA: Sage.

O'Shea, J. (with Ling, R. D.). (2003). *The beast within: Vietnam, the cause and effect of post-traumatic stress disorder*. New York: Vintage.

Park, R. E. (1967). *On social control and collective behavior* (R. H. Turner, Ed.). Chicago: University of Chicago Press.

Patton, M. Q. (1990). *Qualitative research and evaluation methods*. Newbury Park, CA: Sage.

Patton, M. Q. (2002). *Qualitative research and evaluation methods* (2nd ed.). Thousand Oaks, CA: Sage.

Pfaffenberger, B. (1988). *Microcomputer applications in qualitative research*. Newbury Park, CA: Sage.

Pierce, B. N. (1995). The theory of methodology in qualitative research. *TESOL Quarterly, 29*, 569–576.

Piper, H., & Simons, H. (2005). Ethical responsibility in social research. In B. Somekh & C. Lewin (Eds.), *Research methods in the social sciences* (pp. 56–63). London: Sage.

Ragin, C., & Becker, H. (Eds.). (1992). *What is a case? Exploring the foundations of social inquiry*. Cambridge, UK: Cambridge University Press.

Rasimus, E. (2003). *When thunder rolled: An F-105 pilot over North Vietnam*. New York: Ballantine Books.

Reid, C. (2004). *The wounds of exclusion: Poverty, women's health and social justice*. Edmonton, AB: Qualitative Institute Press.

Riemann, G. (2003, September). A joint project against the backdrop of a research tradition: An introduction to "doing biographical research" [Electronic version]. *Forum: Qualitative Social Research, 4*(3). Retrieved September 10, 2006, from http://www.qualitative-research.net/fqs-texte/3-03/3-03hrsg-e.htm

Riemann, G., & Schütze, F. (1991). "Trajectory" as a basic theoretical concept for analyzing suffering disorderly social process. In D. R. Maines (Ed.), *Social organization and social process* (pp. 333–357). New York: Aldine de Gruyter.

Roberts, K. A., & Wilson, R. W. (2002, May). ICT and the research process: Issues around compatibility of technology with qualitative analysis [Electronic version]. *Forum Qualitative Sozialforschung/Forum: Qualitative Social Research, 3*(2). Retrieved December 6, 2005, from http://www.qualitative-research.net/fqs-texte/2-02/2-02robertswilson-e.htm

Rodgers, B. L., & Cowles, K. V. (1993). The qualitative research audit trail: A complex collection of documentation. *Research in Nursing and Health, 16*(3), 219–226.

Rolfe, G. (2006). Validity, trustworthiness and rigour: Quality and the idea of qualitative research. *Journal of Advanced Nursing, 53*(3), 304–310.

Rosenbaum, M. (1981). *Women on heroin*. New Brunswick, NJ: Rutgers University Press.

Rosenthal, G. (1993). Reconstruction of life stories: Principles of selection in generating stories for narrative biographical interviews. *Narrative Study of Lives, 1*(1), 59–91.

Rosenthal, G., & Völter, B. (1998). Three generations within Jewish and non-Jewish German families after the unification of Germany. In Y. Danieli (Ed.), *International handbook of multigenerational legacies of trauma* (pp. 297–313). New York: Plenum.

Saiki-Craighill, S. (2001a). The grieving process of Japanese mothers who have lost a child to cancer: Part I. Adjusting to life after losing a child. *Journal of Pediatric Oncology Nursing, 18*(6), 260–267.

Saiki-Craighill, S. (2001b). The grieving process of Japanese mothers who have lost a child to cancer: Part II. Establishing a new relationship from the memories. *Journal of Pediatric Oncology Nursing, 18*(6), 268–275.

Sallah, M., & Weiss, M. (2006). *Tiger force A true story of men and war.* New York: Little, Brown.

Sandelowski, M. (1993). Theory unmasked: The uses and guises of theory in qualitative research. *Research in Nursing and Health, 16*, 213–218.

Sandelowski, M. (1994). The proof is in the pottery. In J. M. Morse (Ed.), *Critical issues in qualitative research methods* (pp. 46–63). Thousand Oaks, CA: Sage.

Sandelowski, M. (2000). Whatever happened to qualitative description. *Research in Nursing and Health, 23*, 334–340.

Santoli, A. (Ed.). (1981). *Everything we had: An oral history of the Vietnam War by thirty-three American soldiers who fought it.* New York: Ballantine Books.

Santoli, A. (1985). *To bear any burden.* New York: E. P. Dutton.

Sar Desai, D. R. (2005). *Vietnam past and present.* Cambridge, MA: Westview Press.

Schatzman, L. (1986). *The structure of qualitative analysis.* Paper presented (in absentia) at the World Congress of Sociology Session on Issues in Qualitative Interpretation, New Delhi, India.

Schatzman, L. (1991). Dimensional analysis: Notes on an alternative approach to the grounding of theory. In D. R. Maines (Ed.), *Social organization and social process* (pp. 303–314). New York: Aldine de Gruyter.

Schatzman, L., & Strauss, A. (1973). *Field research.* Englewood Cliffs, NJ: Prentice Hall.

Schawndt, T. A. (1998). Constructivist, interpretivist approaches to human inquiry. In N. K. Denzin & Y. S. Lincoln (Eds.), *The landscape of qualitative research theories and issues* (pp. 221–259). Thousand Oaks, CA: Sage.

Schneider J., & Conrad, P. (1983). *Having epilepsy: The experience and control of illness.* Philadelphia: Temple University Press.

Schütze, F. (1992a). Pressure and guilt: War experiences of a young German soldier and their biographical implications, Part 1. *International Sociology, 7*(2), 187–208.

Schütze, F. (1992b). Pressure and guilt: War experiences of a young German soldier and their biographical implications, Part 2. *International Sociology, 7*(3), 347–367.

Seale, C. (1999). *The quality of qualitative research.* London: Sage.

Seale, C. (2002). Qualitative issues in qualitative inquiry. *Qualitative Social Work, 1*(1), 97–110.

Selye, H. (1956). *The stress of life.* New York: McGraw-Hill.

Sheeham, N. (1998). *A bright shining lie: John Paul Vann & America in Vietnam.* New York: Random House.

Shibutani, T. (1966). *Improvised news: A sociological study of rumor*. Indianapolis, IN: Bobbs Merrill.
Shibutani, T. (1991). On the empirical investigation of self-concept. In D. R. Maines (Ed.), *Social organization and social process* (pp. 59–69). New York: Aldine de Gruyter.
Shuval, J. T., & Mizrahi, N. (2004). Changing boundaries: Modes of coexistence of alternative and biomedicine. *Qualitative Health Research, 14*(5), 675–690.
Silverman, D. (2004). *Doing qualitative research* (2nd ed.). Thousand Oaks, CA: Sage.
Smith, W. (1992). *American daughters gone to war*. New York: William Morrow.
Somekh, B., & Lewin, C. (2005). *Research methods in the social sciences*. London: Sage.
Sparkes, A. C. (2001). Myth 94: Qualitative health researchers will agree about validity. *Qualitative Health Journal, 11*(4), 538–552.
Star, S. L. (1989). *Regions of the mind: Brain research and the quest for scientific certainty*. Stanford, CA: Stanford University Press.
Strauss, A. (1978). *Negotiations: Varieties, contexts, processes, and social order*. San Francisco: Jossey-Bass.
Strauss, A. (1987). *Qualitative analysis for social scientists*. Cambridge, UK: Cambridge University Press.
Strauss, A. (1991). The Chicago tradition's ongoing theory of action/interaction. In A. Strauss (Ed.), *Creating sociological awareness* (pp. 3–32). New Brunswick, NJ: Transaction Publishers.
Strauss, A. (1993). *Continual permutations of action*. Hawthorne, NY: Aldine de Gruyter.
Strauss, A. (1995). Notes on the nature and development of general theories. *Qualitative Inquiry, 1*(1), 7–18.
Strauss, A., & Corbin, J. (1988). *Shaping a new health care system*. San Francisco: Jossey-Bass.
Strauss, A., Fagerhaugh, S., Suczek, B., & Wiener, C. (1985). *Social organization of medical work*. Chicago: University of Chicago Press.
Strauss, A., Schatzman, L., Bucher, R., Ehrlich, D., & Sabshin, M. (1964). *Psychiatric ideologies and institutions*. New York: Free Press.
Suczek, B., & Fagerhaugh, S. (1991). AIDS and outreach work. In D. R. Maines (Ed.), *Social organization and social process* (pp. 159–193). New York: Aldine de Gruyter.
Summers, H. G., Jr. (1999). *The Vietnam War almanac*. Novato, CA: Presidio Press.
Terry, W. (1984). *Bloods: An oral history of the Vietnam War by black veterans*. New York: Ballantine Books.
Thomas W. I. (1966). *On social control and collective behavior* (M. Januwitz, Ed.). Chicago: University of Chicago Press.
Trotti, J. (1984). *Phantom over Vietnam*. Novato, CA: Presidio Press.
Tucker, S. C. (1998). *The encyclopedia of the Vietnam War*. New York: Oxford University Press.

Van Devanter, L. (1983). *Home before morning*. New York: Warner Books.

Van Manen, M. (2006). Writing qualitatively, or the demands of writing. *Qualitative Health Research, 16*(5), 713–722.

Vanhook, P. M. (2007). *Comeback of the Appalachian female stroke survivor: The interrelationships of cognition, function, self-concept, interpersonal, and social relationships*. Unpublished doctoral dissertation, East Tennessee State University.

Watson, L. A., & Girad, F. M. (2004). Establishing integrity and avoiding methodological misunderstanding. *Qualitative Health Research, 14*(6), 875–881.

Waugh, B. (with Keown, T.). (2004). *Hunting the Jackal*. New York: HarperCollins.

Weiss, R. S. (1994). *Learning from strangers: The art and method of qualitative interview studies*. New York: Free Press.

Weitzman, E. A., & Miles, M. B. (1995). *Computer programs for qualitative data analysis*. Thousand Oaks, CA: Sage.

Wiener, C. L. (1991). Arenas and careers: The complex interweaving of personal and organizational destiny. In D. R. Maines (Ed.), *Social organization and social process* (pp. 175–188). New York: Aldine de Gruyter.

Wiener, C. L., Fagerhaugh, S., Strauss, A., & Suczek, B. (1979). Trajectories, biographies, and the evolving medical sense: Labor and delivery and the intensive care nursery. *Sociology of Health and Illness, 1*(3), 261–283.

Whittemore, R., Chase, S., & Mandle, C. L. (2001). Validity in qualitative research. *Qualitative Health Research, 11*(4), 522–537.

Whyte, W. (1955). *Street corner society*. Chicago: University of Chicago Press.

Wicker, A. (1985). Getting out of our conceptual ruts: Strategies for expanding conceptual frameworks. *American Psychologist, 40*(10), 1094–1103.

Winter, G. (March 2000). A comparative discussion of the notion of "validity" in qualitative and quantitative research [Electronic version]. *Qualitative Report, 4*(3 & 4). Retrieved December 6, 2005, from http://www.nova.edu/ssss/QR/Qr4-3/winter.html

Wolcott, H. (1994). *Transforming qualitative research*. Thousand Oaks, CA: Sage.

Wolcott, H. (2001). *Writing up qualitative research* (2nd ed.). Thousand Oaks, CA: Sage.

Wolcott, H. (2002). Writing up qualitative research . . . better. *Qualitative Health Research, 12*(1), 91–103.

Yamamoto, N., & Wallhagen, M. I. (1997). The continuation of family caregiving in Japan. *Journal of Health and Social Behavior, 38*(June), 164–176.

Yarborough, T. (2002). *Da Nang diary: A forward air controller's gunsight view of combat in Vietnam*. New York: St. Martin's.

# Index

# About the Authors

**Juliet Corbin** (M.S. in nursing, D.N.Sc. in nursing, family nurse practitioner) was an instructor in the School of Nursing at San Jose State University before her retirement. She was also an adjunct professor at the International Institute of Qualitative Research at the University of Alberta, Alberta, Canada. She has remained active in teaching, consulting, and presenting papers and seminars in various countries throughout the world. She is coauthor (with Anselm Strauss) of the first and second editions of *Basics of Qualitative Research* (1990), *Unending Work and Care* (1988), and *Shaping a New Health Care System* (1988), and is coeditor (with Strauss) of *Grounded Theory in Practice* (1997). Her research interests, teaching, presentations, and publications are in the areas of qualitative methodology, chronic illness, and the sociology of work and the professions.

**Anselm Strauss** was born December 18, 1916, and died September 5, 1996. He was, at the time of his death, Professor Emeritus, Department of Social and Behavioral Sciences, University of California, San Francisco. His main research and teaching activities were in the sociology of health and illness and of work and professions. His approach to doing research was qualitative with the aim of theory building, and with Barney Glaser was co-founder of the method that has come to be known as grounded theory. Over the years, he was asked to be a visiting professor to the universities of Cambridge, Paris, Manchester, Constance, Hagen, and Adelaide. During his lifetime he wrote numerous papers and books, many of which have been translated into other languages. Among his books, written with various coworkers are *Awareness of Dying* (1965), *Mirrors and Masks* (1969), *Professions, Work and Careers* (1971), *Negotiations* (1978), *The Social Organization of Medical Work* (1985), *Unending Work and Care* (1988), and *Continual Permutations of Action* (1993). Though formally retired, he was still actively engaged in writing and research at the time of his death, on topics including work in hospitals and a sociological perspective on body.